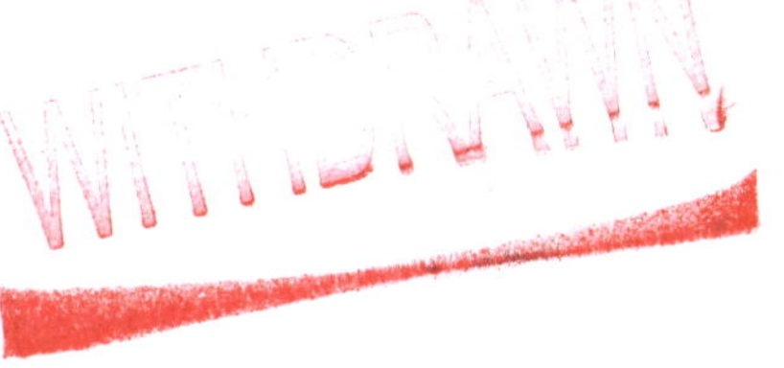

THE LOGIC OF CAUSATION

Definition, Induction and Deduction of Deterministic Causality

PHASES I, II AND III

AVI SION, Ph. D.

Published 1999, 2003, and 2010, Avi Sion, Geneva, Switzerland.
CreateSpace edition 2014

www.TheLogician.net
avi-sion@thelogician.net

Library Cataloguing Information:

Sion, Avi.
The Logic of Causation.
Phases I, II and III.

No index, no bibliography.

ISBN 978-1495221101

Abstract

The Logic of Causation is a treatise of formal logic and of aetiology. It is an original and wide-ranging investigation of the definition of causation (deterministic causality) in all its forms, and of the deduction and induction of such forms. The work was carried out in three phases over a dozen years (1998-2010), each phase introducing more sophisticated methods than the previous to solve outstanding problems. This study was intended as part of a larger work on causal logic, which additionally treats volition and allied cause-effect relations (2004).

The Logic of Causation deals with the main technicalities relating to reasoning about causation. Once all the deductive characteristics of causation in all its forms have been treated, and we have gained an understanding as to how it is induced, we are able to discuss more intelligently its epistemological and ontological status. In this context, past theories of causation are reviewed and evaluated (although some of the issues involved here can only be fully dealt with in a larger perspective, taking volition and other aspects of causality into consideration, as done in ***Volition and Allied Causal Concepts***).

Phase I: Macroanalysis. Starting with the paradigm of causation, its most obvious and strongest form, we can by abstraction of its defining components distinguish four genera of causation, or generic determinations, namely: complete, partial, necessary and contingent causation. When these genera and their negations are combined together in every which way, and tested for consistency, it is found that only four species of causation, or specific determinations, remain conceivable. The concept of causation thus gives rise to a number of positive and negative propositional forms, which can be studied in detail with relative ease because they are compounds of conjunctive and conditional propositions whose properties are already well known to logicians.

The logical relations (oppositions) between the various determinations (and their negations) are investigated, as well as their respective implications (eductions). Thereafter, their interactions (in syllogistic reasoning) are treated in the most rigorous manner. The main question we try to answer here is: *is (or when is) the cause of a cause of something itself a cause of that thing, and if so to what degree?* The figures and moods of positive causative syllogism are listed exhaustively; and the resulting arguments validated or invalidated, as the case may be. In this context, a general and sure method of evaluation called 'matricial analysis' (macroanalysis) is introduced. Because this (initial) method is cumbersome, it is used as little as possible – the remaining cases being evaluated by means of reduction.

Phase II: Microanalysis. Seeing various difficulties encountered in the first phase, and the fact that some issues were left unresolved in it, a more precise method is developed in the second phase, capable of systematically answering most outstanding questions. This improved matricial analysis (microanalysis) is based on tabular prediction of all logically conceivable combinations and permutations of conjunctions between two or more items and their negations (grand matrices). Each such possible combination is called a 'modus' and is assigned a permanent number within the framework concerned (for 2, 3, or more items). This allows us to identify each distinct (causative or other, positive or negative) propositional form with a number of alternative moduses.

This technique greatly facilitates all work with causative and related forms, allowing us to systematically consider their eductions, oppositions, and syllogistic combinations. In fact, it constitutes a most radical approach not only to causative propositions and their derivatives, but perhaps more importantly to their constituent conditional propositions. Moreover, it is not limited to logical conditioning and causation, but is equally applicable to other modes of modality, including extensional, natural, temporal and spatial conditioning and causation. From the results obtained, we are able to settle with formal certainty most of the historically controversial issues relating to causation.

Phase III: Software Assisted Analysis. The approach in the second phase was very 'manual' and time consuming; the third phase is intended to 'mechanize' much of the work involved by means of spreadsheets (to begin with). This increases reliability of calculations (though no errors were found, in fact) – and also allows for a wider scope. Indeed, we are now able to produce a larger, 4-item grand matrix, and on its basis find the moduses of causative and other forms needed to investigate 4-item syllogism. As well, now each modus can be interpreted with greater precision and causation can be more precisely defined and treated.

In this latest phase, ***the research is brought to a successful finish!*** Its main ambition, to obtain a complete and reliable listing of all 3-item and 4-item causative syllogisms, being truly fulfilled. This was made technically feasible, in spite of limitations in computer software and hardware, by cutting up problems into smaller pieces. For every mood of the syllogism, it was thus possible to scan for conclusions 'mechanically' (using spreadsheets), testing all forms of causative and preventive conclusions. Until now, this job could only be done 'manually', and therefore not exhaustively and with certainty. It took over 72'000 pages of spreadsheets to generate the sought for conclusions.

This is a historic breakthrough for causal logic and logic in general. Of course, not all conceivable issues are resolved. There is still some work that needs doing, notably with regard to 5-item causative syllogism. But what has been achieved solves the core problem. The method for the resolution of all outstanding issues has definitely now been found and proven. The only obstacle to solving most of them is the amount of labor needed to produce the remaining (less important) tables. As for 5-item syllogism, bigger computer resources are also needed.

Contents

Phase One: Macroanalysis.

Phase Two: Microanalysis.

Phase Three: Software Assisted Analysis.

AVI SION

THE LOGIC OF CAUSATION:

PHASE ONE: MACROANALYSIS

Phase I: Macroanalysis. Starting with the paradigm of causation, its most obvious and strongest form, we can by abstraction of its defining components distinguish four genera of causation, or generic determinations, namely: complete, partial, necessary and contingent causation. When these genera and their negations are combined together in every which way, and tested for consistency, it is found that only four species of causation, or specific determinations, remain conceivable. The concept of causation thus gives rise to a number of positive and negative propositional forms, which can be studied in detail with relative ease because they are compounds of conjunctive and conditional propositions whose properties are already well known to logicians.

The logical relations (oppositions) between the various determinations (and their negations) are investigated, as well as their respective implications (eductions). Thereafter, their interactions (in syllogistic reasoning) are treated in the most rigorous manner. The main question we try to answer here is: *is (or when is) the cause of a cause of something itself a cause of that thing, and if so to what degree?* The figures and moods of positive causative syllogism are listed exhaustively; and the resulting arguments validated or invalidated, as the case may be. In this context, a general and sure method of evaluation called 'matricial analysis' (macroanalysis) is introduced. Because this (initial) method is cumbersome, it is used as little as possible – the remaining cases being evaluated by means of reduction.

Chapter 1. THE PARADIGM OF CAUSATION

1. Causation.

Causality refers to causal relations, i.e. the relations between causes and effects. This generic term has various, more specific meanings. It may refer to **Causation**, which is *deterministic causality*; or to Volition, which is (roughly put) indeterministic causality; or to Influence, which concerns the interactions between causation and volition or between different volitions.

The term 'causality' may also be used to refer to causal issues: i.e. to negative as well as positive answers to the question "are these things causally related?" In the latter sense, negations of causality (in the positive sense) are also causality (in the broad sense). This allows us to consider Spontaneity (i.e. causelessness, the lack of any causation or volition) as among the 'causal' explanations of things.

A study of the field of causality must also include an investigation of non-causality in all its forms. For, as we shall see, even if we were to consider spontaneity impossible, the existence of causality in one form or other between things in general does not imply that any two things taken at random are necessarily causally related or causally related in a certain way. We need both positive and negative causal propositions to describe the relations between things.

In the present work, *The Logic of Causation*, we shall concentrate on causation, ignoring for now other forms of causality. Causative logic, or the logic of causative propositions, has three major goals, as does the study of any other type of human discourse.

(a) To *define* what we mean by causation (or its absence) and identify and classify the various forms it might take.

(b) To work out the *deductive* properties of causative propositions, i.e. how they are opposed to each other (whether or not they contradict each other, and so forth), what else can be immediately inferred from them individually (eduction), and what can be inferred from them collectively in pairs or larger numbers (syllogism).

(c) To explain how causative propositions are, to start with, *induced* from experience, or constructed from simpler propositions induced from experience.

Once these goals are fulfilled, in a credible manner (i.e. under strict logical supervision), we shall have a clearer perspective on wider issues, such as (d) whether there is a universal law of causation (as some philosophers affirm) or spontaneity is conceivable (as others claim), and (e) whether other forms of causality (notably volition, and its derivative influence) are conceivable.

Note well, we shall to begin with theoretically define and interrelate the various possible forms of causation, *leaving aside for now the epistemological issue as to how they are to be identified and established in practice, as well as discussions of ontological status.*

We shall thus in the present volume primarily deal with the main *technicalities* relating to reasoning about causation, and only later turn our attention to some larger epistemological and ontological issues (insofar as they can be treated prior to further analysis of the other forms of causality). The technical aspect may at times seem tedious, but it is impossible to properly understand causation and its implications without it. Most endless debates about causation (and more generally, causality) in the history of philosophy have arisen due to failure to first deal with technical issues.

2. The Paradigmic Determination.

Causation, or deterministic causality, varies in strength, according to the precise combinations of conditioning found to hold between the predications concerned. We may call the different forms thus identified the **determinations** of causation.

The *paradigm*, or basic pattern, of causation is its strongest determination. This has the form:

> **If the cause is present, the effect is invariably present;**
> **if the cause is absent, the effect is invariably absent.**

Our use, here, of the definite article, as in *the* cause or *the* effect, is only intended to pinpoint the predication under consideration, without meaning to imply that there is only one such cause or effect in the context concerned. Use of an indefinite article, as in *a* cause or *an* effect, becomes more appropriate when discussing a multiplicity of causes or effects, which as we shall later see may take various forms.

We may rewrite the above *static* formula in the following more *dynamic* expression:

> If the cause shifts from absent to present, the effect invariably shifts from absent to present;
> if the cause shifts from present to absent, the effect invariably shifts from present to absent;

We shall presently see how this model is variously reproduced in lesser determinations. For now, it is important to grasp the underlying principle it reflects.

The essence of causation (or 'effectuation') is that ***when some change is invariably accompanied by another, we say that the first phenomenon that has changed has "caused" (or "effected") the second phenomenon that has changed.*** In the above model, the changes involved are respectively from the absence to the presence of the first phenomenon (called the cause) and from the absence to the presence of the second phenomenon (called the effect); or vice versa. We may, incidentally, commute this statement and say that the effect has been caused (or effected) by the cause.

Now, some comments about our terminology here:

The term "**change**", here, must be understood in a very broad sense, as referring to any event of difference, whatever its modality.

- Its primary meaning is, of course, *natural change*, with reference to *time* or more to the point with respect to broader changes in surrounding *circumstances*[1]. Here, the meaning is

---

1 The difference between time and circumstance as concepts of reference seems very slim. How do we *pinpoint* an undefined 'circumstance' other than with reference to time? Yet the distinction seems important, since we construct two different types of modality or modes on its basis. The only answer I can think of for now is that whereas times (e.g. "on 17 August 1999, I wrote this footnote") are unrepeatable, circumstances (e.g. "at the time Turkey experienced an earthquake, I wrote this footnote") are in principle repeatable. A circumstance is loosely specified by *describing some events* in

that some object or characteristic of an object which initially existed or appeared, later did not exist or disappeared (ceasing to be), or vice-versa (coming to be); or something existed or appeared at one place and time and recurred or reappeared at another place, at another time (mutation, alteration or movement). This gives rise to temporal and natural modalities of causation.

- Another, secondary sense is *diversity* in *individuals* or *groups*. This signifies that an individual object has different properties in different parts of its being[2]; or that a kind of object has some characteristic in some of its instances and lacks that characteristic (and possibly has another characteristic, instead) in some other of its instances. This gives rise to spatial and extensional modalities of causation.
- Tertiary senses are *epistemic* or *logical change*, which focus respectively on the underlying acts of consciousness or the status granted them: something is at first noticed and later ignored, or believed and later doubted, or vice-versa, by someone. This gives rise to epistemic and logical modalities of causation.

Regarding the terms "**present**" and "**absent**" (i.e. not present), they may be understood variously, with reference to the situations just mentioned. They may signify existence or appearance or instancing (i.e. occurrence in some indicated cases) or being seen or being accredited true – or the negations of these.

The term "**phenomenon**" is here, likewise, intended very broadly, to include physical, mental or spiritual phenomena (things, appearances, objects), concrete or abstract. Also, a phenomenon may be static or dynamic: that is, the changing cause and effect need not be a quality or quantity or state or position, though some such static phenomena are always ultimately involved; the cause and effect may themselves be changes or events or movements. For instance, motion is change of place, acceleration is change in the speed or direction of motion. What matters is the switch from presence to absence, or vice-versa, of that thing, whatever its nature (be it static or dynamic). The cause and effect need not even be of similar nature; for example, a change of quality may cause a change of quantity.

Another term to clarify in the above principle is "**accompanied**". Here again, our intent is very large. The cause and effect may be in or of the same object or different objects, adjacent or apart in space, contemporaneous or in a temporal sequence. The definition of causation contains no prejudice in these respects, though we may eventually find fit to postulate relatively non-formal rules, such as that in natural causation the effect cannot precede the cause in time or that all causation at a distance implies intermediate contiguous causations[3].

---

a time (without always intending that reference item to be more than coincidental – i.e. the earthquake did not cause me to write these comments).

[2] This is the basis for a concept of *spatial modality*, which I did not treat in *Future Logic*. At the time I wrote that book, I did not take time to think about it. However, I can predict that the properties of this mode should be very similar to those of extensional modality, just as temporal modality is akin to natural (or circumstantial) modality. Spatial and temporal modality should behave in similar ways in various respects.

[3] Be it said in passing, these specific rules, mentioned here for purposes of illustration, though seemingly true for natural causation, are certainly not relevant in the extensional or logical modes of causation. Indeed, it is no longer sure that a 'contiguity principle' applies universally even to natural causation: recent discoveries by physicists may suggest the existence of 'instant action at a distance' between pairs of particles, which seemingly goes against Relativity Theory prediction since the limit of the speed of light is not maintained. Whatever the theoretical physics outcome of such discoveries, the current question mark demonstrates that logic theory must remain open in such issues; i.e. principles

Indeed, it is in some cases difficult for us, if not impossible, to say which of the two phenomena is the cause and which is the effect. And this often is not only an epistemological issue, but more deeply an ontological one. For, though there is sometimes a **direction of causation** to specify, there is often in fact no basis for such a specification. The phenomena named 'cause' and 'effect' are in a reciprocal relation of causation; the terms cause and effect are in such cases merely verbal distinctions. All that we can say is that the phenomena are bound together, and either can be accessed through the other; the labels applied to them become a matter of convenience for purposes of discourse.

Finally, the term "**invariably**" has to be stressed. How such constancy is established is not the issue here; we shall consider that elsewhere. In the paradigm of causation given above, it would not do for the conjunction of the cause and effect, or the conjunction of their negations, to be merely occasional. We would not regard such varying conjunctions as signifying genuine causation, but quite the opposite as signs of mere coincidence, happenstance of togetherness. *Post hoc ergo propter hoc.* The problem is complicated in lesser determinations of causation; but as we shall see it can be overcome, a constancy of conjunction or of non-conjunction is always ultimately involved.

In this context, a warning is in order. When something is invariably accompanied by another, we say that the first (the presence or absence of the cause) "is followed by" the second (the presence or absence of the effect). This refers to causal sequence and should not be confused with temporal sequence; the term "followed" is ambivalent (indeed, it is also used in relation to spatial or numerical series). Even though causal and temporal sequence are often both involved (which is why the term "to follow" is equivocal), causal sequence may occur without temporal sequence (even in natural causation) or in a direction opposite to temporal sequence (though supposedly not in natural causation, certainly in logical causation, and by abstraction of the time factor also in extensional causation). The context usually makes the intent clear, of course.

Now, for some formal analysis:

In our present treatment of causation, we shall focus principally on the logical 'mode' of causation, note well. There are (as we shall later discuss) other modes, notably the natural, the temporal, the spatial and the extensional, whose definitions differ with respect to the type of modality considered. Having investigated modality and conditioning in detail in a previous treatise (*Future Logic*, 1990), I can predict that most of the behavior patterns of logical causation are likely to be found again in the other modes of causation; but also, that some significant differences are bound to arise.

Returning now to the paradigm of causation, it may be expressed more symbolically as follows, using the language of logical conditioning (as developed in my *Future Logic*, Part III):

> **If C, then E; and**
> **if notC, then notE.**

A sentence of the form "If P, then Q" means "the conjunction of P and the negation of Q is impossible", i.e. there are no knowledge-contexts where this conjunction (P + notQ) credibly occurs. Such a proposition can be recast in the contraposite form "If notQ, then notP", which

---

like that of contiguity must be regarded as generalizations which might be abandoned if the need to do so is found overwhelming.

means "the conjunction of notQ and the negation of notP is impossible" – the same thing in other words.

Such a proposition, note, does not formally imply that P is possible or that notQ is possible. Normally, we do take it for granted that such a proposition may be realized, i.e. that P is possible, and therefore (by apodosis) Q is possible and the conjunction "P and Q" is possible; and likewise that notQ is possible, and therefore (by apodosis) notP is possible and the conjunction "notQ and notP" is possible.

However, in some cases such assumption is unjustified. It may happen that, though "If P, then Q" is true, P is impossible, in which case "If P, then notQ" must also be true; or it may happen that, though "if P, then Q" is true, notQ is impossible, in which case "If notP, then Q" must also be true. These results are paradoxical, yet quite logical. I will not go into this matter in detail here, having dealt with it elsewhere (see *Future Logic*, ch. 31). It is not directly relevant to the topic under discussion, except that it must be mentioned to stress that such paradox cannot occur in the context of causation (except to deny causation, of course).

Therefore, when discussing causation, it is tacitly understood that:

C is contingent and E is contingent[4].

That is, each of C, E is possible but unnecessary; likewise, by obversion, for their negations, each of notC, notE is possible but unnecessary. If any of these positive or negative terms is by itself necessary or impossible, it is an antecedent or consequent in valid (and possibly true) propositions, but it is not a cause or effect within the causation specified. This is, by the way, one difference in meaning between the expressions cause/effect, and the expressions antecedent/consequent. We shall see, as we deal with lesser determinations of causation, that their meanings diverge further. All the more so, when the terms cause/effect are used in other forms of causality.

Furthermore, as above shown with reference to "P" and "Q", granting the contingencies of C and E, each of the propositions "If C, then E" and "If notC, then notE" implies the following possibilities:

The conjunction (C + E) is possible; and
the conjunction (notC + notE) is possible.

All this is hopefully clear to the reader. But we must eventually consider its implications with reference to statements dealing with lesser determinations of causation or statements denying causation.

---

4 To avoid any confusion, we should add "in the type of modality characterizing the causal relation". But this specification would be incomprehensible to most readers, as the issue of mode of causation is dealt with in a later chapter.

Chapter 2. THE GENERIC DETERMINATIONS

1. Strong Determinations.

The strongest determination of causation, which we identified as the paradigm of causation, may be called complete and necessary causation. We shall now repeat the three constituent propositions of this form and their implications, all of which must be true to qualify:

(i) If C, then E;
(ii) if notC, then notE;
(iii) where: C is contingent and E is contingent.

As we saw, these propositions together imply the following:

The conjunction (C + E) is possible;
the conjunction (notC + notE) is possible.

Clauses (i) and (iii) signify complete causation. With reference to this positive component, we may call C a **complete cause** of E and E a **necessary effect** of C. Where there is complete causation, the cause is said to ***make necessary*** (or necessitate) the effect[5]. This signifies that the presence of C is *sufficient* (or enough) for the presence E.
Clause (ii) and (iii) signify necessary causation. With reference to this negative component, we may call C a **necessary cause** of E and E a **dependent effect** of C. Where there is necessary causation, the cause is said to ***make possible*** (or be necessitated by) the effect. This signifies that the presence of C is *requisite* (or indispensable) for the presence E[6].
Clause (iii) is commonly left tacit, though as we saw it is essential to ensure that the first two clauses do not lead to paradox. Strictly speaking, it would suffice, given (i), to stipulate that C is possible (in which case so is E) and E is unnecessary (in which case so is C). Or equally well, given (ii), that C is unnecessary (in which case so is E) and E is possible (in which case so is C). The possibilities of the conjunctions (C + E) and (notC + notE), logically follow, and so need not be included in the definition.

Looking at the paradigm, we can identify two distinct lesser determinations of causation, which as it were split the paradigm in two components, *each of which by itself conforms to the paradigm* through an ingenuous nuance, as shown below.
Also below, I list the various clauses of each definition, renumbering them for purposes of reference. Then a table is built up, including all the causal and effectual items involved (positive and negative) and all their conceivable combinations[7]. The *modus* of each item or

5 The expression "X makes Y *impossible*" means that X makes notY necessary, incidentally.

6 We commonly say, in such case, that C is a ***sine qua non*** (Latin for 'without which not') or *proviso* of E.

7 I use the word 'item' to refer to a cause or effect (or the negation of a cause or effect), indifferently. An item is, thus, for the logician, primarily a *thesis* (in the largest sense), i.e. a categorical

combination, i.e. whether it is defined or implied as possible or impossible, or left open, is then identified. In each case, the *source* of such modus is noted, i.e. whether it is given or derivable from given(s).

Complete causation:

(i) **If C, then E;**
(ii) **if notC, not-then E**;
(iii) where: **C is possible.**

Table 2.1. Complete causation.				
No.	Element/compound		Modus	Source/relationship
1	**C**		**possible**	**(iii)**
2	notC		possible	implied by (ii)
3		E	possible	implied by (i) + (iii)
4		notE	possible	implied by (ii)
5	C	E	possible	implied by (i) + (iii)
6	**C**	**notE**	**impossible**	**(i)**
7	notC	E	*open*	
8	**notC**	**notE**	**possible**	**(ii)**

Complete causation conforms to the paradigm of causation by means of the same main clause (i); whereas its clause (ii), note well, concerning what happens in the absence of C, substitutes for the invariable absence of E (i.e. "then notE"), the *not*-invariable *presence* of E (i.e. "not-then E"). However, remember, contraposition of (i) implies that "If notE, then notC", meaning that in the absence of E we can be sure that C is also absent[8].

Clause (ii) means that (notC + notE) is possible, so we are sure from it that C is unnecessary and E is unnecessary; also it teaches us that C and E cannot be exhaustive. Technically, it would suffice for us to know that notE is possible, for we could then infer clause (ii) from (i); but it is best to specify clause (ii) to fit the paradigm of causation. As for clause (iii), we need only specify that C is possible; it follows from this and clause (i) that (C + E) is possible and so that E is also possible.

Note well the nuance that, to establish such causation, the effect has to be found *invariably* present in the presence of the cause, otherwise we would commit the fallacy of *post hoc ergo propter hoc*; but the effect need not be invariably absent in the absence of the cause: it suffices for the effect *not to be* invariably present.

The segment of the above table numbered 5-8 (shaded) may be referred to as the *matrix* of complete causation. It considers the possibility or impossibility of all conceivable conjunctions of all the items involved in the defining clauses or the negations of these items.

Necessary causation:

---

or other form of proposition. But an item may also signify a *term*, since theses are ultimately predications. An item, then, is a thesis, or term within a thesis, involved in a causal proposition.

8 In some but not all cases, notE not only implies but causes notC, note.

(i) **If notC, then notE;**
(ii) **if C, not-then notE;**
(iii) where: **C is unnecessary**.

Table 2.2. Necessary causation				
No.	Element/compound		Modus	Source/relationship
1	C		possible	implied by (ii)
2	**notC**		**possible**	**(iii)**
3		E	possible	implied by (ii)
4		notE	possible	implied by (i) + (iii)
5	**C**	**E**	**possible**	**(ii)**
6	C	notE	*open*	
7	**notC**	**E**	**impossible**	**(i)**
8	notC	notE	possible	implied by (i) + (iii)

Necessary causation conforms to the paradigm of causation by means of the same main clause (i)[9]; whereas its clause (ii), note well, concerning what happens in the presence of C, substitutes for the invariable presence of E (i.e. "then E"), the *not*-invariable *absence* of E (i.e. "not-then notE"). However, remember, contraposition of (i) implies that "If E, then C", meaning that in the presence of E we can be sure that C is also present[10].
Clause (ii) means that (C + E) is possible, so we are sure from it that C is possible and E is possible; also it teaches us that C and E cannot be incompatible. Technically, it would suffice for us to know that E is possible, for we could then infer clause (ii) from (i); but it is best to specify clause (ii) to fit the paradigm of causation. As for clause (iii), we need only specify that C is unnecessary; it follows from this and clause (i) that (notC + notE) is possible and so that E is also unnecessary.
Note well the nuance that, to establish such causation, the effect has to be found *invariably* absent in the absence of the cause, otherwise we would commit the fallacy of *post hoc ergo propter hoc*; but the effect need not be invariably present in the presence of the cause: it suffices for the effect *not to be* invariably absent.
Note the *matrix* of necessary causation, i.e. the segment of the above table numbered 5-8 (shaded).

Lastly, notice that complete and necessary causation are *'mirror images'* of each other. All their characteristics are identical, except that the polarities of their respective cause and effect opposite: C is replaced by notC, and E by notE, or vice-versa. The one represents the positive aspect of strong causation; the other, the negative aspect. Accordingly, their logical properties correspond, *mutatis mutandis* (i.e. if we make all the appropriate changes).

---

9 Notice that clause (i), here, in necessary causation, was labeled as clause (ii) in complete and necessary causation. The numbering is independent.
10 In some but not all cases, E not only implies but causes C, note.

Following the preceding analysis of necessary and complete causation into two distinct components each of which independently conforms to the paradigm, we can conceive of complete causation *without* necessary causation and necessary causation *without* complete causation. These two additional determinations of causation are conceivable, note well, only because they do not infringe logical laws; that is, we already know that the various propositions that define them are individually and collectively logically compatible.

2. Parallelism of Strongs.

Before looking into weaker determinations of causation, we must deal with the phenomenon of *parallelism*.

The definition of complete causation does not exclude that there be some cause(s) other than C – such as say C_1 – having the same relation to E. In such case, C and C_1 may be called *parallel complete causes* of E. The minimal relation between such causes is given by the following normally valid 2nd figure syllogism (see *Future Logic*, p.162):

> If C, then E (and if notC, not-then E / and C is possible);
> and if C_1, then E (and if notC_1, not-then E / and C_1 is possible);
> therefore, if notC_1 not-then C (= if notC, not-then C_1 – by contraposition).

The possibility of parallel complete causes is clear from the logical compatibility of these premises, which together merely imply that in the absence of E both C and C_1 are absent. The main clauses of the premises can be merged in a compound proposition of the form "If notE, then neither C nor C_1", which by contraposition yields "If C or C_1, then E". Thus, such parallel causes may be referred to as 'alternative' complete causes (in a large sense of the term 'alternative').

Since the conclusion of the above syllogism is subaltern to each of the propositions "if notC_1, then notC" and "if notC, then notC_1", it may happen that C implies C_1 and/or C_1 implies C – but they need not do so. Likewise, since the conclusion is compatible with the proposition "if C_1, then notC" or "if C, then notC_1", it may happen that C and C_1 are incompatible with each other – but they do not have to be. The conclusion merely specifies that C and C_1 not be exhaustive (i.e. be neither contradictory nor subcontrary; this is the sole formal specification of the disjunction in "If C or C_1, then E").

Similarly, still in complete causation, E need not be the exclusive necessary effect of C; there may be some other thing(s) – such as say E_1 – which invariably follow C, too. In such case, E and E_1 may be called *parallel necessary effects* of C. The minimal relation between such effects is given by the following normally valid 3rd figure syllogism (see *Future Logic*, pp. 162-164):

> If C, then E (and if notC, not-then E_1 / and C is possible);
> and if C, then E_1 (and if notC, not-then E_1 / and C is possible);
> therefore, if E_1, not-then notE (= if E, not-then notE_1 – by contraposition).

The possibility of parallel necessary effects is clear from the logical compatibility of these premises, which together merely imply that in the presence of C both E and E_1 are present.

The main clauses of the premises can be merged in a compound proposition of the form "If C, then both E and E_1". Thus, such parallel effects may be said to be 'composite' necessary effects.
Since the conclusion of the above syllogism is subaltern to each of the propositions "if E_1, then E" and "if E, then E_1", it may happen that E_1 implies E and/or E implies E_1 – but they need not do so. Likewise, since the conclusion is compatible with the proposition "if notE_1, then E" or "if notE, then E_1", it may happen that E and E_1 are exhaustive – but they do not have to be. The conclusion merely specifies that E and E_1 not be incompatible (i.e. be neither contradictory nor contrary).

Again, *mutatis mutandis*, the definition of necessary causation does not exclude that there be some cause(s) other than C – such as say C_1 – having the same relation to E. In such case, C and C_1 may be called *parallel necessary causes* of E. The minimal relation between such causes is given by the following normally valid 2nd figure syllogism (see *Future Logic*, p. 162):

> If notC, then notE (and if C, not-then notE / and notC is possible);
> and if notC_1, then notE (and if C_1, not-then notE / and notC_1 is possible);
> therefore, if C_1, not-then notC (= if C, not-then notC_1 by contraposition).

The possibility of parallel necessary causes is clear from the logical compatibility of these premises, which together merely imply that in the presence of E both C and C_1 are present. The main clauses of the two premises can be merged in a compound proposition of the form "If E, then both C and C_1", which by contraposition yields "If notC or notC_1, then notE". Thus, such parallel causes may be referred to as 'alternative' necessary causes (in a large sense of the term 'alternative').
Since the conclusion of the above syllogism is subaltern to each of the propositions "if C_1, then C" and "if C, then C_1", it may happen that C_1 implies C and/or C implies C_1 – but they need not do so. Likewise, since the conclusion is compatible with the proposition "if notC_1, then C" or "if notC, then C_1", it may happen that C and C_1 are exhaustive – but they do not have to be. The conclusion merely specifies that C and C_1 not be incompatible (i.e. be neither contradictory nor contrary; this is the sole formal specification of the disjunction in "If notC or notC_1, then notE").
Similarly, still in necessary causation, E need not be the exclusive dependent effect of C; there may be some other thing(s) – such as say E_1 – which are invariably preceded by C, too. In such case, E and E_1 may be called *parallel dependent effects* of C. The minimal relation between such effects is given by the following normally valid 3rd figure syllogism (see *Future Logic*, p. 162-164):

> If notC, then notE (and if C, not-then notE / and notC is possible);
> and if notC, then notE_1 (and if C, not-then notE_1 / and notC is possible);
> therefore, if notE_1, not-then E (= if notE, not-then E_1 by contraposition).

The possibility of parallel dependent effects is clear from the logical compatibility of these premises, which together merely imply that in the absence of C both E and E_1 are absent. The main clauses of the premises can be merged in a compound proposition of the form "If notC,

then neither E nor E_1". Thus, such parallel effects may be said to be 'composite' dependent effects.
Since the conclusion of the above syllogism is subaltern to each of the propositions "if notE_1, then notE" and "if notE, then notE_1", it may happen that E implies E_1 and/or E_1 implies E – but they need not do so. Likewise, since the conclusion is compatible with the proposition "if E_1, then notE" or "if E, then notE_1", it may happen that E and E_1 are incompatible with each other – but they do not have to be. The conclusion merely specifies that E and E_1 not be exhaustive (i.e. be neither contradictory nor subcontrary).

It happens that parallel causes or parallel effects are themselves causally related. That this is possible, is implied by what we have seen above. Since each of the following pairs of items may have any formal relation with one exception, namely:

- parallel complete causes cannot be exhaustive (since "if notC, not-then C_1" is true for them); and parallel necessary effects cannot be incompatible (since "if E, not-then notE_1" is true for them);
- parallel necessary causes cannot be incompatible (since "if C, not-then notC_1" is true for them); and parallel dependent effects cannot be exhaustive (since "if notE, not-then E_1" is true for them);

... it follows that either one of parallel causes C and C_1 may be a complete or necessary cause of the other; and likewise, either one of parallel effects E and E_1 may be a complete or necessary cause of the other.
In certain situations, as we shall see in a later chapter, it is possible to infer such causal relations between parallels. But, it must be stressed, the mere fact of parallelism does not in itself imply such causal relations.

In sum, complete and/or necessary causation should not be taken to imply *exclusiveness* (i.e. that a unique cause and a unique effect are involved); such relation(s) allow for *plurality* of causes or effects in the sense of parallelism as just elucidated.
Indeed, it is very improbable that we come across exclusive relations in practice, since every existent has many facets, each of which might be selected as cause or effect. Our focusing on this or that aspect as most significant or essential, is often arbitrary, a matter of convenience; though often, too, it is guided by broader considerations, which may be based on intuition of priorities or complicated reasoning.
In any case, it is important to distinguish plurality arising in strong causation, which signifies alternation of causes or composition of effects, as above, from plurality arising in weak causation, which signifies composition of causes or alternation of effects, which we shall consider in the next section.

3. Weak Determinations.

Having clarified the complete and necessary forms of causation, as well as parallelism, we are now in a position to deal with lesser determinations of causation. Let us first examine partial causation; contingent causation will be dealt with further on.

Partial causation:

(i) **If (C1 + C2), then E;**
(ii) **if (notC1 + C2), not-then E;**
(iii) **if (C1 + notC2), not-then E;**
(iv) where: **(C1 + C2) is possible.**

Table 2.3. Partial causation.

No.	Element/compound			Modus	Source/relationship
1	C1			possible	implied by (iii) or (iv)
2	notC1			possible	implied by (ii)
3		C2		possible	implied by (ii) or (iv)
4		notC2		possible	implied by (iii)
5			E	possible	implied by (i) + (iv)
6			notE	possible	implied by (ii) or (iii)
7	C1		E	possible	implied by (i) + (iv)
8	C1		notE	possible	implied by (iii)
9	notC1		E	*open*	
10	notC1		notE	possible	implied by (ii)
11		C2	E	possible	implied by (i) + (iv)
12		C2	notE	possible	implied by (ii)
13		notC2	E	*open*	
14		notC2	notE	possible	implied by (iii)
15	**C1**	**C2**		**possible**	**(iv)**
16	C1	notC2		possible	implied by (iii)
17	notC1	C2		possible	implied by (ii)
18	notC1	notC2		*open*	
19	C1	C2	E	possible	implied by (i) + (iv)
20	**C1**	**C2**	**notE**	**impossible**	**(i)**
21	C1	notC2	E	*open*	
22	**C1**	**notC2**	**notE**	**possible**	**(iii)**
23	notC1	C2	E	*open*	
24	**notC1**	**C2**	**notE**	**possible**	**(ii)**
25	notC1	notC2	E	*open*	
26	notC1	notC2	notE	*open*	

Two phenomena C1, C2 may be called **partial causes** of some other phenomenon E, only if *all* the above conditions (i.e. the four defining clauses) are satisfied. In such case, we may call E a **contingent effect** of each of C1, C2. Of course, the *compound* (C1 + C2) is a complete cause of E, since in its presence, E follows (as given in clause (i)); and in its absence, i.e. if not(C1 + C2), E does not invariably follow (as evidenced by clauses (ii) and (iii)). Rows 19-26 of the above table (shaded) constitute the matrix of partial causation.

We may thus speak of this phenomenon as a composition of partial causes; and stress that C1 and C2 belong in that particular causation of E by calling them *complementary* partial causes of it. Indeed, instead of saying "C1 and C2 are complementary partial causes of E", we may equally well formulate our sentence as "C1 (complemented by C2) is a partial cause of E" or as "C2 (complemented by C1) is a partial cause of E". These three forms are identical, except for that the first treats C1 and C2 with equal attention, whereas the latter two lay stress on one or the other cause. Such reformatting, as will be seen, is useful in some contexts.

We may make a distinction between *absolute* and *relative* partial causation, as follows. The 'absolute' form specifies one partial cause without mentioning the complement(s) concerned; it just says: "C1 is a partial cause of E", meaning "C1 (*with some unspecified complement*) is a partial cause of E". This is in contrast to the 'relative' form, which does specify a complement, as in the above example of "C1 (complemented by C2) is a partial cause of E". This distinction reflects common discourse. Its importance will become evident when we consider negations of such forms.

One way to see the appropriateness of our definition of partial causation, *its conformity to the paradigm of causation*, is by resorting to nesting (see *Future Logic*, p. 148). We may rewrite it as follows:

From (i) if C2, then (if C1, then E);
from (ii) if C2, then (if notC1, not-then E);
from (iii) if notC2, not-then (if C1, then E).[11]

Clause (i) tells us that given C2, C1 implies E. Clause (ii) tells us that given C2, notC1 does not imply E. Thus, under condition C2, C1 behaves like a complete cause of E. Moreover, clause (iii) shows that under condition notC2, C1 ceases to so behave. Similarly, *mutatis mutandis*, C2 behaves conditionally like a complete cause of E.[12]

Let us now examine the definition of partial causation more closely. The terminology adopted for it is obviously intended to contrast with that for complete causation.

Clause (i) informs us that in the presence of the two elements C1 and C2 together, the effect is invariably also present. However, that clause alone would not ensure that both C1 and C2 are *relevant* to E, *participants* in its causation. We need clause (ii) to establish that without C1, C2 would not by itself have the same result. And, likewise, we need clause (ii) to establish that without C2, C1 would not by itself have the same result.

Suppose, for instance, clause (ii) were false; then, combining it with (i), we would obtain the following simple dilemma:

If (C1 + C2), then E – and – if (notC1 + C2), then E;
therefore, if C2, then E.

That is, C2 would be a complete cause of E, without need of C1, which would in such case be *an accident* in the relation "If (C1 + C2), then E", note well. Similarly, if clause (iii) were false, it would follow that C1 is sufficient by itself for E, irrespective of C2. In the special case where both (ii) and (iii) are denied, C1 and C2 would be parallel complete causes of E

11 These three forms are implied, respectively, by our first givens; but they do not imply them unconditionally.

12 We can also, incidentally, view the matter as follows, by focusing on the nested clauses. Clauses (i) and (iii) mean that the partial cause C2 of E may be regarded as a complete cause of the new effect "if C1, then E". Similarly, *mutatis mutandis*, clauses (i) and (ii) can be taken to mean that C1 is a complete cause of "if C2, then E".

(compatible ones, since they are conjoined in the antecedent of clause (i)). Therefore, as well as clause (i), clauses (ii) and (iii) have to specified for partial causation.
Furthermore, our definition of partial causation thus mentions three combinations of C1, C2 and their respective negations, namely:

- C1 + C2
- notC1 + C2
- C1 + not C2

And it tells us what happens in relation to E in each of these situations: in the first, E follows; in the next two, it does not. One might reasonably ask, what about the fourth combination, namely:

- notC1 + notC2?[13]

Well, for that, there are only two possibilities: either E follows or it does not. Note first that both these possibilities are logically compatible with clauses (i), (ii) and (iii).
Suppose that "If (notC1 + notC2), then E" is true. In that case, notC1 and notC2 would each have the same relation to E that C1 and C2 have by virtue of clauses (i), (ii), (iii). For if we combine this supposed additional clause with clauses (ii) and (iii), we see that, whereas E follows the conjunction of notC1 and notC2, E does not follow the conjunction of not(notC1) with notC2 or that of notC1 with not(notC2). In that case, we would simply have two, instead of just one, compound causes of E, namely (C1 + C2) and (notC1 + notC2), sharing the same clauses (ii) and (iii) which establish the relevance of each of the elements. Though at first sight surprising, such a state of affairs is quite conceivable, being but a special case of parallel causation! Thus, the proposition "If (notC1 + notC2), then E" may well be true. But may it be false? Suppose that its contradictory "If (notC1 + notC2), not-then E" is true, instead. Here again, the causal significance of the first three clauses remains unaffected. We can thus conclude that what happens in the situation "notC1 + notC2", i.e. whether E follows or not, is irrelevant to the roles played by C1 and C2. Our definition of partial causation through the said three clauses is thus satisfactory.
Lastly the following should be noted. If we replaced clauses (ii) and (iii) by "If not(C1 + C2), not-then E", to conform with clause (i) to the definition of complete causation, we would only be sure that the compound (C1 + C2) causes E. It does not suffice to establish that both its elements are involved in that causation, since it could be adequately realized by the eventuality that "If (notC1 + notC2), not-then E". For this reason, too, clauses (ii) and (iii) are unavoidable.
Regarding clause (iv), which serves to ensure that the first three clauses do not lead to paradox, it is easy to show that the possibility of the conjunction (C1 + C2) is the minimal requirement. For this through clause (i) implies that E is possible and (C1 + C2 + E) is possible. Additionally, clause (ii) means that (notC1 + C2 + notE) is possible, and therefore implies that (notC1 + C2) is possible and each of notC1, C2, notE is possible. Similarly, clause

13 Note that the combination "notC1 + notC2" may occasionally be impossible. In such case, notC1 implies C2 and notC2 implies C1. But according to syllogistic theory (see *Future Logic*, pp. 158-160), this would *not* allow us to abbreviate clauses (ii) and (iii) of the definition to "If notC1, not-then E" and "If notC2, not-then E". Thus, even in such case, the definition remains unaffected.

(iii) means that (C1 + notC2 + notE) is possible, and therefore implies that (C1 + notC2) is possible and each of C1, notC2, notE is possible. It is thus redundant to specify these various contingencies.

The **methodological principle** underlying the definition of partial causation is well known to scientists and oft-used. It is that *to establish the causal role of any element such as C1, of a compound (C1 + C2...) in whose presence a phenomenon E is invariably present, we must find out what happens to E when the element C1 is absent while all other elements like C2 remain present*. That is, we observe how the putative effect is affected by removal of the putative cause ***while keeping all other things equal***[14]. Only if a change in status occurs (minimally from "then E" to "not-then E"), may the element be considered as participating in the causation, i.e. as a relevant *factor*.
Once this is understood, it is easy to generalize our definition of partial causation from two factors (C1, C2) to any number of them (C1, C2, C3...), as follows:

(i) If (C1 + C2 + C3...), then E;
(ii) if (notC1 + C2 + C3...), not-then E;
(iii) if (C1 + notC2 + C3...), not-then E;
(iv) if (C1 + C2 + notC3...), not-then E;
...etc. (if more than three factors);
and (C1 + C2 + C3...) is possible.

Clause (i) establishes the complete causation of the effect E by the compound (C1 + C2 + C3...). But additionally there has to be for each element proof that its absence would be felt: this is the role of clauses (ii), (iii), (iv)..., each of which negates *one and only one* of the elements concerned. Thus, the number of additional clauses is equal to the number of factors involved.
Whatever the relation to E of other possible combinations of the elements and their negations, the partial causation of E by elements C1, C2, C3... is settled by the minimum number of clauses specified in our definition. As we saw, with two factors the combination "notC1 + notC2" is not significant. Similarly, we can show that with three factors the following combinations are not significant:

- notC1 + notC2 +C3
- notC1 + C2 + notC3
- C1 + notC2 + notC3
- notC1 + notC2 + notC3

And so forth. Generally put, if the number of elements is n, the number of insignificant combinations will be is $2^n - (1 + n)$. Whether any of these further combinations implies or does not imply E does not affect the role of partial causation signified by the defining clauses for the factors C1, C2, C3... per se. Other causations may be involved in certain cases, but they do not disqualify or diminish those so established.

[14] This phrase "keeping all other things equal" is not mine – but a consecrated phrase often found in textbooks. I do not know who coined it first.

The very last clause, that (C1 + C2 + C3...) is possible, is required and sufficient, for reasons already seen.
Clearly, we can say that *the more factors are involved, the weaker the causal bond.* If C is a complete cause of E, it plays a big role in the causation of E. If C1 is a partial cause of E, with one complement C2, it obviously plays a lesser role than C. Similarly, the more complements C1 has, like C2, C3..., the less part it plays in the whole causation of E. We may thus view the degree of determination involved as inversely proportional to the number of causes involved, though we may (note well) be able to assign different *weights* to the various partial causes[15].
Note finally that we can facilitate mental assimilation of multiple (i.e. more than two) partial causes through successive reductions to pairs of partial causes, one of which is compound. Thus, (C1 + C2 + C3 + ...) may be viewed as (C1 + (C2 + C3 +...)), *provided* all the above mentioned conditions are entirely satisfied.

Let us now turn our attention to contingent causation.

Contingent causation:

(i) **If (notC1 + notC2), then notE;**
(ii) **if (C1 + notC2), not-then notE;**
(iii) **if (notC1 + C2), not-then notE;**
(iv) where: **(notC1 + notC2) is possible.**

Table 2.4. Contingent causation.

No	Element/compound			Modus	Source/relationship
1	C1			possible	implied by (ii)
2	notC1			possible	implied by (iii) or (iv)
3		C2		possible	implied by (iii)
4		notC2		possible	implied by (ii) or (iv)
5			E	possible	implied by (ii) or (iii)
6			notE	possible	implied by (i) + (iv)
7	C1		E	possible	implied by (ii)
8	C1		notE	*open*	
9	notC1		E	possible	implied by (iii)
10	notC1		notE	possible	implied by (i) + (iv)
11		C2	E	possible	implied by (iii)
12		C2	notE	*open*	
13		notC2	E	possible	implied by (ii)
14		notC2	notE	possible	implied by (i) + (iv)
15	C1	C2		*open*	
16	C1	notC2		possible	implied by (ii)
17	notC1	C2		possible	implied by (iii)
18	**notC1**	**notC2**		**possible**	**(iv)**

15 For instance, with reference to concomitant variations (see Appendix on J. S. Mill's Methods); if the C1 and C2 enter in a mathematical formula like, say, $E = C1^2 + C2$, C1 has less weight than C2.

19	C1	C2	E	*open*	
20	C1	C2	notE	*open*	
21	**C1**	**notC2**	**E**	**possible**	**(ii)**
22	C1	notC2	notE	*open*	
23	**notC1**	**C2**	**E**	**possible**	**(iii)**
24	notC1	C2	notE	*open*	
25	**notC1**	**notC2**	**E**	**impossible**	**(i)**
26	notC1	notC2	notE	possible	implied by (i) + (iv)

Two phenomena C1, C2 may be called **contingent causes** of some other phenomenon E, only if *all* the above conditions (i.e. the four defining clauses) are satisfied. In such case, we may call E a **tenuous effect**[16] of each of C1, C2. Of course, the *compound* (notC1 + notC2) is a necessary cause of E, since in its presence, notE follows (as given in clause (i)); and in its absence, i.e. if not(notC1 + notC2), notE does not invariably follow (as evidenced by clauses (ii) and (iii)). Rows 19-26 of the above table (shaded) constitute the matrix of contingent causation.

We may thus speak of this phenomenon as a composition of contingent causes; and stress that that C1 and C2 belong in that particular causation of E by calling them *complementary* contingent causes of it. Indeed, instead of saying "C1 and C2 are complementary contingent causes of E", we may equally well formulate our sentence as "C1 (complemented by C2) is a contingent cause of E" or as "C2 (complemented by C1) is a contingent cause of E". These three forms are identical, except for that the first treats C1 and C2 with equal attention, whereas the latter two lay stress on one or the other cause. Such reformatting, as will be seen, is useful in some contexts.

We may make a distinction between *absolute* and *relative* contingent causation, as follows. The 'absolute' form specifies one contingent cause without mentioning the complement(s) concerned; it just says: "C1 is a contingent cause of E", meaning "C1 (*with some unspecified complement*) is a contingent cause of E". This is in contrast to the 'relative' form, which does specify a complement, as in the above example of "C1 (complemented by C2) is a contingent cause of E". This distinction reflects common discourse. Its importance will become evident when we consider negations of such forms.

Here again, we can demonstrate that our definition of contingent causation *conforms to the paradigm of causation* through nesting. We may rewrite it as follows:

From (i) if notC2, then (if notC1, then notE);
from (ii) if notC2, then (if C1, not-then notE);
from (iii) if C2, not-then (if notC1, then notE).

Clause (i) tells us that given notC2, notC1 implies notE. Clause (ii) tells us that given notC2, C1 does not imply notE. Thus, under condition notC2, C1 behaves like a necessary cause of E.

---

[16] I use the name "tenuous effect" for lack of a better one, to signify a lesser degree of non-independence than a "dependent effect". Alternatively, broadening the connotation of dependence, we might say that the effect of a necessary cause is strongly dependent (it depends on that one cause) and the effect of a contingent cause is weakly dependent (it depends on that cause, if no other is available).

Moreover, clause (iii) shows that under condition C2, C1 ceases to so behave. Similarly, *mutatis mutandis*, C2 behaves conditionally like a necessary cause of E.

Note well that the main clause of contingent causation is not "If not(C1 + C2), then notE"[17], but more specifically "If (notC1 + notC2), then notE". Considering that in partial causation the antecedent is (C1 + C2) and that this compound behaves as a complete cause, one might think that in contingent causation the antecedent would be a negation of the same compound, i.e. not(C1 + C2), which would symmetrically behave as a necessary cause. But the above demonstration of conformity to paradigm shows us that this is not the case. The explanation is simply that two of the alternative expressions of "If not(C1 + C2), then notE", namely "If (C1 + notC2), then notE" and "If (notC1 + C2), then notE" are contradictory to clauses (ii) and (iii), respectively. Therefore, only "If (notC1 + notC2), then notE" is a formally appropriate expression in this context. Our definition of contingent causation is thus correct.

We need not repeat our further analysis of partial causation for contingent causation; all that has been said for the former can be restated, *mutatis mutandis*, for the latter. For partial and contingent causation are *'mirror images'* of each other. The one represents the positive aspect of weak causation; the other, the negative aspect. All their characteristics are identical, except that the polarities of their respective causes and effect are opposite: C1 is replaced by notC1, C2 by notC2, and E by notE, or vice-versa.

Note that partial and contingent causation each involves a *plurality* of causes, though *in a different sense* from that found in parallelism.

We should also mention that partial causation often underlies *alternation* or plurality of effects.

Consider the form "If C, then (E or E_1)", which may be interpreted as "the conjunction (C + notE + notE1) is impossible", and therefore implies "If (C + notE), then E_1" and "If (C + notE_1), then E". Take the latter, for instance, and you have a type (i) clause. If additionally it is true that (notC + notE_1 + notE), (C + E_1 + notE), (C + notE_1) are possible conjunctions, you have clauses of types (ii), (iii) and (iv), respectively. In such case[18], C is a partial cause of E (the other partial cause being notE_1 or, more precisely, some complete and necessary cause of notE_1).

Just as we may have plurality of effects in partial causation, so we may have it in contingent causation.

Note, concerning the term '*occasional*'. When parallel complete causes may occur separately (i.e. neither implies the other), they are often called occasional causes; however, note well, the same term is often used to refer to partial causes, in the sense that each of them is effective only when the other(s) is/are present. The term occasional effect is used with reference to alternation of effects; i.e. when a cause has alternative effects, each of the latter is occasional; but the term is also applicable more generally, to any effect of a partial cause as such, i.e. to contingent effects.

Partial and contingent causation may conceivably occur in tandem or separately; i.e. no formal inconsistency arises in such cases.

---

17 This form, note well, does not specify which of the three alternative combinations (C1 + notC2), (notC1 + C2) or (notC1 + notC2) implies notE; it means only that *at least one of them* does.

18 As can be seen by renaming C as "C1" and notE1 as "C2".

4. Parallelism of Weaks.

Before going further let us here deal with parallelism in relation to the weaker determinations of causation.

In partial causation, this would mean, that there are two (or more) sets of two (or more) partial causes, viz. C1, C2... and C3, C4... (and so forth), with the same effect E:

If (C1 + C2...), then E; etc.
If (C3 + C4...), then E; etc.
...

Clearly, we have 'plurality of causes' in *both* senses of the term at once, here. By "etc.", I refer to the further clauses involved in partial causation, such as "if (C1 + notC2), not-then E" and so on, here left unsaid to avoid repetitions. Such statements may be merged; thus, the above two become a single statement in which each bracketed conjunction constitutes an alternative complete cause:

If (C1 + C2...) or (C3 + C4...) or..., then E; etc.

The bracketed conjunctions, as we have seen when dealing with parallel complete causes, may be interrelated in various ways except be exhaustive. These interrelations would be expressed in additional statements. The resulting information, including the above statement where all the conjunctions are disjoined in a single antecedent and all statements not explicitated[19] here, can then be analyzed in great detail by tabulating all the items and their negations, and considering the modus of each combination. We can, in this way, have a clear picture of all eventualities, and avoid all ambiguity.

Similarly, *mutatis mutandis*, for contingent causation:

> If (notC1 + notC2...), then notE; etc.
> If (notC3 + notC4...), then notE; etc.
> ...

We may merge these complex causal statements, consider additional specifications regarding the opposition of alternatives, and analyze the mass of information through a table.

Note the following special cases of the above parallelisms.

A partial cause may be found *common* to two (or more) such causations with the same effect; if say C3 is identical with C1, C1 would have C2... as complement(s) in the first relation and C4... as complement(s) in the second, without problem. But may something (say C1) be a partial cause in one relation and its negation (say, notC1 = C3) a partial cause in the other? Yes, since the negation of E would imply both not(C1 + C2...) and not(notC1 + C4...), which is consistent; except that in such case the two compounds could not occur together.

Similarly, a contingent cause may be found *common* to two (or more) such causations with the same effect; if say C3 is identical with C1, notC1 would have notC2... as complement(s) in the

---

19 i.e. made explicit.

first relation and notC4... as complement(s) in the second, without problem. But may something (say C1) be a contingent cause in one relation and its negation (say, notC1 = C3) a contingent cause in the other? Yes, since the negation of notE would imply both not(notC1 + notC2...) and not(C1 + notC4...), which is consistent; except that in such case the two compounds could not occur together.

5. The Four Genera of Causation.

We have found the minimal formal definitions of, respectively, complete, necessary, partial and contingent causation. We are now in a position to begin synthesizing our accumulated findings concerning these determinations of causation. Remember how we developed these four concepts....

We started with the paradigm of causation (later named complete and necessary causation). From this we *abstracted two constituent forms*, or (strong) determinations, which we called complete causation and necessary causation. Then we *derived by means of an analogy two additional forms*, or (weak) determinations, which we called by way of contrast partial causation and contingent causation.

These four constructs apparently *exhaust* what we mean by causation, in view of their respective conceptual derivations from the paradigm of causation, and of their symmetry in relation to each other and the whole. No further expressions of the concept of causation, direct or indirect, seem conceivable.

The four forms thus identified can thus be referred to as the genera of causation, or as its **generic determinations**. And we can safely postulate that:

> ***Nothing can be said to be a cause or effect of something else (in the causative sense), if it is not related to it in the way of at least one of these four genera of causation.***

We shall need **symbols** for these four genera, to facilitate their discussion. I propose (remember them well) the following letters, simply:

n for Necessary causation,
m for coMplete causation (to rhyme with **n**),
p for Partial causation, and
q for 'Qontingent' causation (to rhyme with **p**)[20].

This notation will be found particularly useful when we deal with causative syllogism. We will also occasionally distinguish between absolute and relative partial or contingent causation, by means of the symbols: $\mathbf{p}_{abs}$ and $\mathbf{q}_{abs}$ for absolutes (i.e. those not mentioning any complement)

---

20 I have previously used, in my work *Future Logic*, the letters **n** and **p** for the *modalities* of necessity and possibility (or more specifically, particularity or potentiality). These should not be confused, note well. In any case, their relations are very different. In modality, **n** implies **p** (i.e. if something is necessary, it is possible). But here, in causation, as we shall soon see, **n** and **p** are merely compatible (i.e. a necessary cause need not be a partial cause, though something may be both a necessary and partial cause).

and $\mathbf{p}_{rel}$ and $\mathbf{q}_{rel}$ for relatives (i.e. those specifying some complement). Unless specified as relative, **p** and **q** may always be considered absolute.
It follows from what we have just said that we may interpret the causative proposition "**P is a cause of Q**" as "P is a complete or necessary or partial or contingent cause of Q (or a consistent combination of these alternatives)".
It is easy to demonstrate that any compounds of the four genera involving both **m** and **p**, and/or both **n** and **q**, are inconsistent, i.e. formally excluded. That is, one and the same thing cannot be both a complete and partial cause of the same effect; for if clause (i) of **m**, namely "if C1, then E", is true, then clause (iii) of **p**, namely "if (C1 + notC2), not-then E", cannot be true, and vice-versa. Similarly, necessary and contingent causation, i.e. **n** and **q**, are incompatible. We shall see at a later stage that certain other combinations are also formally impossible.
We shall consider the remaining, *consistent compounds* involving the four generic determinations, which we shall call the **specific determinations**, in the next chapter.

We may, as already suggested, refer to something as a *strong cause*, if it is a complete and/or necessary cause; and to something as a *weak cause*, if it is a partial and/or contingent cause. Conversely, a necessary and/or dependent effect may be said to be a *strong effect*; and a contingent and/or tenuous effect, it may be said to be a *weak effect*. Mixtures of these characters are conceivable, as we shall see.
Another classification based on common characters: if something is known to be a complete or partial cause, it may be called a '*contributing cause*'[21]; and if something is known to be a necessary or contingent cause, it may be called a '*possible cause*'. Likewise, if something is known to be a necessary or contingent effect, it may be called a '*possible effect*'; and if something is known to be a dependent or tenuous effect, it may be called (say) a '*subject effect*'.
Moreover: we have characterized complete and partial causation as positive aspects of causation; and necessary and contingent causation as its negative aspects, comparatively. We may in this sense, relative to a given set of items, speak of 'positive' or 'negative' causation. The latter, of course, should not be confused with negations of causation. Accordingly, we may refer to *positive or negative causes or effects*.

The reader is referred to the **Appendix** on J. S. Mill's Methods, for comparison of our treatment of causation in this chapter (and the next).

6. Negations of Causation.

So far, we have only considered in detail *positive* causative propositions, i.e. statements affirming causation of some determination. We must now look at *negative* causative propositions, i.e. statements denying causation of some determination or any causation whatever. For this purpose, to avoid the causal connotations implied by use of symbols like C and E for the items involved, we shall rather use neutral symbols like P and Q.
Statements denying causation may be better understood by studying the negations of conditional propositions.

---

21 In the sense that it is a cause *to some extent*, sufficient or not.

'hypothetical' proposition has the form "If X, then Y" (which may be read as X ... X is logically followed by Y); it means by definition "the conjunction (X + ...) is impossible". Its contradictory is a *'negative hypothetical'* proposition of the form "If X, not-then Y"[22] (which may be read as X does not imply Y, or X is not logically followed by Y); it means by definition "the conjunction (X + notY) is possible".
In the positive form, though X and notY are together impossible, they are not implied (or denied) to be individually impossible. In the negative form, since X and notY are possible together, *each of* X, notY is also formally implied as possible. In either form, there is no formal implication that notX be possible or impossible, or that Y be possible or impossible. As for the remaining conjunctions (X + Y), (notX + Y), (notX + notY) – nothing can be inferred concerning them, either.
However, as we have seen, when such statements appear as implicit clauses of causation, the interactions between clauses will inevitably further specify the situation for many of the items concerned.

The negation of complete causation or necessary causation, through statements like "P is not a complete cause of Q" or "P is not a necessary cause of Q", is feasible if *any one or more* of the three constituent clauses of such causation is deniable. That is, such negation consists of a disjunctive proposition saying "not(i) and/or not(ii) and/or not(iii)", which may signify non-causation or another determination of causation (necessary instead of complete, or vice-versa, or a weaker form of causation).
To give an example: the denial of "P is a complete cause of Q" means "if P, not-then Q" *and/or* "if notP, then Q" *and/or* "P is impossible". These alternatives may give rise to different outcomes; in particular note that if "P is impossible" is true, then P cannot be a cause at all, and if "if P, then Q" and "if notP, then Q" are both true, then Q is necessary, in which case Q cannot be an effect at all.
The negation of strong causation as such means the negation of both complete and necessary causation.

With regard to negation of partial or contingent causation, we must distinguish two degrees, according as *a given* complement is intended or *any* complement whatever.
The more restricted form of negation of partial causation or contingent causation mentions a complement, as in statements like "P1 (complemented by P2) is not a partial cause of Q" or "P1 (complemented by P2) is not a contingent cause of Q". Such negation is feasible if *any one or more* of the four constituent clauses of such causation is deniable. That is, such negation consists of a disjunctive proposition saying "not(i) and/or not(ii) and/or not(iii) and/or not(iv)".
In contrast, note well, the negation of partial causation or contingent causation through statements like "P1 is not a partial cause of Q" or "P1 is not a contingent cause of Q", is more radical. "P1 is not a partial cause of Q" means "P1 (*with whatever complement*) is not a partial cause of Q" – it may thus be viewed as a conjunction of an infinite number of more restricted statements, viz. "P1 (complemented by P2) is not a partial cause of Q, and P1 (complemented by P3) is not a partial cause of Q, and... etc.", where P2, P3, etc. are *all* conceivable complements. Similarly with regard to "P1 is not a contingent cause of Q".

---

22 The proposition "If X, not-then Y" is not to be confused with "If X, then notY", note well. The latter implies but is not implied by the former.

A restricted negative statement is very broad in its possible outcomes: it may signify that P1 is not a cause of Q at all, or that P1 is instead a complete or necessary cause of Q, or that P1 is a weak cause of Q but a contingent rather than partial one or a partial rather than contingent one, or that P1 is a partial or contingent cause (as the case may be) of Q but with some complement *other than* P2.

A radical negative statement comprises many restricted ones, and is therefore less broad in its possible outcomes, specifically excluding that P1 be involved in a partial or contingent causation (as the case may be) with *any* complement(s) whatsoever. A restricted negation is *relative* to a complement (say, P2); a radical negation is a generality comprising all similar restricted negations for the items concerned (P1, Q), and is therefore relative to *no* complement (neither P2, nor P3, etc.).

The negation of weak causation as such means the negation of both partial and contingent causation, either in a restricted sense (i.e. relative to some complement) or in a radical sense (i.e. irrespective of complement).

This brings us to the relation of **non-causation**, which is also very complex.

As we saw, the positive causative proposition "P is a cause of Q" may be interpreted as "P is a complete or necessary or partial or contingent cause of Q". Accordingly, we may interpret the negative causative proposition "P is not a cause of Q" as "P is not a complete *and* not a necessary *and* not a partial *and* not a contingent cause of Q", i.e. as a denial of all four genera of causation in relation to P and Q (with whatever complement).

It is noteworthy that we cannot theoretically *define* non-causation except through negation of all the concepts of causation, which have to be defined first[23]. In contrast, on a practical level, we proceed in the opposite direction: in accord with general rules of induction, we presume any two items P and Q to be without causative relation, until if ever we can establish inductively or deductively that a causative relation obtains between them.[24]

Nevertheless, 'non-causation' refers to denial of causation, and is not to be confused with ignorance of causation; it is an ontological, not an epistemological concept.

Note well that non-causation is *not* defined by the propositions "if P, not-then Q, and if notP, not-then notQ". Such a statement, though suggestive of non-causation, is equally compatible with partial and/or contingent causation; so it cannot suffice to distinguish non-causation. *To specify a relation of non-causation, we have to deny every determination of causation.*

Furthermore, "P is not a cause of Q" refers to *relative* non-causation – it is relative to the items P and Q specifically, and does not exclude that Q may have *some other* cause P_1, or that P may have *some other* effect Q_1. Two items, say P and Q, taken at random, need not be causatively related at all (even in cases where they happen to be respectively causatively related to some third item, as will be seen when we study syllogism in later chapters). In such case, P and Q are called *accidents* of each other; their eventual conjunction is called a *coincidence*.

Relative non-causation is an integral part of the formal system of deterministic causality. We have to acknowledge the possibility, indeed inevitability, of such a relation. If I say "the position of stars does not affect[25] people's destinies", I mean that there is no causal relation

23 We shall later see that this truism is ignored by some philosophers.

24 The philosophical problems of defining causation (its forms) and identifying specific cases of causation (its contents), are distinct, as we shall see.

25 To 'affect' some thing is to cause a change in it.

specifically between stars and people; yet I may go on to say that stars affect other things or that people are affected by other things, without contradicting myself.

Relative non-causation should not be confused with *absolute* non-causation. The *causelessness* of some item A would be expressed as "nothing causes A", a proposition summarizing innumerable statements of the form "B does not cause A; C does not cause A;...etc.", where B, C,... are *all existents other than* A. Similarly, the *effectlessness* of some item A would be expressed as "nothing is caused by A", a proposition summarizing innumerable statements of the form "B is not caused by A; C is not caused by A;... etc.", where B, C,... are *all existents other than* A.

We thus see that whereas positive causative propositions are defined by conjunctions of clauses, negative ones are far more complex in view of their involving disjunctions.

The negations of determinations, or the negation altogether of causation, should not themselves be regarded as further determinations, since they by their breadth allow for non-causation (between the items concerned), note.

Chapter 3. THE SPECIFIC DETERMINATIONS

1. The Species of Causation.

We shall now look into the consistent combinations of the four genera of causation, symbolized as **m**, **n**, **p**, **q**, with each other or their negations. Implicit in our gradual development of these concepts of causation from a common paradigm, was the idea that they are abstractions, indefinite concepts that are eventually concretized in the more specific and definite compounds.

We have already found some of their combinations, namely **mp** and **nq** to be **inconsistent**. This was due to incompatibilities between clauses of their definitions, or in other words, certain rows of their matrices. Thus, row 6 of **m** (C + notE is impossible) is in conflict with row 22 (C1 + notE is possible) of **p**; similarly, row 7 of **n** (notC + E is impossible) is in conflict with row 23 (notC1 + E is possible) of **q**.

It is also possible to prove certain other combinations to be logically impossible. This can be done formally, but not at the present stage of development, because we do not yet have the technical means at this stage to treat *negations* of generic determinations. To define *not***m**, *not***n**, *not***p**, *not***q** in verbal terms would be extremely arduous and confusing. I will therefore for now merely affirm to you that **combinations of *any one positive* generic determination with *the negations of the three other* generic determinations, for the very same terms, are inconsistent**.

By elimination, we are left with only *four* consistent compounds, i.e. remaining combinations give rise to no inconsistency, i.e. whose respective clauses do not contradict each other. This means that, from the logical point of view, they are conceivable, and therefore worthy of further formal treatment. We may refer to them as the specific determinations, or species of causation.

The following table (where + and – signify, respectively, affirmation and denial of a determination) lists all combinations of the generics and identifies the logically possible specifics among them:

Table 3.1. Possible specifications of the 4 generic determinations.

No. of genera	Compound	m	n	p	q	Modus
Four	**mnpq**	+	+	+	+	**mp**, **nq** impossible
Three	**mnp**	+	+	+	-	**mp** impossible
	mnq	+	+	-	+	**nq** impossible
	mpq	+	-	+	+	**mp** impossible
	npq	-	+	+	+	**nq** impossible
Two	**mp**	+	-	+	-	**mp** impossible
	nq	-	+	-	+	**nq** impossible
	mn	+	+	-	-	**possible**
	mq	+	-	-	+	**possible**
	np	-	+	+	-	**possible**
	pq	-	-	+	+	**possible**
Only one	**m-alone**	+	-	-	-	*will be proved impossible*
	n-alone	-	+	-	-	*will be proved impossible*
	p-alone	-	-	+	-	*will be proved impossible*
	q-alone	-	-	-	+	*will be proved impossible*
None	**non causation**	-	-	-	-	**possible**

The formulae given in the above table for each specific determination is as brief as possible. For instance, since **m** implies the negation of **p** and **n** implies the negation of **q**, '**mn**' (meaning both complete and necessary causation) tacitly implies 'not**p** and not**q**' (neither partial nor contingent causation, *with whatever complement*); the latter negations need not therefore be mentioned. Similarly, an expression like **m-alone** signifies the affirmation of one generic determination (here, **m**) *and* the denial of all three others (i.e. not**n** and not**q**, as well as not**p**). This notation is far from ideal, but suffices for our current needs, since many combinations are eliminated at the outset.

We see that four specific determinations, namely **mn**, **mq**, **np**, **pq**, are formed by conjunction of positive causative propositions; these we shall call (following J. S. Mill's nomenclature) **joint determinations**. It follows from the above table that each generic determination has only *two* species. Each generic determination may therefore be interpreted as a disjunction of its two possible embodiments; thus, **m** means **mn** or **mq**; **n** means **mn** or **np**; **p** means **np** or **pq**; and **q** means **mq** or **pq**. Also note, we could refer to **mn** as 'only-strong causation' and to **pq** 'only-weak causation', while **mq** and **np** are 'mixtures of strong and weak'.

The four specific determinations formed by composing positive causative propositions with negative ones, namely **m-alone**, **n-alone**, **p-alone**, **q-alone**, will be called **lone determinations**. This expression is introduced at this stage to contrast it with generic and joint determinations. Clearly, one should not confuse an isolated generic symbol such as **m** with the corresponding specific symbol **m-alone**; I use this heavy notation to ensure no confusion arises. Moreover, *nota bene*: In the above table, these forms are eliminated at the outset, because they concern *absolute* partial or contingent causation, i.e. they are irrespective of complement and mean **m-alone**$_{abs}$ etc. But as we shall later see, when they involve *relative* partial or contingent causation, i.e. when some complement is specified (in $\mathbf{p}_{rel}$ or $\mathbf{q}_{rel}$ or their negations), so that they mean **m-alone**$_{rel}$ etc., they remain possible forms. This need not concern us at the moment, but is said to explain why these forms need to be named.

We would label as, simply, **causation** (or 'any causation'), the disjunctive proposition "**m** or **n** or **p** or **q**", or the more specific "**mn** or **mq** or **np** or **pq**". Such positive propositions merely *imply* causation, if they involve less disjuncts or an isolated generic or joint determination. The contradictory of causation, **non-causation**, is the only remaining allowable combination, our table being exhaustive. This last possible combination involves negation of *all four* generic or joint determinations, note well. That is, it means "neither **m** nor **n** nor **p** nor **q**" or equally "neither **mn** nor **mq** nor **np** nor **pq**".

The above table also allows us to somewhat interpret complex negations. The negation of any compound is equivalent to the disjunction of all remaining four compounds (three of causation *and* one of non-causation). For instance "**not(mn)**" means **mq**, **np**, **pq**, or **non-causation**. Similarly with any other formula.

Note that where one of the weak determinations is denied by reason of the affirmation of the contrary strong determination (**m** in the case of **p**, or **n** in the case of **q**), any and all proposed complements are denied. Where one of the weaks is affirmed (even if the other is radically denied), at least one complement is implied; and of course, the contrary strong determination is denied. In all other cases, we must remember to be careful and distinguish between restricted and radical negations of **p** or **q**, as already explained in the previous chapter.

2. The Joint Determinations.

We shall now examine in detail the four joint determinations, symbolized by **mn**, **mq**, **np**, and **pq**, each of which is obtained by consistent conjunction of two generic determinations. Each is thus a species shared by the two genera constituting it. Thus, **mn** is a specific case of **m** and a specific case of **n**; and so forth.

We have already encountered one of these joint determinations, viz. complete and necessary causation, the paradigm of causation. We shall now examine it in further detail, and also treat the other three joint determinations.

Complete and Necessary causation by C of E:

(i) **If C, then E**;
(ii) if notC, not-then E (may be left tacit);
(iii) where: **C is possible**.
And:
(iv) **if notC, then notE**;
(v) if C, not-then notE (may be left tacit);
(vi) where: **C is unnecessary**.

Table 3.2. Complete necessary causation.				
No.	Element/compound		Modus	Source/relationship
1	**C**		**Possible**	**(iii)**
2	**notC**		**Possible**	**(vi)**
3		E	Possible	implied by (v)
4		notE	possible	implied by (ii)
5	C	E	possible	(v) or implied by (i) + (iii)
6	**C**	**notE**	**impossible**	**(i)**
7	**notC**	**E**	**impossible**	**(iv)**
8	notC	notE	possible	(ii) or implied by (iv) + (vi)

Notice how the merger of clauses (i), (ii) and (iii) with (iv), (v) and (vi) renders clauses (ii) and (v) redundant (though still implicit). Rows 5-8 of the above table (shaded) constitute the matrix of complete-necessary causation.

Complete but Contingent causation by C1 of E:

(i) **If C1, then E**;
(ii) if notC1, not-then E (may be left tacit);
(iii) where: C1 is possible (may be left tacit).
And:
(iv) **if (notC1 + notC2), then notE**;
(v) **if (C1 + notC2), not-then notE**;
(vi) **if (notC1 + C2), not-then notE**;
(vii) where: **(notC1 + notC2) is possible**.

Table 3.3. Complete contingent causation.

No.	Element/compound			Modus	Source/relationship
1	C1			possible	(iii) or implied by (v)
2	notC1			possible	implied by (vi) or (vii)
3		C2		possible	implied by (vi)
4		notC2		possible	implied by (v) or (vii)
5			E	possible	implied by (v) or (vi)
6			notE	possible	implied by (iv) + (vii)
7	C1		E	possible	implied by (v)
8	**C1**		**notE**	**impossible**	**(i)**
9	notC1		E	possible	implied by (vi)
10	notC1		notE	possible	(ii) or implied by (iv) + (vii)
11		C2	E	possible	implied by (vi)
12		C2	notE	*open*	if #12 is impossible, so is #24; and in view of (i): if #12 is possible, so is #24
13		notC2	E	possible	implied by (v)
14		notC2	notE	possible	implied by (iv) + (vii)
15	**C1**	**C2**		*open*	if #15 is impossible, so is #19; and in view of (i): if #15 is possible, so is #19
16	C1	notC2		possible	implied by (v)
17	notC1	C2		possible	implied by (vi)
18	**notC1**	**notC2**		**possible**	**(vii)**
19	C1	C2	E	*open*	if #19 is possible, so is #15; and in view of (i): if #19 is impossible, so is #15
20	C1	C2	notE	impossible	implied by (i)
21	**C1**	**notC2**	**E**	**possible**	**(v)**
22	C1	notC2	notE	impossible	implied by (i)
23	**notC1**	**C2**	**E**	**possible**	**(vi)**
24	**notC1**	**C2**	**notE**	*open*	if #24 is possible, so is #12; and in view of (i): if #24 is impossible, so is #12
25	**notC1**	**notC2**	**E**	**impossible**	**(iv)**
26	notC1	notC2	notE	possible	implied by (iv) + (vii)

Notice how the merger of clauses (i), (ii) and (iii) with (iv), (v), (vi) and (vii) renders clauses (ii) and (iii) redundant (though still implicit). Rows 19-26 of the above table constitute the matrix of complete-contingent causation.

Concerning the four positions labeled *open* in the above table, note that the moduses of Nos. 12 and 24 are tied and likewise those of Nos. 15 and 19. Proof for the first two: if #12 (C2 + notE) is impossible, #24 (notC1 + C2 + notE) must also be impossible; if #24 (notC1 + C2 + notE) is impossible, then knowing #20 (C1 + C2 + notE) to be impossible, #12 (C2 + notE) must also be impossible; the rest follows by contraposition. Proof for the other two: if #15 (C1 + C2) is impossible, #19 (C1 + C2 + E) must also be impossible; if #19 (C1 + C2 + E) is impossible, then knowing from (i) that #20 (C1 + C2 + notE) is impossible, #15 (C1 + C2) must also be impossible; the rest follows by contraposition. The interpretation of these open cases is as follows.

(a) Suppose #12 is impossible; this means that "If C2, then E". We know from #14 that "If notC2, not-then E"; and from #3 that "C2 is possible". Whence, C2 satisfies the definition for being a *complete* cause of E, just like C1. Thus, in such case, C1 and C2 are simply *parallel* complete (and contingent) causes of E. This is quite conceivable, and as we have seen in an earlier section such causes may be compatible or incompatible. If #15 is possible, they are compatible; and if #15 is impossible, they are incompatible.

(b) Suppose #12 is possible; this means that "If C2, not-then E", in which case C2 is not a complete cause of E. This is quite conceivable, covering situations where one of the contingent causes (namely, C1) is also complete, while the other (C2) is not complete. Additionally, we can say: if #15 is possible, they are compatible; and if #15 is impossible, they are incompatible; there is no problem of consistency either way.

However, a very interesting question arises in such case: is a contingent but not complete cause (like C2, here) bound to be a partial cause? C2 is certainly not a partial cause of E in conjunction with C1, since C1 is a complete cause of E. Therefore, *if* C2 is a partial cause of E, it will be so in conjunction with *some other* partial cause of E, say C3. But since C3 is unmentioned in our original givens, its existence is not formally demonstrable. We thus have no certainty that an *incomplete* contingent cause is implicitly a *partial* contingent cause! We will return to this issue later.

Partial yet Necessary causation by C1 of E:

(i) **If notC1, then notE;**
(ii) if C1, not-then notE (may be left tacit);
(iii) where: C1 is unnecessary (may be left tacit).

And:

(iv) **if (C1 + C2), then E;**
(v) **if (notC1 + C2), not-then E;**
(vi) **if (C1 + notC2), not-then E;**
(vii) where: **(C1 + C2) is possible**.

Table 3.4. Partial necessary causation.					
No.	Element/compound			Modus	Source/relationship
1	C1			possible	implied by (vi) or (vii)
2	notC1			possible	(iii) or implied by (v)
3		C2		possible	implied by (v) or (vii)
4		notC2		possible	implied by (vi)
5			E	possible	implied by (iv) + (vii)
6			notE	possible	implied by (v) or (vi)
7	C1		E	possible	(ii) or implied by (iv) + (vii)
8	C1		notE	possible	implied by (vi)
9	**notC1**		**E**	**impossible**	**(i)**
10	notC1		notE	possible	implied by (v)
11		C2	E	possible	implied by (iv) + (vii)
12		C2	notE	possible	implied by (v)
13		notC2	E	*open*	if #13 is impossible, so is #21; and in view of (i): if #13 is possible, so is #21
14		notC2	notE	possible	implied by (vi)
15	**C1**	**C2**		**possible**	**(vii)**
16	C1	notC2		possible	implied by (vi)
17	notC1	C2		possible	implied by (v)
18	notC1	notC2		*open*	if #18 is impossible, so is #26; and in view of (i): if #18 is possible, so is #26
19	C1	C2	E	possible	implied by (iv) + (vii)
20	**C1**	**C2**	**notE**	**impossible**	**(iv)**
21	C1	notC2	E	*open*	if #21 is possible, so is #13; and in view of (i): if #21 is impossible, so is #13
22	**C1**	**notC2**	**notE**	**possible**	**(vi)**
23	notC1	C2	E	impossible	implied by (i)
24	**notC1**	**C2**	**notE**	**possible**	**(v)**
25	notC1	notC2	E	impossible	implied by (i)
26	notC1	notC2	notE	*open*	if #26 is possible, so is #18; and in view of (i): if #26 is impossible, so is #18

Notice here again how the merger of clauses (i), (ii) and (iii) with (iv), (v), (vi) and (vii) renders clauses (ii) and (iii) redundant (though still implicit). Rows 19-26 of the above table (shaded) constitute the matrix of partial-necessary causation.

Concerning the four positions labeled *open* in the above table, note that the moduses of Nos. 13 and 21 are tied and likewise those of Nos. 18 and 21. These statements may be proved in the same manner as done for the preceding table; this is left to the reader as an exercise. We can also interpret these situations in similar ways. If #13 is impossible, C2 is a partial and necessary cause of E, parallel to C1; and notC2 is either compatible or incompatible with notC1 according to whether #18 is possible or impossible. If #13 is possible, C2 is a partial but not necessary cause of E, and notC2 is either compatible or not with notC1, according to whether #18 is possible or not.

However, it is not formally demonstrable that an *unnecessary* partial cause is implicitly a *contingent* partial cause; and the implications of this finding (or absence of finding) will have to be considered later.

Partial and Contingent causation by C1 of E:

(i) **If (C1 + C2), then E;**
(ii) **if (notC1 + C2), not-then E;**
(iii) **if (C1 + notC2), not-then E;**
(iv) where: **(C1 + C2) is possible.**
And:
(v) **if (notC1 + notC2), then notE;**
(vi) **if (C1 + notC2), not-then notE;**
(vii) **if (notC1 + C2), not-then notE;**
(viii) where: **(notC1 + notC2) is possible.**

Table 3.5. Partial contingent causation.

No.	Element/compound			Modus	Source/relationship
1	C1			possible	implied by (iii) or (iv) or (vi)
2	notC1			possible	implied by (ii) or (vii) or (viii)
3		C2		possible	implied by (ii) or (iv) or (vii)
4		notC2		possible	implied by (iii) or (vi) or (viii)
5			E	possible	implied by (vi) or (vii)
6			notE	possible	implied by (ii) or (iii)
7	C1		E	possible	implied by (vi)
8	C1		notE	possible	implied by (iii)
9	notC1		E	possible	implied by (vii)
10	notC1		notE	possible	implied by (ii)
11		C2	E	possible	implied by (vii)
12		C2	notE	possible	implied by (ii)
13		notC2	E	possible	implied by (vi)
14		notC2	notE	possible	implied by (iii)
15	**C1**	**C2**		**possible**	**(iv)**
16	C1	notC2		possible	implied by (iii) or (vi)
17	notC1	C2		possible	implied by (ii) or (vii)
18	**notC1**	**notC2**		**possible**	**(viii)**
19	C1	C2	E	possible	implied by (i) + (iv)
20	**C1**	**C2**	**notE**	**impossible**	**(i)**
21	**C1**	**notC2**	**E**	**possible**	**(vi)**
22	**C1**	**notC2**	**notE**	**possible**	**(iii)**
23	**notC1**	**C2**	**E**	**possible**	**(vii)**
24	**notC1**	**C2**	**notE**	**possible**	**(ii)**
25	**notC1**	**notC2**	**E**	**impossible**	**(v)**
26	notC1	notC2	notE	possible	implied by (v) + (viii)

Rows 19-26 of the above table (shaded) constitute the matrix of partial-contingent causation. We note that here none of the original clauses are made redundant by the combination of partial and contingent causation. Furthermore, no position in the above table is left open, with regard to the possibility or impossibility of the item or combination concerned.
Additionally we can say that if C1 and C2 are, as here, *complementary* partial contingent causes of E, then they have the same set of relations to each other and to E. But this does not mean that if C1 and C2 are complementary partial causes of E, they are bound to be complementary contingent causes of E, since as we have seen both or just one of them may be necessary cause(s) of E. Similarly, we cannot say that if C1 and C2 are complementary contingent causes of E, they are bound to be complementary partial causes of E, since as we have seen both or just one of them may be complete cause(s) of E.

There may, of course, be more than one complement to C1 (i.e. complements C3, C4..., in addition to C2) in the last three joint determinations, **mq**, **np** or **pq**. Such cases may be similarly treated, as we have explained when considering the weaker generic determinations separately.

It is with reference to the joint determinations **mq** and **np** that the utility of reformatting sentences about partial or contingent causation becomes apparent. An **mq** proposition is best stated as "C1 is a complete and (complemented by C2) a contingent cause of E", and an **np** proposition is best stated as "C1 is a necessary and (complemented by C2) a partial cause of E".

We must now consider the **hierarchy** between the above four forms, since there are clearly differences in degree in the 'bond' between cause(s) and effect. Causation is obviously at its *strongest* when both complete and necessary (**mn**). It is difficult to say which of the next two forms (**mq** or **np**) is the stronger and which the weaker, they are not really comparable to each other; all we can say is that they are both less determining than the first and more determining than the last; let us call them *middling* determinations. Causation is *weakest* for each factor involved in partial and contingent causation (**pq**).

With regard to **parallelism**, we can infer that it is conditionally possible with reference to our previous findings in the matter.

Two complete-necessary causes, C, C_1, of the same effect E, may be parallel, provided they are neither exhaustive nor incompatible with each other, i.e. provided "if C, not-then notC_1 and if notC, not-then C_1" is true.

For complete-contingent causation, it is conceivable that C1, C2 have this relation to E and C3, C4 have this same relation to E, provided the complete causes C1 and C3 are not exhaustive and the compounds (notC1 + notC2) and (notC3 + notC4) are not exhaustive. An interesting special case is when C2 = C4, i.e. when the two complete causes have the same complement in the contingent causation of E.

For partial-necessary causation, it is conceivable that C1, C2 have this relation to E and C3, C4 have this same relation to E, provided the necessary causes C1 and C3 are not incompatible and the compounds (C1 + C2) and (C3 + C4) are not exhaustive. An interesting special case is when C2 = C4, i.e. when the two necessary causes have the same complement in the partial causation of E.

For partial-contingent causation, the same condition of non-exhaustiveness between the parallel compounds involved applies. And here, too, note the special case when C2 = C4 as interesting.

Tables involving all the items concerned and their negations in all combinations may be constructed to analyze the implications of such parallelisms in detail.

The **negations** of the four joint determinations may be reduced to the denial of one or both of their constituent generic determinations. That is, **not(mn)** means 'not-**m** and/or not-**n**'; **not(mq)** means 'not-**m** and/or not-**q**'; **not(np)** means 'not-**p** and/or not-**n**'; and **not(pq)** means 'not-**p** and/or not-**q**'. Each of these alternative denials in turn implies denial of one or more of the constituent clauses, obviously.

3. The Significance of Certain Findings.

Let us review how we have proceeded so far. We started with the paradigm of causation, namely, complete necessary causation. We then abstracted its constituent "determinations", the complete and the necessary aspects of it, and by negation formulated another two generic determinations, namely partial and contingent causation. We then recombined these abstractions, to obtain all initially conceivable formulas. Some of these formulas (**mp**, **nq**) could be eliminated as logically impossible by inspecting their definitions and finding contradictory elements in them. Others (the *lone* determinations, obtained by conjunction of only one generic determination and the negations of all three others) were eliminated on the basis of later findings not yet presented here. This left us with only five logically tenable specific causative relations between any two items, namely the four *joint* determinations (the consistent conjunctions generic determinations) and non-causation (the negation of all four generic determinations).

When I personally first engaged in the present research, I was not sure whether or not the (absolute) lone determinations were consistent or not. Because each lone determination involves three negative causative propositions in conjunction, and each of these is defined by disjunction of the negations of the defining clauses of the corresponding positive form, it seemed very difficult to reliably develop matrixes for them. I therefore, as a logician[26], had to assume as a working hypothesis that they were logically possible. It is only in a later phase, when I developed "matricial microanalysis" that I discovered that they can be formally eliminated. Take my word on this for now. This discovery was very instructive and important, because it signified that *causation is more "deterministic" than would otherwise have been the case*.

If lone determinations had been logically possible, causation would have been moderately deterministic. For two items might be causatively related on the positive side, but not on the negative side, or vice-versa. Something could be *only* a complete cause (or *only* a partial cause) of another without having to also be a necessary or contingent one; or it could be *only* a necessary cause (or only a contingent cause) of another without having to also be a complete or partial one. But as it turned out there is logically no such degree of freedom in the causative realm.

If two things are causatively related at all, they *have to be* ultimately related in one (and indeed only one) of the four ways described as the joint determinations[27], i.e. in the way of **mn**, **mq**, **np**, or **pq**. The concepts **m**, **n**, **p**, **q** are common aspects of these four relations and no others. There is no "softer" causative relation. Causation is "full" or it is not at all; no "holes" are allowed in it. We can formulate the following "laws of causation" in consequence:

- If something is a complete or partial cause of something, it must also be either a necessary or (with some complement or other) a contingent cause of it.

---

26 The logician must keep an open mind so long as an issue remains unresolved. Logic cannot at the outset, without good reason, close doors to alternatives. Where formal considerations leave spaces, we cannot impose prejudices or speculations. The reason being that the aim of the science of logic is *to prepare the ground for discourse and debate*. If it takes arbitrary 'metaphysical' positions at the outset, it deprives us of *a language* with which to even consider opposite views. So long as formal grounds for some thesis is lacking, its antithesis must remain utterable.

27 It is interesting to note that, although J. S. Mill did not (to my knowledge) consider the issue of lone determinations, he turned out to be right in acknowledging only the four joint determinations.

- If something is a necessary or contingent cause of something, it must also be either a complete or (with some complement or other) a partial cause of it.
- In short, since a lone determination is impossible, if something is at all a causative of anything, it must be related in the way of a joint determination with it.

These laws have the following corollaries:

- If something is neither a necessary nor contingent cause of something, it must also be neither a complete nor (with whatever complement) a partial cause of it.
- If something is neither a complete nor partial cause of something, it must also be either neither a necessary nor (with whatever complement) a contingent cause of it.
- In short, since a lone determination is impossible, if two things are known not to be related in the way of either pair of contrary generic determinations (i.e. **m** and **p**, or **n** and **q**), they can be inferred to be not causatively related at all.

Also:

- The *complement* of a partial cause of something, being also itself a partial cause of that thing, must either be a necessary or (with some complement or other) a contingent cause of that thing.
- The *complement* of a contingent cause of something, being also itself a contingent cause of that thing, must either be a complete or (with some complement or other) a partial cause of that thing.

With regard to the epistemological question, as to how these causative relations are to be established, we may say that they are ultimately based on ***induction*** (including deduction from induced propositions): we have no other credible way to knowledge. Causative propositions may of course be built up gradually, clause by clause (see definitions in the previous chapter).

As I showed in my work *Future Logic*, the *positive* hypothetical (i.e. if/then) forms, from which causatives are constructed, result from generalizations from experience of conjunctions between the items concerned (which generalizations are of course revised by particularization, when and if they lead to inconsistency with new information). The *negative* hypothetical (i.e. if/not-then) forms are assumed true if no positive forms have been thus established, or are derived by the demands of consistency from positive forms thus established. In their case, an epistemological quandary may be translated into an ontological *fait accompli* (at least until if ever reason is found to prefer a positive conclusion).

We may first, by such induction (or deduction thereafter), propose one of the four generic determinations in isolation. The proposed generic determination is effectively treated as a joint determination "in-waiting", a convenient abstraction that does not really occur separately, but only within conjunctions. We are of course encouraged by methodology to subsequently **vigorously research** which of the four joint determinations can be affirmed between the items concerned. In cases where all such research efforts prove fruitless, we are simply left with a *problematic* statement, such as (to give an instance) "P is a complete cause, and either a necessary or a contingent cause, of Q".

But, since lone determination does not exist, we can never opt for a *negative* conclusion, like "P is a complete cause, but neither a necessary nor a contingent cause, of Q". We may *not* in this context effectively generalize from "I did not find" to "there is not" (a further causative relation). We may not interpret a structural doubt as a negative structure, an uncertainty as an indeterminacy.

In the history of Western philosophy, until recent times, the dominant hypothesis concerning causation has been that it is applicable *universally*. Some philosophers mitigated this principle, reserving it for 'purely physical' objects, *excepting* beings with volition (humans, presumably G-d, and even perhaps higher animals). A few, notably David Hume, denied any such "law of causation" as it has been called.

But in the 20th Century, the idea that there might, even in Nature (i.e. among entities without volition), be 'spontaneous' events gained credence, due to unexpected developments in Physics. That idea tended to be supported by the Uncertainty Principle of Werner Heisenberg for quantum phenomena, interpreted by Niels Bohr as an ontological (and not merely epistemological) principle of indeterminacy, and the Big-Bang theory of the beginning of the universe, which Stephen Hawking considered as possibly implying an *ex nihilo* and non-creationist beginning.

We shall not here try to debate the matter. All I want to do at this stage is stress the following nuances, which are now brought to the fore. The primary thesis of determinism is that ***there is causation*** in the world; i.e. that causal relations of the kind identified in the previous chapter (the four generic determinations) ***do occur*** in it. Our above-mentioned discovery that such causation has to fit in one of the four specific determinations may be viewed as a corollary of this thesis, or a logically consistent definition of it.

This is distinct from various universal causation theses, such as that nothing can occur except through causation (implying that causation is the only existing form of causality), or that at least nothing in Nature can do so (though for conscious beings other forms of causality may apply, notably volition), among others.

We shall analyze such so-called ***laws*** of causation in a later chapter; suffices for now to realize that they are extensions, attempted generalizations, of the apparent ***fact*** of causation, and not identical with it. Many philosophers seem to be unaware of this nuance, effectively regarding the issue as either '*causation everywhere*' or '*no causation anywhere*'.

The idea that causation is present ***somewhere*** in this world is logically quite compatible with the idea that there may be ***pockets*** or ***borders*** where it is absent, a thesis we may call 'particular (i.e. non-universal) causation'. We may even, more extremely, consider that causation is poorly scattered, in a world moved principally by spontaneity and/or volition.

The existence of causation thus does *not* in itself exclude the spontaneity envisaged by physicists (in the subatomic or astronomical domains); and it does *not* conflict with the psychological theory of volition or the creationist theory of matter[28].

Apparently, then, though determinism may be the major relation between things in this world, it leaves some room, however minor (in the midst or at the edges of the universe), for indeterminism.

We will give further consideration to these issues later, for we cannot deal with them adequately until we have clarified the different modes of causation.

---

28 Note incidentally that to say that G-d created the world does not imply that He did so specifically as and when the Bible seems to describe it; He may equally well have created the first concentration of matter and initiated the Big-Bang. Note also, that Creationism implies the pre-existence of G-d, a 'spiritual' entity; it is therefore a theory concerning the beginning of 'matter', *but not of existence as such*. G-d is in it posited as Eternal and Transcendental, or prior to or beyond time and space, but still 'existent'. With regard to such issues, including the compatibility of spontaneity and volition with Creation, see my *Buddhist Illogic*, chapter 10.

Chapter 4. IMMEDIATE INFERENCES

1. Oppositions.

The logical interrelations between the truths and falsehoods of propositions involving the same items are referred to as their 'oppositions'. This expression is unfortunate, because in everyday speech (and often in logical discourse) it connotes more specifically a 'conflict' between propositions; whereas in the science of logic, the term is intended more broadly as a 'face-off'. Thus, the possible oppositions between two propositions are:

- *contradiction* (provided they cannot be both true and they cannot be both false);
- *contrariety* (provided they cannot be both true);
- *subcontrariety* (provided they cannot be both false);
- *subalternation* (provided one, called the *subalternant*, cannot be true if the other, called the *subaltern*, is false – though the latter may be true if the former is false);
- *equivalence* (provided neither can be true if the other is false, and neither can be false if the other is true);
- *neutrality* (provided either can be true or false without the other being true or false).

Contradictory or contrary propositions are *incompatible* or mutually exclusive; propositions otherwise related are *compatible* or conjoinable. Equivalents *mutually imply* each other; among subalternatives, the subalternant *implies but is not implied by* the subaltern. Propositions are neutral to each other if they are not related in *any* of the other ways above listed.

Let us now consider the oppositions between the four *generic* determinations. We can show, with reference to the definitions in the preceding chapter, that:

- **If "P is a complete cause of Q", then "(*whatever* P is complemented by) P is *not* a partial cause of Q";**
- **if "P (with *whatever* complement) is a partial cause of Q", then "P is *not* a complete cause of Q".**
- **If "P is a necessary cause of Q", then (*whatever* P is complemented by) "P is *not* a contingent cause of Q";**
- **if "P (with *whatever* complement) is a contingent cause of Q", then "P is *not* a necessary cause of Q".**

For clause (i) of complete causation, viz. "if P, then Q", implies both "if (P + R), then Q" and "if (P + notR), then Q", for *any* item R whatsoever; whereas clause (iii) of partial causation implies that there is an item R such that "if (P + R), then Q". Similarly, necessary and contingent causation have conflicting implications, and therefore cannot both be true.

More briefly put, **m** and **p** are incompatible and **n** and **q** are incompatible. Other than that, no incompatibilities or implications exist between the four generic determinations. P may have no causative relation to Q at all, without any inconsistency ensuing. Thus, **m** and **p** are contrary (but not contradictory) and **n** and **q** are contrary (but not contradictory). As for the pairs **m** and **n**, **m** and **q**, **n** and **p**, **p** and **q** – they are all neutral to each other.

With regard to the negations of generic determinations, it follows that they are all neutral to each other. Of course, by definition, **not-m** is the contradictory of **m**, **not-n** is the contradictory of **n**, **not-p** is the contradictory of **p**, and **not-q** is the contradictory of **q**. Also, as above seen, **m** implies **not-p**, and **n** implies **not-q**. But between the four negations themselves, no incompatibilities or implications exist.
It should be stressed that partial causation is not to be considered as identical with the negation of complete causation, but only as one of the possible outcomes of such negation. That is, it would be illogical to infer from "P is not a complete cause of Q" that "P is a partial cause of Q", or from "P is not a partial cause of Q" that "P is complete a cause of Q". The labels 'complete' and 'partial' could be misleading, connoting a relation of inclusion between whole and part; here, note well, 'complete' excludes 'partial', and vice-versa. Similarly, of course, contingent causation is not equivalent to the negation of necessary causation.
Most importantly, keep in mind the inferences already mentioned in the last section of the preceding chapter, namely:

- If **m** is true, then **n** or **q** must be true.
- If **n** is true, then **m** or **p** must be true.
- If **p** is true, then **n** or **q** must be true.
- If **q** is true, then **m** or **p** must be true.
- If neither **m** nor **p** is true, then neither **n** nor **q** can be true.
- If neither **n** nor **q** is true, then neither **m** nor **p** can be true.

With regard to the oppositions between the four *joint* determinations.
Each of the four joint determinations obviously implies but is not implied by its constituent generic determinations. That is, **m** and **n** are subalterns of **mn**, **m** and **q** are subalterns of **mq**, and so forth. It follows that each joint determination is contrary to the negations of its constituent generic determinations. That is, **mn** is contrary to **not-m** and to **not-n**; and so forth. Or in other words, if either or both of its constituent generic determinations is/are denied, the joint determination as a whole must be denied.
Furthermore, the four joint determinations are all mutually exclusive. That is, if any one of them is true, the three others have to be false. For if **mn** is true, **mq** cannot be true (since **n** and **q** are incompatible), and **np** cannot be true (since **m** and **p** are incompatible), and **pq** cannot be true (for both reasons). Similarly, if we affirm **mq**, we must deny the combinations **mn**, **np**, **pq**; and so forth. On the other hand, the negation of any joint determination has no consequence on the others; they may all be false without resulting inconsistency.

2. Eductions.

Immediate inference is inference of a conclusion from one premise, in contrast to syllogistic (or mediate) inference. 'Opposition' is one form of it, in which the items concerned retain the same position and polarity. 'Eduction' is another form of it, involving some change in position and/or polarity of the items occurs.
Let us now look into the feasibility of eductions from causative propositions, with reference to their definitions. We shall *for now ignore the issue of direction of causation*, dealt with further

on. All the usual eductive processes[29], namely inversion, conversion, and contraposition, obversion, obverted-inversion, obverted-conversion, and obverted-contraposition, can be used in the ways shown below. First however, we must consider eduction by negation of the complement, a process applicable to the weak determinations.

a. **Negations of the complement** (from R to notR) for the same items (P-Q). This concerns the weak determinations, and results in a negative conclusion.

- **"P (complemented by R) is a partial cause of Q" implies "P (complemented by *not*R) is *not* a partial cause of Q".**

Proof: Clause (i) of "P (complemented by R) is a partial cause of Q" and clause (iii) of "P (complemented by notR) is a partial cause of Q" contradict each other; therefore they are incompatible propositions.

In contrast, note, "P (complemented by R) is a partial cause of Q" is compatible with "P (complemented by notR) is a contingent cause of Q".

- **"P (complemented by R) is a contingent cause of Q" implies "P (complemented by *not*R) is *not* a contingent cause of Q".**

Proof: In a similar manner, *mutatis mutandis*.

In contrast, note, "P (complemented by R) is a contingent cause of Q" is compatible with "P (complemented by notR) is a partial cause of Q".

Negation of the complement for the joint determination **pq** follows by conjunction:

- If P (complemented by R) is a partial and contingent cause of Q, then P (complemented by *not*R) is *neither* a partial nor a contingent cause of Q.

b. **Inversions** (changes from P-Q to **notP-notQ**); the conclusion is called the inverse of the premise.

All four generic determinations are invertible to a positive causative proposition, simply by substituting not{notP} for P, not{notQ} for Q. In the case of weak determinations, additionally, not{notR} replaces R; and moreover, eduction by negation of the complement of the positive conclusion yields a further negative conclusion. Thus,

- **"P is a complete cause of Q" implies "notP is a necessary cause of notQ".**

And vice-versa. In contrast, note, "P is a complete cause of Q" and "notP is a complete cause of notQ" are merely compatible.

- **"P is a necessary cause of Q" implies "notP is a complete cause of notQ".**

And vice-versa. In contrast, note, "P is a necessary cause of Q" and "notP is a necessary cause of notQ" are merely compatible.

- **"P (complemented by R) is a partial cause of Q" implies "notP (complemented by notR) is a contingent cause of notQ".**

And vice-versa. It follows by negation of the complement that:

- **"P (complemented by R) is a partial cause of Q" implies "notP (complemented by R) is *not* a contingent cause of notQ".**

In contrast, note, "P (complemented by R) is a partial cause of Q" is compatible with "notP (complemented by R) is a partial cause of notQ" and with "notP (complemented by notR) is a partial cause of notQ".

---

29 The terminology here used is the same as that traditionally used in other fields of logic, except for "negation of complement".

- **"P (complemented by R) is a contingent cause of Q" implies "notP (complemented by notR) is a partial cause of notQ".**

And vice-versa. It follows by negation of the complement that:

- **"P (complemented by R) is a contingent cause of Q" implies "notP (complemented by R) is *not* a partial cause of notQ".**

In contrast, note, "P (complemented by R) is a contingent cause of Q" is compatible with "notP (complemented by R) is a contingent cause of notQ" and with "notP (complemented by notR) is a contingent cause of notQ".

Notice, with regard to the positive implications of the weak determinations, that P, Q, and R all change polarity. Evidently, inversions involve a change of determination from positive (complete or partial) to negative (necessary or contingent, respectively), or vice-versa.

With regard to the *joint* determinations, their inversions follow from those relative to the generic determinations.

Inversion of **mn** or **pq** is possible, without change of determination (i.e. to **mn** or **pq**, respectively), since the changes for each constituent determination balance each other out; and all items change polarity. Thus:

- If P is a complete and necessary cause of Q, then notP is a complete and necessary cause of notQ.
- If P (complemented by R) is a partial and contingent cause of Q, then notP (complemented by notR) is a partial and contingent cause of notQ; also, notP (complemented by R) is *not* a partial or contingent cause of notQ.

Inversion of **mq** or **np** is possible, though with changes of determination (i.e. to **np** or **mq**, respectively); and all items change polarity. Thus:

- If P is a complete and (complemented by R) a contingent cause of Q, then notP is a necessary and (complemented by notR) a partial cause of notQ; also, notP (complemented by R) is *not* a partial cause of notQ.
- if P is a necessary and (complemented by R) a partial cause of Q, then notP is a complete and (complemented by notR) a contingent cause of notQ; also, notP (complemented by R) is *not* a contingent cause of notQ.

With regard to *negative* causative propositions, we can easily derive analogous inversions on the basis of[30] the above findings:

- "P is not a complete cause of Q" implies "notP is not a necessary cause of notQ".
- "P is not a necessary cause of Q" implies "notP is not a complete cause of notQ".
- "P (complemented by R) is not a partial cause of Q" implies "notP (complemented by notR) is not a contingent cause of notQ".
- "P (complemented by R) is not a contingent cause of Q" implies "notP (complemented by notR) is not a partial cause of notQ".

Similarly for the negations of joint determinations.

c. **Conversions** (changes from P-Q to **Q-P**); the conclusion is called the converse of the premise.

The strong generic determinations are convertible, as follows:

- **"P is a complete cause of Q" implies "Q is a necessary cause of P".**

---

30 Specifically, by contraposition of the positive implications.

Proof: Clause (i) of the given proposition may be contraposed to "if notQ, then notP"; clauses (i) and (iii) together imply that (P + Q) is possible, which means that "if Q, not-then notP"; and clause (ii) implies "notQ is possible". Thus, the conditions for the said conclusion are satisfied, and conversion is feasible.

And vice-versa. In contrast, note, "P is a complete cause of Q" and "Q is a complete cause of P" are merely compatible.

- **"P is a necessary cause of Q" implies "Q is a complete cause of P".**

Proof: In a similar manner, *mutatis mutandis*.

And vice-versa. In contrast, note, "P is a necessary cause of Q" and "Q is a necessary cause of P" are merely compatible.

The weak generic determinations are also convertible, as follows:

- **"P (complemented by R) is a partial cause of Q" implies "Q (complemented by *not*R) is a contingent cause of P".**

Proof: Clause (i) of the given proposition means that (P + R + notQ) is impossible, which may be restated as "if (notQ + R), then notP"; clauses (i) and (iv) together imply that (P + R + Q) is possible, which means that "if (Q + R), not-then notP"; clause (iii) means that (P + notR + notQ) is possible, which may be restated as "if (notQ + notR), not-then notP"; and clause (ii) implies "(notQ + R) is possible". Thus, the conditions for the said conclusion are satisfied (reading not{notR} instead of R), and conversion is feasible.

And vice-versa. It follows by negation of the complement that:

- **"P (complemented by R) is a partial cause of Q" implies "Q (complemented by R) is *not* a contingent cause of P".**

In contrast, note, "P (complemented by R) is a partial cause of Q" is compatible with "Q (complemented by R) is a partial cause of P" and with "Q (complemented by notR) is a partial cause of P".

- **"P (complemented by R) is a contingent cause of Q" implies "Q (complemented by *not*R) is a partial cause of P".**

Proof: In a similar manner, *mutatis mutandis*.

And vice-versa. It follows by negation of the complement that:

- **"P (complemented by R) is a contingent cause of Q" implies "Q (complemented by R) is *not* a partial cause of P".**

In contrast, note, "P (complemented by R) is a contingent cause of Q" is compatible with "Q (complemented by R) is a contingent cause of P" and with "Q (complemented by notR) is a contingent cause of P".

Note well, with regard to the positive implications of the weak determinations, that R changes polarity, while P, Q do not; in this sense, their conversion may be qualified as imperfect. Evidently, conversions involve a change of determination from positive (complete or partial) to negative (necessary or contingent, respectively), or vice-versa.

With regard to the *joint* determinations, their conversions follow from those relative to the generic determinations.

Conversion of **mn** is possible, without change of determination (i.e. to **mn**), since the changes for each constituent determination balance each other out. Thus:

- If P is a complete and necessary cause of Q, then Q is a complete and necessary cause of P.

Conversion of **pq** is possible, without change of determination (i.e. to **pq**), for the same reason; but the subsidiary item (R) changes polarity in the positive implication. Thus:

- If P (complemented by R) is a partial and contingent cause of Q, then Q (complemented by notR) is a partial and contingent cause of P; also, Q (complemented by R) is *not* a partial or contingent cause of P.

Conversion of **mq** or **np** is possible, though with changes of determination (i.e. to **np** or **mq**, respectively); also, the subsidiary item (R) changes polarity in the positive implication. Thus:

- If P is a complete and (complemented by R) a contingent cause of Q, then Q is a necessary and (complemented by notR) a partial cause of P; also, Q (complemented by R) is *not* a partial cause of P.
- If P is a necessary and (complemented by R) a partial cause of Q, then Q is a complete and (complemented by notR) a contingent cause of P; also, Q (complemented by R) is *not* a contingent cause of P.

With regard to *negative* causative propositions, we can easily derive analogous conversions on the basis of the above findings:

- "P is not a complete cause of Q" implies "Q is not a necessary cause of P".
- "P is not a necessary cause of Q" implies "Q is not a complete cause of P".
- "P (complemented by R) is not a partial cause of Q" implies "Q (complemented by notR) is not a contingent cause of P".
- "P (complemented by R) is not a contingent cause of Q" implies "Q (complemented by notR) is not a partial cause of P".

Similarly for the negations of joint determinations.

d. **Contrapositions** (changes from P-Q to **notQ-notP**); the conclusion is called the contraposite of the premise.

All four generic determinations are contraposable, simply by conversion of their inverses:

- **"P is a complete cause of Q" implies "notQ is a complete cause of notP".**

And vice-versa. In contrast, note, "P is a complete cause of Q" and "notQ is a necessary cause of notP" are merely compatible.

- **"P is a necessary cause of Q" implies "notQ is a necessary cause of notP".**

And vice-versa. In contrast, note, "P is a necessary cause of Q" and "notQ is a complete cause of notP" are merely compatible.

- **"P (complemented by R) is a partial cause of Q" implies "notQ (complemented by R) is a partial cause of notP".**

And vice-versa. It follows by negation of the complement that:

- **"P (complemented by R) is a partial cause of Q" implies "notQ (complemented by notR) is *not* a partial cause of notP".**

In contrast, note, "P (complemented by R) is a partial cause of Q" is compatible with "notQ (complemented by R) is a contingent cause of notP" and with "notQ (complemented by notR) is a contingent cause of notP".

- **"P (complemented by R) is a contingent cause of Q" implies "notQ (complemented by R) is a contingent cause of notP".**

And vice-versa. It follows by negation of the complement that:

- **"P (complemented by R) is a contingent cause of Q" implies "notQ (complemented by notR) is *not* a contingent cause of notP".**

In contrast, note, "P (complemented by R) is a contingent cause of Q" is compatible with "notQ (complemented by R) is a partial cause of notP" and with "notQ (complemented by notR) is a partial cause of notP".

Notice, with regard to the positive implications of the weak determinations, that while P, Q change polarity, R does not; in this sense, their contraposition may be qualified as imperfect. Evidently, contrapositions distinctively do *not* involve changes of determination.

With regard to the joint determinations, their contrapositions follow from those relative to the generic determinations.

Contraposition of **mn** is possible, without change of determination (i.e. to **mn**). Thus:

- If P is a complete and necessary cause of Q, then notQ is a complete and necessary cause of notP.

Contraposition of **pq**, **mq** or **np** is possible, without change of determination (i.e. to **pq**, **mq** or **np**, respectively); and the subsidiary item (R) does not change polarity in the positive implication. Thus:

- If P (complemented by R) is a partial and contingent cause of Q, then notQ (complemented by R) is a partial and contingent cause of notP; also, notQ (complemented by notR) is *not* a partial or contingent cause of notP.
- If P is a complete and (complemented by R) a contingent cause of Q, then notQ is a necessary and (complemented by R) a partial cause of notP; also, notQ (complemented by notR) is *not* a partial cause of notP.
- If P is a necessary and (complemented by R) a partial cause of Q, then notQ is a complete and (complemented by R) a contingent cause of notP; also, notQ (complemented by notR) is *not* a contingent cause of notP.

With regard to *negative* causative propositions, we can easily derive analogous contrapositions on the basis of the above findings:

- "P is not a complete cause of Q" implies "notQ is not a complete cause of notP".
- "P is not a necessary cause of Q" implies "notQ is not a necessary cause of notP".
- "P (complemented by R) is not a partial cause of Q" implies "notQ (complemented by R) is not a partial cause of notP".
- "P (complemented by R) is not a contingent cause of Q" implies "notQ (complemented by R) is not a contingent cause of notP".

Similarly for the negations of joint determinations.

e. **Obversions** (changes from P-Q to **P-notQ**); the conclusions are called obverses of the premise.[31]

All four generic determinations are obvertible in various ways, though the obverses are negative causative propositions.

- **"P is a complete cause of Q" implies "P is *not* a complete cause of notQ" and "P is *not* a necessary cause of notQ".**

Proof: Clauses (i) and (iii) of "P is a complete cause of Q" together imply that (P + Q) is possible, whereas clause (i) of "P is a complete cause of notQ" implies that conjunction impossible; therefore they are incompatible propositions. Also, clause (i) of "P is a complete cause of Q" and clause (ii) of "P is a necessary cause of notQ" contradict each other; therefore they are incompatible.

---

[31] The form "P causes notQ" is often reworded in everyday speech as "**P prevents Q**" (or other similar words – see your thesaurus). We could treat the latter expression as a form in its own right, and look into all the logic of its four genera: complete, necessary, partial and contingent prevention. But we do not need to do so, for all that logic is implicit in the work here in process.

- **"P is a necessary cause of Q" implies "P is *not* a necessary cause of notQ" and "P is *not* a complete cause of notQ".**

Proof: In a similar manner, *mutatis mutandis*.

- **"P (complemented by R) is a partial cause of Q" implies "P (complemented by R) is *not* a partial cause of notQ" and "P (complemented by *not*R) is *not* a contingent cause of notQ".**

Proof: Clauses (i) and (iv) of "P (complemented by R) is a partial cause of Q" together imply that (P + R + Q) is possible, whereas clause (i) of "P (complemented by R) is a partial cause of notQ" implies that conjunction impossible; therefore they are incompatible propositions. Also, clause (ii) of "P (complemented by R) is a partial cause of Q" and clause (i) of "P (complemented by notR) is a contingent cause of notQ" contradict each other; therefore they are incompatible. Notice in the latter case, the change in polarity of the complement (from R to notR), as well as the change in determination (from **p** to **q**).

In contrast, *note well*, "P (complemented by R) is a partial cause of Q" is compatible with "P (complemented by R) is a contingent cause of notQ", and with "P (complemented by notR) is a partial cause of notQ".

- **"P (complemented by R) is a contingent cause of Q" implies "P (complemented by R) is *not* a contingent cause of notQ" and "P (complemented by *not*R) is *not* a partial cause of notQ".**

Proof: In a similar manner, *mutatis mutandis*. Notice in the latter case, the change in polarity of the complement (from R to notR), as well as the change in determination (from **q** to **p**).

In contrast, *note well*, "P (complemented by R) is a contingent cause of Q" is compatible with "P (complemented by R) is a partial cause of notQ", and with "P (complemented by notR) is a contingent cause of notQ".

With regard to the *joint* determinations, their obversions follow from those relative to the generic determinations.

- If P is a complete and necessary cause of Q, then P is neither a complete nor a necessary cause of notQ.
- If P is a complete and contingent cause of Q, then P is neither a complete, nor (complemented by R) a contingent, cause of notQ, and P is neither a necessary, nor (complemented by *not*R) a partial, cause of notQ.
- If P is a necessary and partial cause of Q, then P is neither a necessary, nor (complemented by R) a partial, cause of notQ, and P is neither a complete, nor (complemented by *not*R) a contingent, cause of notQ.
- If P (complemented by R) is a partial and contingent cause of Q, then P (whether complemented by R or notR) is neither a partial nor a contingent cause of notQ.

f. **Obverted inversions** (changes from P-Q to **notP-Q**); the conclusions are called obverted-inverses of the premise.

All four generic determinations may be subjected to obverted-inversion, by successive inversion then obversion. The conclusions are therefore negative causative propositions.

- **"P is a complete cause of Q" implies "notP is *not* a complete cause of Q" and "notP is *not* a necessary cause of Q".**
- **"P is a necessary cause of Q" implies "notP is *not* a necessary cause of Q" and "notP is *not* a complete cause of Q".**

- **"P (complemented by R) is a partial cause of Q" implies "notP (complemented by R) is *not* a partial cause of Q" and "notP (complemented by *not*R) is *not* a contingent cause of Q".**

In contrast, *note well*, "P (complemented by R) is a partial cause of Q" is compatible with "notP (complemented by notR) is a partial cause of Q", and with "notP (complemented by R) is a contingent cause of Q".

- **"P (complemented by R) is a contingent cause of Q" implies "notP (complemented by R) is *not* a contingent cause of Q" and "notP (complemented by *not*R) is *not* a partial cause of Q".**

In contrast, *note well*, "P (complemented by R) is a contingent cause of Q" is compatible with "notP (complemented by notR) is a contingent cause of Q", and with "notP (complemented by R) is a partial cause of Q".

With regard to the *joint* determinations, their obverted-inversions follow from those relative to the generic determinations, as usual.

g. **Obverted conversions** (changes from P-Q to **Q-notP**); the conclusions are called obverted-converses of the premise.

All four generic determinations may be subjected to obverted-conversion, by successive conversion then obversion. The conclusions are therefore negative causative propositions.

- **"P is a complete cause of Q" implies "Q is *not* a complete cause of notP" and "Q is *not* a necessary cause of notP".**
- **"P is a necessary cause of Q" implies "Q is *not* a necessary cause of notP" and "Q is *not* a complete cause of notP".**
- **"P (complemented by R) is a partial cause of Q" implies "Q (complemented by R) is *not* a partial cause of notP" and "Q (complemented by notR) is *not* a contingent cause of notP".**

In contrast, *note well*, "P (complemented by R) is a partial cause of Q" is compatible with "Q (complemented by notR) is a partial cause of notP", and with "Q (complemented by R) is a contingent cause of notP".

- **"P (complemented by R) is a contingent cause of Q" implies "Q (complemented by R) is *not* a contingent cause of notP" and "Q (complemented by notR) is *not* a partial cause of notP".**

In contrast, *note well*, "P (complemented by R) is a contingent cause of Q" is compatible with "Q (complemented by notR) is a contingent cause of notP", and with "Q (complemented by R) is a partial cause of notP".

With regard to the *joint* determinations, their obverted-conversions follow from those relative to the generic determinations, as usual.

h. **Obverted contrapositions**, also known as conversions by negation (changes from P-Q to **notQ-P**); the conclusions are called obverted-contraposites of the premise.

All four generic determinations may be subjected to obverted-contraposition, by successive contraposition then obversion. The conclusions are therefore negative causative propositions.

- **"P is a complete cause of Q" implies "notQ is *not* a complete cause of P" and "notQ is *not* a necessary cause of P".**
- **"P is a necessary cause of Q" implies "notQ is *not* a necessary cause of P" and "notQ is *not* a complete cause of P".**

- **"P (complemented by R) is a partial cause of Q" implies "notQ (complemented by R) is *not* a partial cause of P" and "notQ (complemented by *not*R) is *not* a contingent cause of P".**

In contrast, *note well*, "P (complemented by R) is a partial cause of Q" is compatible with "notQ (complemented by notR) is a partial cause of P", and with "notQ (complemented by R) is a contingent cause of P".

- **"P (complemented by R) is a contingent cause of Q" implies "notQ (complemented by R) is *not* a contingent cause of P" and "notQ (complemented by *not*R) is *not* a partial cause of P".**

In contrast, *note well*, "P (complemented by R) is a contingent cause of Q" is compatible with "notQ (complemented by notR) is a contingent cause of P", and with "notQ (complemented by R) is a partial cause of P".

With regard to the *joint* determinations, their obverted-contrapositions follow from those relative to the generic determinations, as usual.

We may finally note the following derivative eductions, though they are virtually useless except that they partly summarize the preceding findings:

- If P is a *strong* cause of Q, then notP is a strong cause of notQ (inversion), and Q is a strong cause of P (conversion), and notQ is a strong cause of notP (contraposition).
- If P (complemented by R) is a *weak* cause of Q, then notP (complemented by notR) is a weak cause of notQ (inversion), and Q (complemented by notR) is a weak cause of P (conversion), and notQ (complemented by R) is a weak cause of notP (contraposition).
- If P is a cause of Q, then notP is a cause of notQ (inversion), and Q is a cause of P (conversion), and notQ is a cause of notP (contraposition).

Moreover, we can say:

- If P is a *strong* cause of Q, then P is not a strong cause of notQ (obversion), and notP is not a strong cause of Q (obverted inversion), and Q is not a strong cause of notP (obverted conversion), and notQ is not a strong cause of P (obverted contraposition).

But similar negative implications are not possible for "P (complemented by R) is a weak cause of Q", in view of variations in the complement in such cases. It follows that similar negative implications are not possible for "P is a cause of Q".

Finally, concerning the weak determinations, it should be noted that wherever the inference results in no change of complement, i.e. wherever the premise and conclusion concern *the same* complement, the complement need not be mentioned at all. That is, we can in some cases simply say: if "P is a partial (or contingent) cause of Q", then "(the new cause) is (or is not) a partial (or contingent) cause of (the new effect)" (as the case may be), on the tacit understanding that the complement, whatever it happens to be, has not been altered.

More broadly, whether or not the complement changes polarity, it is clear that *we do not need to specify or even remember its precise content*, in order to perform the inference. When the complement is unchanged, we need not mention it at all (or, to be sure, we can say in the conclusion "with the same complement, whatever it be"); and when it is changed, we can add in the conclusion "with the negation of the initial complement, whatever it be, as complement". It is good to know this, because it allows us to proceed with inferences without immediately having to or being able to pin-point the complement involved.

Note lastly, that all immediate inferences could also be validated or invalidated, as the case may be, by means of matricial analysis (see later). I have here preferred the less systematic, but also less voluminous, method of reduction to conditional arguments.
All these inferences add to our knowledge and understanding of causative propositions, of course. Some of them will prove useful for validations or invalidations of causative syllogisms by direct reduction to others.

3. The Directions of Causation.

Now, the implications between different forms of causative propositions identified above, such as that "P is a complete cause of Q" implies "Q is a necessary cause of P", demonstrate that our definitions of causation were incomplete. For we well know that causation has a direction! However, bear with me – we deal with this issue.
Strictly-speaking, when we utter a statement of the form "P is a (complete, necessary, partial, contingent) cause of Q", we imply a tacit clause specifying the direction (or ordering of items), in addition to the various clauses (treated in the preceding chapter) concerning determination. This means that denial of the tacit clause on direction would suffice to deny the causation concerned, even if all the other clauses are affirmed. However, there are good reasons why in our formal treatment we are wise to keep the issue of direction separate.
First of these is the epistemological fact that *the direction of causation is not always known*. We may by inductive or deductive means arrive at knowledge of all the other clauses, and yet be hard put to immediately specify the direction. If we wished to summarize our position in such case, and were not permitted to use the language of causation, we would have to introduce a relational expression *other than* "is a ... cause of" (say, "is a ... determinant of") to allow us to verbalize the situation. Causation would then be defined as the combination of this relation ("determination") with a directional clause. This is feasible, but in my view redundant; we can manage without such an artifice.
Secondly, we have to consider the ontological fact that *causation does not always occur in only one direction: it may occur in both*. Sometimes, the direction is exclusively from P to Q, or from Q to P; but sometimes, the causal relation is two-way or **reversible**. Moreover, reversible causation is not always **reciprocal**: there may be one determination in one direction, and another in the opposite sense; or there may, in the case of weak causations, be different complements in each direction. For this reason, too, we are wise to handle the issue of direction flexibly, considering it expressed in an additional clause, but left 'hidden' or ignored until specifically dealt with. This is the course adopted in the present work.

The directional clause for a causative proposition can be a phrase qualifying the sentence "P is a ... cause of Q", a phrase of the form "**in the direction from P to Q**" (which is identical with 'notP to notQ') *and/or* "**in the direction from Q to P**" (which is identical with 'notQ to notP'). We must additionally allow for (hopefully temporary) ignorance with the phrase "**direction unknown**".
We allow for only two directions, not four, note well. "P to Q" and its inverse "notP to notQ" are one and the same direction; likewise "Q to P" and "notQ to notP" are identical in direction. In this manner, causative statements remain *always* or formally invertible – but strictly-speaking only *sometimes* or conditionally convertible or contraposable, specifically when the causation is known to be reversible. That is, whereas inversion is ontologically universal,

conversion and contraposition have the status of formal artifices until and unless their ontological applicability is established in a given case. The latter two eductions, of course, go together; if either is applicable, so is the other (since the contraposite is the inverse of the converse).

As we shall see, consideration of direction of causation affects other deductive processes in a similar manner, i.e. making them conditional instead of universal. Thus, in causative syllogism, arguments in the first figure guarantee the direction implicit in the conclusion (given the directions implied by the premises), whereas arguments in the other two figures cannot do so.

However, this is not a great difficulty, because we know that wherever a causative conclusion is drawn, the direction of causation *has to be either* as implied by that conclusion *or* as implied by its formal converse *or both.* Thus, the issue of direction is relatively minor. It is without impact on the inferred 'bond', on the fact *that there is* a certain (strong or weak) causative relationship between the items concerned; the only problem it sets for us is *in which form* this relationship is expressed, as 'P-Q' or its converse 'Q-P'.

The real problem with direction of causation is identifying how it is to be induced in the first place. We shall try to solve this problem later, in the chapter on induction of causation. For now, suffices to say the following. In *de dicta* (logical) causation, theses are hierarchized by their epistemological roles (an axiom causes but is not caused by a resulting theorem, even if the latter implies the former, for instance); in *de re* (natural, temporal, spatial or extensional) causation, the order of things is often dictated by temporal or spatial sequences, for instances (logical issues also come into play).

There are cases in practice where deciding which item is the cause and which is the effect is virtually a matter of convention. This may occur in reciprocal causation, as well as in causation with permanently unknown (i.e. practically unknowable) direction. In such cases, the expressions "the cause" and "the effect" merge into one, becoming mere verbal differentiations. This is often true in the logical mode, and in the spatial and extensional modes; it occurs more rarely in the temporal and natural modes. The reason being that the only really absolute rule of direction we know is temporal sequence; other rules, though credible, are open to debate.

It should be stressed that the concept of direction (or orientation) concerns not only causation, but more broadly space and time in a variety of guises. It is therefore an issue in a wider and deeper ontological and epistemological context, not one reserved to causation. It might be viewed as one of the fundamental building-blocks of knowledge, and therefore not entirely definable with reference to other concepts.

It may be exemplified concretely by drawing a line on paper (this expresses its spatial component), and running a finger along it first one way, then the other (this expresses its temporal component, since the movement takes time to cover space); and saying "though the path covered is the same in both instances, the first movement is to be distinguished from the second – and this difference will be called one of direction". In this manner, the words 'from' and 'to', though very abstract[32], are shown to be meaningful, i.e. to symbolize a communicable distinction, which can by analogy be applied in other contexts.

---

[32] We cannot physically point to cases of 'direction', we can only point a finger *in the direction of* concrete objects (dogs, trees) – which is not the same act. A special mental capability and effort is

Such visual and mechanical demonstration merely aids the intuition[33] in focusing upon the intention of verbal expressions of direction. It does not, of course, by itself suffice to clearly define directionality in the context of causation, or to establish the direction of causation in particular cases. We must search for more precise means to achieve these ends. But we at least have a sort of ostensive-procedural definition of directionality in general, which gives *some* meaning to clauses like "from P to Q" and "from Q to P".

The propositions "P causes Q" and "Q causes P" are simply *declared unequal*. Causation in general is symbolized by a string of words, namely "P"-"causation"-"Q", analogous to a line; this line of relation is, however, to be taken as two-fold, i.e. as occasionally different in the senses P-Q and Q-P. *What* this difference signifies more deeply in formal terms, we cannot yet say; but we do believe *that* it exists, and wish to prepare for its linguistic expression by such declaration.

---

required to transcend the object pointed to and shift attention to the act and significance of pointing, of which direction is an abstract aspect.

33 Incidentally, even animals seem to intuitively (in the sense of wordlessly, effectively) understand direction. This is suggested, for instance, by their homing abilities, based on visual, auditory, and other sensory data (such as olfactory or gustative, in the case of ants, or magnetic, in the case of migrating birds). They can evidently even grasp direction of causation, knowing and remembering who or what hurt or pleased them. In my experience, however, dogs do not seem to understand finger-pointing; but some people claim that they do.

Chapter 5. CAUSATIVE SYLLOGISM

1. Causal or Effectual Chains.

The topic of concatenation of causations is an important field of research, though a tedious one. It is important not only to the natural sciences, which need to monitor or trace causal or effectual chains, but also to law and ethics.

To grasp its practical value in legal or ethical discourse, consider this example[34]: *a motorist overruns a pedestrian, who in the hospital where he is rushed is additionally the victim of some medical mishap – can the motorist be blamed for the poor pedestrian's subsequent misfortunes?* Such questions can only be convincingly answered through a systematic and wide-ranging reflection on causal logic.

The concept of concatenation refers primarily to 'chain reactions': P causes Q, which causes R, and so on; or conversely, R is effected by Q, which is effected by P, and so forth.

Clearly, the concepts of cause and effect here are relative to each other. In the context of deterministic causality, nothing is absolutely a cause or absolutely an effect; it is always the cause or effect *of* something.

All we wish to point out here is the obvious: that a phenomenon Q which is a cause in relation to another phenomenon R may itself stand as effect in relation to yet a third phenomenon P. Similarly, a phenomenon Q which is an effect in relation to another phenomenon P may itself stand as cause in relation to yet a third phenomenon R.

When we speak in terms of chains like P-Q-R, we stand back from the underlying bipolar relations of cause and effect and focus on the wider picture. The items P, Q, R may then be referred to, more indifferently, as successive *links* in the chain.

Needless to say, concatenation of events implies but is not implied by the seriality of events (in whatever appropriate sense of the term 'series'). Furthermore, even knowing that P causes Q and that Q causes R, we cannot presume concatenation. A series P-Q-R may be said to really form a chain, only if we can demonstrate that P, through the intermediary of Q, indeed causes R. This is not always feasible, for as we have seen the verb "causes" has a large variety of meaning.

You cannot just say "P causes Q and Q causes R, therefore P causes R" indiscriminately. This is one reason why a theoretical treatment of causal logic is essential to scientific thinking.

The search for concatenations varies in motive. Sometimes we are looking for the cause(s) of a cause, sometimes for the effect(s) of an effect, sometimes for some intermediary between a cause and an effect. We need not assume at the outset that all phenomena are bound to have causes and effects *ad infinitum*, nor that there has to be an infinity of intermediaries between any two given items.

---

[34] This example is based on one given by H. L. A. Hart and A. M. Honoré in *Causation in the Law* (Oxford: Oxford, 1959).

A cause without apparent prior cause would be called a primary cause; an effect without apparent posterior effect would be called an ultimate effect. A cause and effect without apparent intermediary would be referred to as immediate or contiguous; if they have an intermediary, they would be referred to as mediated.
If we speculate that Existence as a whole has a Beginning and/or an End, then of course we may speak of that as a First Cause and/or a Last Effect. Likewise, we need not *ab initio* prejudice the issue concerning specific events within Existence, be it infinite or finite, and at least to start with make allowances for (in some sense) causeless or effectless phenomena.

We have so far mentioned what may be called *orderly concatenation*. We also search for chains in the context of *parallelism of causes, or of effects*. We may need to know whether parallel causes or parallel effects are themselves causally related, and thus order them in relation to the initial cause or terminal effect concerned. In such case, we are identifying one of the two causes or two effects (as the case may be) as an intermediary between the other two items.
It should be stressed, however, that the arguments about parallelism considered here cannot strictly-speaking tell us which one of the two causes (or two effects) causes the other; for as we have mentioned in the preceding chapters, sometimes there is a hidden issue of direction of causation to consider. This issue has to be resolved separately, with reference to spatial, temporal, or other conceptual or logical considerations[35]. We shall simply ignore this problem of ordering for now, and regard the tacit condition as always satisfied.

We should, additionally, in passing, mention the phenomenon of **spiraling causation**, which we commonly refer to as *vicious circles*. This phenomenon is a special case of concomitant variation[36]. It occurs when an increase or decrease in a cause C ($C\pm x_1$) causes an increase or decrease in an effect E ($E\pm y_1$), which in turn causes another increase or decrease in C ($C\pm x_1\pm x_2$), which in turn causes another increase or decrease in E ($E\pm y_1\pm y_2$), and so forth.
The spiral need not constitute an infinite chain, even if complete causation is involved at each step, because each of the causations involved is independent of (i.e. not formally implied by) its predecessors, note well. Even so, a spiral may come to a halt because it is in fact implicitly conditional, i.e. partial causation is involved at each step. But we can also conceive of infinite spirals, in the case of ongoing processes continuing as long as the universe lasts.

The problem of causal or effectual chains is, as we shall see, essentially *syllogistic*. We need to identify which syllogisms involving causative propositions as premises yield such propositions as conclusion. In this research, it is as important to expose the invalidity of certain syllogisms as to identify the valid syllogisms, for inappropriate reasoning is common[37].

---

35 Meaning, ultimately, with reference to our insights and hypotheses concerning the phenomena of nature in question, and more radically to our philosophical ordering of knowledge on a grand scale.

36 See Appendix on J. S. Mill's Methods.

37 As a prescriptive science, Logic is ultimately only interested in valid argument. But as a descriptive one, it is very interested in knowing how (and how often) people tend to err in their reasoning processes. In this context we might apply the rule, if it can happen it will happen!

valid[39], i.e. logically acceptable, albeit their having premises of different (though equally strong) determinations. The conclusion, notice, has the same determination as the major premise. On the other hand, as we will show later, if the premises (with P, Q, R in a similar arrangement) have the same determination, i.e. both concern complete causation or both necessary causation, we are not permitted to draw any causative conclusion.

These examples reveal some of the complexities of causative argument.

We see from them that the ordering of the items involved in the premises affects the logical possibility of drawing a conclusion. In the first figure, two identical strong determinations yield a valid conclusion (of the same determination), whereas a mixture of such determinations is fruitless. In the second and third figures, the opposite is true; and furthermore, these differ from each other, in that a valid conclusion in the second figure follows the determination of the minor premise, whereas one in the third figure follows that of the major premise.

The problem becomes even more complicated when we investigate weak causations, which involve at least three items each (instead of two, as with strong causations). We discover, to give an extreme example, that whatever the figure considered, no conclusion can be drawn from two premises each of which concerns partial or contingent causation only. We then wonder what combinations of premises may be used to draw a conclusion about weak causation.

More broadly, considering that we have to deal with three figures, and eight possible determinations of causation for each premise, we have to examine 3*64=192 combinations, or 'moods' (as logicians say). What conclusion, if any, can be drawn from each one of those arguments; and how do we go about demonstrating it? Furthermore, we have so far mentioned syllogisms with only affirmative causative propositions; what of syllogisms involving propositions denying causation or a particular determination of causation?

Clearly, we cannot hope to reason correctly about causation without first dealing with causative syllogism in a thorough and systematic manner, so that we know precisely when an argument is valid and when it is not. If we limit our research to a few frequently used arguments, like those above shown, we will miss many opportunities for valid inference and risk making some invalid inferences. And in view of the volume of the problem, it has to be treated in as global a manner as possible.

This is our task in the next few chapters.

The research is tedious, because causative propositions are, as we have seen, very complex; they are each composed of two or more clauses, and most of these clauses are positive or negative conditional propositions, i.e. themselves complex.

In the simplest syllogisms, those involving strong determinations of causation only, and therefore the minimum number of (i.e. three) items, we can readily reduce causative reasoning to syllogism involving conditional propositions. The latter are reasonably well-known to logicians and to the public at large; a full treatment of them may be found in my work *Future Logic*.

---

39 Strictly speaking, the conclusion is permitted only if we can separately establish that the ordering of P and R as respectively cause and effect is acceptable. We shall deal with such details eventually.

But soon we find such simple methods inadequate. Syllogisms involving weak determinations or mixtures of strong and weak determinations are too complicated for us to feel secure with the results obtained by means of reduction. For certainty, I have had to develop a more complex method, called matricial analysis.

3. Figures and Moods.

A syllogism, we know thanks to Aristotle, consists of at least two premises and a conclusion. The premises together contain at least three items (terms or theses), at least one of which they have in common, and the conclusion contains at least two items, each of which was contained in a premise not containing the other.

Our job in syllogistic reasoning is to obtain from the premises, i.e. the given data, the information we need to construct the putative conclusion. If the premises, together and without reference to unstated assumptions, justify the conclusion, the syllogism proposed is *valid* deduction; otherwise it is *invalid*.

Validation (i.e. showing valid) justifies a form of reasoning; removing any uncertainty we may have about it or teaching us a new way of inference. Invalidation (i.e. showing invalid) is just as important, to contrast valid with invalid moods and thus set the limits of validity, and most of all to prevent us making mistakes in our thinking.

A putative conclusion may be invalid in the way of a *non-sequitur*, meaning that the conclusion does not conflict with anything in the premises, but just does not logically follow from them. Or, worse, it may be invalid in the way of *antinomy*, meaning that the conclusion is inconsistent (contradictory or contrary) with something in the premises[40].

In the case of a non-sequitur, we may be able to save the situation by stipulating some condition(s) under which the conclusion would follow; in that event, we may call the conclusion conditionally valid, or add the condition(s) to be satisfied to our premises as an additional premise to obtain an unconditionally valid conclusion, or again consider that we have a disjunctive conclusion whose alternatives include the satisfaction or non-satisfaction of the said condition(s).

In the case of an antinomy, we can redeem things by proposing the contradictory or a contrary of our invalid conclusion as a valid conclusion; if the invalid conclusion is a compound, we may be able to obtain a valid conclusion of the kind desired by negating some element(s) in it.

We are usually able to infer *some* information from the premises; but if this information does not add up to a *causative* proposition of some kind, we here consider the conjunction of the premises as a failure. For our task, in the present context, is not an investigation of deduction in the broadest sense, but specifically deduction of causative propositions from other causative propositions. Thus, do not be surprised if a syllogism is declared invalid even though some elements of a putative compound conclusion were inferable.

The evaluations of some moods may seem immediately or intuitively obvious; but some moods are too complicated for that and require careful examination. Some causative syllogisms, as already mentioned, can be validated by *direct reduction* to already established, non-causative syllogisms. Others are too complex for that, and can only be validated through *matricial analysis*, i.e. with painstaking reference to their corresponding matrix; this method

40 Non-sequitur is a generic term, including both antinomy and non-antinomic non-sequitur (or 'merely' non-sequitur) as its species.

will be described in detail later. Still others, though complex, can be validated by *direct and/or indirect reduction* (also known as reduction *ad absurdum*) to causative syllogisms already validated by other means (namely by matricial analysis).

Aristotle taught us that a syllogism may have one of three **figures**, according to the placement of the three items (terms or theses) in its premises and conclusion, as follows:

Table 5.1. The figures of (three-item) syllogism.

	Figure 1	Figure 2	Figure 3
Major premise:	Q – R	R – Q	Q – R
Minor premise:	P – Q	P – Q	Q – P
Conclusion:	P – R	P – R	P – R

Notice, in each of the figures, the positions of the item found in both premises but not in the conclusion (namely Q; this is known as the middle item). Notice also the various positions of the other two items, one of which (R, the major item) is found only in the major premise (traditionally stated first) and conclusion (traditionally stated last), and the other of which (P, the minor item) is found only in the minor premise (traditionally stated second) and conclusion. The positions of the items tell us which 'figure' the reasoning is in.[41]
Each Aristotelian figure refers to three items (P, Q, R). But in the present context we are also dealing with some four-item (P, Q, R, S) arguments, which as we shall see can be combined in three different ways (and many more, which we shall deal with in a later chapter). Thus, we shall have to refer to **subfigures**. We can call Aristotle's primary arrangement subfigure (a), and the three additional arrangements subfigures (b), (c), (d).

Table 5.2. Subfigures of each figure.

Subfigures	a	b	c	d
Definitions	both premises strong only	major premise strong only	minor premise strong only	neither premise strong only
Figure 1	QR PQ PR	QR P(S)Q P(S)R	Q(S)R PQ P(S)R	Q(P)R P(S)Q P(S)R
Figure 2	RQ PQ PR	RQ P(S)Q P(S)R	R(S)Q PQ P(S)R	R(P)Q P(S)Q P(S)R
Figure 3	QR QP PR	QR Q(S)P P(S)R	Q(S)R QP P(S)R	Q(P)R Q(S)P P(S)R

41 There is, in fact a fourth figure, viz. Z-Y/Y-X/X-Z, in which the major and minor premises of the first figure are effectively transposed or whose conclusion is converse compared to a first figure conclusion. But as Aristotle argued, this is not a natural movement of thought, even though we can occasionally make some interesting inferences through it. Considering the matter insignificant, nothing more will be said about it, here. See *Future Logic*, p. 38.

In (a), both premises involve strong determinations only; that is why there are only two items per premise (and in the conclusion). In (b) and (c), one premise (the major or minor, respectively) has only two items (implying the presence of only strong determination) and the other premise (and conclusion) has three items (implying the presence of joint strong and weak, or of only weak, determination). In (d), each premise (and the conclusion) involves three items (implying the absence of only-strong determination).

It is seen that the three-item symbolism (P, Q, R for the minor, middle and major items, respectively) is retained in four-item figures, except that we have an additional item (call it the **subsidiary** item, symbol S) appearing in a premise and the conclusion: S represents 'outside interference', as it were, in relation to the triad P-Q-R.

The important thing to note about this subsidiary item is that though it has to be mentioned in theoretical exposition and evaluation, as here, to place it and judge its impact, **it need not be mentioned**[42] **in practice**, because the conclusion follows the premises whatever its content happen to be. That is, the premise concerned and the conclusion need not specify "(complemented by S)".

On the other hand, the clause "(complemented by P)" in the major premise of subfigures 1d, 2d and 3d, **cannot be ignored in practice**, since the middle item Q might well cause the major item R with some complement(s) other than the minor item P, rather than with P. Even though P is mentioned in association with Q in the minor premise, that in itself does not imply the causation in the major premise to be true with it: this knowledge must be obtained by other means to enable the inference of the conclusion.

If, in our present context, we specify as well as the figure the precise determination and polarity involved in each of the premises and in a putative conclusion, we have pinpointed the precise *mood* under discussion. This expression refers to the formal aspects of a syllogism, which distinguish it from all others. Thus, for each figure of syllogism, there are many conceivable moods.

The **mood determinations** (numbered 1-9 for reference) found in each subfigure are given in the table below. These tell us the determinations of the premises involved, which may be 'strong only' (abbreviation, **so**), a 'mix of strong and weak' (**sw/ws**), or 'weak only' (**wo**). Due to the numbers of items allowed for a premise in each subfigure, the number of determinations found in each subfigure varies.

Table 5.3. Determinations found in each subfigure.

Subfig.	a	b		c		d			
Determ.	1	2	3	4	5	6	7	8	9
Major	so	so	so	sw/ws	wo	sw/ws	sw/ws	wo	wo
Minor	so	sw/ws	wo	so	so	sw/ws	wo	sw/ws	wo

42 The content of S has to be ultimately known, otherwise the clauses involving it cannot be claimed to be known true. Nevertheless, if the premise involving S is a product of previous inductive and deductive arguments, and thus considered reasonably settled, the content of S can be ignored contextually. The same applies, of course, when there are a plurality of complements, i.e. when S itself stands for a composite of partial causes.

There are 64 positive moods per figure, a total of 192 moods in all three of them. The system proposed now is to use three-digit identification numbers, or **mood numbers**. The hundreds will identify the figure 1, 2 or 3. The tens (#s 1-8, no 0 or 9) will tell us the major premise's determination. The units (#s 1-8, no 0 or 9) will specify the minor premise's determination. The subfigures (above labeled a-d) and modes (above labeled 1-9), shown in the preceding tables, are not explicitly mentioned in the mood number, but are tacitly implied by it.

Table 5.4. Subfigures, modes and moods.

Subfig.	Determ.	Premises		Moods Nos.	Qty
a	1	major minor	strong only strong only	tens 1, 4, 5 units 1, 4, 5	9
b	2	major minor	strong only sw, ws	tens 1, 4, 5 units 2, 3	6
	3	major minor	strong only weak only	tens 1, 4, 5 units 6, 7, 8	9
c	4	major minor	sw, ws strong only	tens 2, 3 units 1, 4, 5	6
	5	major minor	weak only strong only	tens 6, 7, 8 units 1, 4, 5	9
d	6	major minor	sw, ws sw, ws	tens 2, 3 units 2, 3	4
	7	major minor	sw, ws weak only	tens 2, 3 units 6, 7, 8	6
	8	major minor	weak only sw, ws	tens 6, 7, 8 units 2, 3	6
	9	major minor	weak only weak only	tens 6, 7, 8 units 6, 7, 8	9

I could of course have used letters instead of numbers to symbolize the different moods, but fearing to confuse the reader with yet more letter-symbols (the science of logic abounds with them) I have preferred number-symbols. Note that there are no moods numbered 01-10, 19-20, 29-30, 39-40, 49-50, 59-60, 69-70, 79-80, or 89+. The table below clarifies the meaning of each mood number within any given figure.

Table 5.5. Mood numbers in each figure.

Minor premise	Major premise mn=1	mq=2	np=3	pq=4	m=5	n=6	p=7	q=8
mn=1	11	21	31	41	51	61	71	81
mq=2	12	22	32	42	52	62	72	82
np=3	13	23	33	43	53	63	73	83
pq=4	14	24	34	44	54	64	74	84
m=5	15	25	35	45	55	65	75	85
n=6	16	26	36	46	56	66	76	86
p=7	17	27	37	47	57	67	77	87
n=8	18	28	38	48	58	68	78	88

It is useful to expand the above table as done below, to show precisely what combination of determinations in the premises each mood number refers to.

Table 5.6. For each figure, mood numbers and determinations of major and minor premises.

Minor premise	Major premise **mn=1**	**mq=2**	**np=3**	**pq=4**	**m=5**	**n=6**	**p=7**	**q=8**
mn=1	11	21	31	41	51	61	71	81
major	mn	mq	np	pq	m	n	p	q
minor	mn	mn	mn	mn	mn	mn	mn	mn
mq=2	12	22	32	42	52	62	72	82
major	mn	mq	np	pq	m	n	p	q
minor	mq	mq	mq	mq	mq	mq	mq	mq
np=3	13	23	33	43	53	63	73	83
major	mn	mq	np	pq	m	n	p	q
minor	np	np	np	np	np	np	np	np
pq=4	14	24	34	44	54	64	74	84
major	mn	mq	np	pq	m	n	p	q
minor	pq	pq	pq	pq	pq	pq	pq	pq
m=5	15	25	35	45	55	65	75	85
major	mn	mq	np	pq	m	n	p	q
minor	m	m	m	m	m	m	m	m
n=6	16	26	36	46	56	66	76	86
major	mn	mq	np	pq	m	n	p	q
minor	n	n	n	n	n	n	n	n
p=7	17	27	37	47	57	67	77	87
major	mn	mq	np	pq	m	n	p	q
minor	p	p	p	p	p	p	p	p

q=8	18	28	38	48	58	68	78	88
major minor	mn q	mq q	np q	pq q	m q	n q	p q	q q

Note that if you divide the above table in four equal squares, the top left square involves premises with only joint determinations, the bottom left one a joint major premise with a generic minor premise, the top right square involves a generic major premise with a joint minor premise, and finally the bottom right square premises with only generic determinations.

In my listing of moods in the next chapter, I do not follow their numerical order. Rather, I present the moods in a diagonal order with reference to the above table, starting with the strongest (top left hand corner) and ending with the weakest (bottom right hand corner).
Four moods, involving only strong determinations (namely, Nos. 11, 14, 41 and 44), have no 'mirror images'; the remaining sixty moods may be treated in pairs, for each has a mirror image (thus, 12 and 13 are essentially the same, as are 21 and 31, and so forth). I present explicitly the more positive mood of each pair (e.g. 12), and only mention its mirror image (e.g. 13).
Moods with a stronger major premise are listed before moods with a stronger minor premise (e.g. 12, 13 before 21, 31). Moods with premises of uniform determination are listed before moods of mixed determination (e.g. 22, 33 before 23, 32). And so forth, the goal being to present all moods in a natural order.

Chapter 6. LIST OF POSITIVE MOODS

1. Valid and Invalid Moods.

As we have seen in the preceding chapter, causative syllogism with both premises affirmative has 64 conceivable moods in each of three figures. In the present chapter, we shall list all these moods, and for each mood specify whether it is valid or invalid, and briefly the basis of this evaluation.

For any positive mood, there are four initially conceivable, putative conclusions, corresponding to the four generic determinations, which we have symbolized as **m** (for complete causation), **n** (for necessary causation), **p** (for partial causation) and **q** (for contingent causation). However, at most two such conclusions may be valid for any given mood, since the determinations **m** and **p** are contrary and **n** and **q** are contrary. Thus, there are eight logically possible conclusions for any positive causative syllogism, namely:

mn, mq, np, pq, m, n, p, q.

A putative conclusion is **valid** – if it logically follows from the given premises, i.e. if its contradictory is logically incompatible with them or any of their implications. A putative conclusion is declared **invalid** – if it is not valid, for whatever reason; the reason may be that the premises themselves are inconsistent, or that the contradictory of the putative conclusion is compatible with them (in which case the putative conclusion is a non-sequitur), or that the putative conclusion is incompatible with the premises (in which case the putative conclusion is an antinomy and its contradictory is valid).

If one of the eight joint or generic determinations is demonstrably inferable from the premises concerned, the mood is valid. If none of them can be legitimately drawn from the premises, the mood is invalid. Additionally, some moods are invalid at the outset because the premises concerned are in fact incompatible in some respect(s); i.e. at least one clause of each is implicitly denied by at least one clause of the other.

We shall, to repeat, in the present chapter only list the moods and their valid conclusion(s) if any, and state succinctly the basis of these results. In the next two chapters, we will show how these results were obtained, systematically and in detail; i.e. we will justify our claims.

Note that, in accord with the tradition in logic, if a mood is valid, only the correct conclusion(s) is/are mentioned in the listing; other conclusions, not mentioned, are tacitly implied to be incorrect. But it is well to keep *both* the valid and invalid conclusions in mind; for the purpose of the whole exercise is not only to instruct us in proper reasoning, but also to save us from improper reasoning!

As will be seen, some conclusions have to be validated or invalidated by matricial analysis; moods with at least one conclusion treated by matricial analysis may be called primary. The remaining conclusions may be validated or invalidated by reduction to the primary moods; moods all of whose conceivable conclusions have been treated by reduction may be called secondary or derived.

As for moods invalid due to inconsistency between the premises, they need not of course be subjected to matricial analysis or reduction. Note that it may be possible to affirm or deny some conclusion(s) from *some* of their clauses, if the inherent contradiction is disregarded; but that would be nonsensical, for if *all* the clauses are taken into consideration, we have to admit that the premises in question cannot in fact come together to yield such conclusion(s).

All evaluations could be performed by matricial analysis; but this process is long-winded, so we try and avoid it as much as possible. Such avoidance is anyway not sheer laziness on our part, for it is instructive to be aware of the interrelations between moods which reduction reveals. We learn, in this way, that causative syllogisms together constitute a close-knit totality, a system.

It should be stressed that the issue of **direction of causation** is ignored throughout the present formal treatment. In figure 1, this is no problem; i.e. given the directions of causation implied in the premises (namely, from P to Q and from Q to R), the direction of causation implied in an eventual valid conclusion (viz. from P to R) follows necessarily. But in figures 2 and 3, any eventual valid conclusions must be regarded as *conditionally* valid, i.e. on the proviso that the implied direction of causation (viz. from P to R) is established by other means.

However, if it turns out that a figure 2 or 3 conclusion is found *not to* satisfy this condition, the underlying implications between the items concerned (P and R) may still *in certain cases* result in a causative conclusion in the reverse direction. Such cases are formally predictable, simply by *transposition of the premises* concerned. If such transposition has some causative conclusion, then the direction of causation implied by that conclusion (i.e. from R to P) will be unconditionally valid. For if there is causation between P and R, it is bound to be in one direction or the other.

a. Strong determinations. If two premises yield the conclusion 'P is a complete cause of R', then their transposition will yield the converse conclusion 'R is a necessary cause of P'. If we do not know the direction of causation, we cannot know which of these conclusions is the correct one, but we *do* know that at least one of them must be. If we know that it is not this one, then we know it must be that one. Similarly, with the eventual conclusions 'P is a necessary cause of R' and 'R is a complete cause of P'.[43]
b. Weak determinations. If two premises yield the conclusion 'P (complemented by S) is a partial cause of R', and this conclusion is found unjustified with regard to the issue of direction of causation, then its converse has to be admitted as valid, viz. 'R (complemented by notS) is a contingent cause of P' (note well the change of polarity of the complement). Similarly, if we know that an eventual conclusion of the form 'P (complemented by S) is a contingent cause of R' is inapplicable with respect to the issue of direction of causation, then we may affirm 'R (complemented by notS) is a partial cause of P' instead.

The following **statistics**, based on the listings below, are of interest:

- In figure 1, out of 64 conceivable positive moods, 30 are valid and 34 are invalid (of which 10, due to inconsistency in the premises).
- In figure 2, out of 64 conceivable positive moods, 18 are valid and 46 are invalid (of which 6, due to inconsistency in the premises).

[43] We can on this basis anticipate, in figures 2 and 3, the validity or invalidity of some moods on the basis of others. For the order of the premises in these figures is arbitrary.

- In figure 1, out of 64 conceivable positive moods, 18 are valid and 46 are invalid (of which 10, due to inconsistency in the premises).

Thus, out of the 192 positive moods considered, **66 (34%) are valid and 126 (66%) are invalid**. Obviously, in view of this validity rate, such reasoning cannot be left to chance!

2. Moods in Figure 1.

§1. Mood No. 111 = **mn/mn/mn**.

Q is a complete and necessary cause of R;
P is a complete and necessary cause of Q;
so, P is a complete and necessary cause of R.

VALID by reduction to moods 155, 166.

No mirror mood.

§2. Mood No. 112 = **mn/mq/mq**.

Q is a complete and necessary cause of R;
P (complemented by S) is a complete and contingent cause of Q;
so, P (complemented by S) is a complete and contingent cause of R.

VALID by reduction to moods 118, 155.

No. 113 = **mn/np/np** (similarly, through 117, 166).

§3. Mood No. 121 = **mq/mn/mq**.

Q (complemented by S) is a complete and contingent cause of R;
P is a complete and necessary cause of Q;
so, P (complemented by S) is a complete and contingent cause of R.

VALID by reduction to moods 155, 181.

No. 131 = **np/mn/np** (similarly, through 166, 171).

§4. Mood No. 122 = **mq/mq**.

Q (complemented by P) is a complete and contingent cause of R;
P (complemented by S) is a complete and contingent cause of Q;
does it follow that P is (complemented by S) a cause of R? No!

INVALID due to inconsistency of premises.

No. 133 = **np/np** (similarly).

§5. Mood No. 123 = **mq/np**.

Q (complemented by P) is a complete and contingent cause of R;
P (complemented by S) is a partial and necessary cause of Q;
does it follow that P is (complemented by S) a cause of R? No!

INVALID due to inconsistency of premises.

No. 132 = **np/mq** (similarly).

§6. Mood No. 114 = **mn/pq/pq**.

Q is a complete and necessary cause of R;
P (complemented by S) is a partial and contingent cause of Q;
so, P (complemented by S) is a partial and contingent cause of R.

VALID by reduction to moods 117, 118.

No mirror mood.

§7. Mood No. 141 = **pq/mn/pq.**

Q (complemented by S) is a partial and contingent cause of R; P is a complete and necessary cause of Q; so, P (complemented by S) is a partial and contingent cause of R.	**VALID** by reduction to moods 171, 181.

No mirror mood.

§8. Mood No. 124 = **mq/pq/q.**

Q (complemented by P) is a complete and contingent cause of R; P (complemented by S) is a partial and contingent cause of Q; so, P (complemented by S) is a contingent cause of R.	**VALID** by reduction to 128 or 184 and by matricial analysis.

No. 134 = **np/pq/p** (similarly, through 137 or 174 and MA).

§9. Mood No. 142. **pq/mq.**

Q (complemented by P) is a partial and contingent cause of R; P (complemented by S) is a complete and contingent cause of Q; does it follow that P is (complemented by S) a cause of R? No!	INVALID due to inconsistency of premises.

No. 143 = **pq/np** (similarly).

§10. Mood No. 144 = **pq/pq/pq.**

Q (complemented by P) is a partial and contingent cause of R; P (complemented by S) is a partial and contingent cause of Q; so, P (complemented by S) is a partial and contingent cause of R.	**VALID** by reduction to moods 147+148, or 174+184.

No mirror mood.

§11. Mood No. 115 = **mn/m/m.**

Q is a complete and necessary cause of R; P is a complete cause of Q; so, P is a complete cause of R.	**VALID** by reduction to moods 111, 112, 155.

No. 116 = **mn/n/n** (similarly, through 111, 113, 166).

§12. Mood No. 151 = **m/mn/m.**

Q is a complete cause of R; P is a complete and necessary cause of Q; so, P is a complete cause of R.	**VALID** by reduction to moods 111, 121, 155.

No. 161 = **n/mn/n** (similarly, through 111, 131, 166).

§13. Mood No. 125 = **mq/m/m.**

Q (complemented by S) is a complete and contingent cause of R; P is a complete cause of Q; so, P is a complete cause of R.	**VALID** by reduction to 121, 155 and by matricial analysis.

No. 136 = **np/n/n** (similarly, through 131, 166 and MA).

§14. Mood No. 126 = **mq/n**. — INVALID by reduction to mood 121 and by matricial analysis.

Q (complemented by S) is a complete and contingent cause of R;
P is a necessary cause of Q;
does it follow that P is (complemented by S) a cause of R? No!

No. 135 = **np/m** (similarly, through 131 and MA).

§15. Mood No. 152 = **m/mq/m**. — **VALID** by reduction to 112, 155 and by matricial analysis.

Q is a complete cause of R;
P (complemented by S) is a complete and contingent cause of Q;
so, P is a complete cause of R.

No. 163 = **n/np/n** (similarly, through 113, 166 and MA).

§16. Mood No. 153 = **m/np**. — INVALID by reduction to mood 113 and by matricial analysis.

Q is a complete cause of R;
P (complemented by S) is a partial and necessary cause of Q;
does it follow that P is (complemented by S) a cause of R? No!

No. 162 = **n/mq** (similarly, through 112 and MA).

§17. Mood No. 117 = **mn/p/p**. — **VALID** by reduction to 113, 114 and by matricial analysis.

Q is a complete and necessary cause of R;
P (complemented by S) is a partial cause of Q;
so, P (complemented by S) is a partial cause of R.

No. 118 = **mn/q/q** (similarly, through 112, 114 and MA).

§18. Mood No. 171 = **p/mn/p**. — **VALID** by reduction to 131, 141 and by matricial analysis.

Q (complemented by S) is a partial cause of R;
P is a complete and necessary cause of Q;
so, P (complemented by S) is a partial cause of R.

No. 181 = **q/mn/q** (similarly, through 121, 141 and MA).

§19. Mood No. 127 = **mq/p**. — INVALID by reduction to mood 124 and by matricial analysis.

Q (complemented by P) is a complete and contingent cause of R;
P (complemented by S) is a partial cause of Q;
does it follow that P is (complemented by S) a cause of R? No!

No. 138 = **np/q** (similarly, through 134 and MA).

§20. Mood No. 128 = **mq/q/q**. — **VALID** by reduction to 122, 124 and by matricial analysis.

Q (complemented by P) is a complete and contingent cause of R;
P (complemented by S) is a contingent cause of Q;
so, P (complemented by S) is a contingent cause of R.

No. 137 = **np/p/p** (similarly, through 133, 134 and MA).

§21. Mood No. 172 = **p/mq**. INVALID due to inconsistency of premises.

Q (complemented by P) is a partial cause of R;
P (complemented by S) is a complete and contingent cause of Q;
does it follow that P is (complemented by S) a cause of R? No!

No. 183 = **q/np** (similarly).

§22. Mood No. 173 = **p/np**. INVALID due to inconsistency of premises.

Q (complemented by P) is a partial cause of R;
P (complemented by S) is a partial and necessary cause of Q;
does it follow that P is (complemented by S) a cause of R? No!

No. 182 = **q/mq** (similarly).

§23. Mood No. 145 = **pq/m**. INVALID by reduction to mood 141 and by matricial analysis.

Q (complemented by S) is a partial and contingent cause of R;
P is a complete cause of Q;
does it follow that P is (complemented by S) a cause of R? No!

No. 146 = **pq/n** (similarly, through 141 and MA).

§24. Mood No. 154 = **m/pq**. INVALID by reduction to 114, 124 and by matricial analysis.

Q is a complete cause of R;
P (complemented by S) is a partial and contingent cause of Q;
does it follow that P is (complemented by S) a cause of R? No!

No. 164 = **n/pq** (similarly, through 114, 134 and MA).

§25. Mood No. 147 = **pq/p/p**. **VALID** by reduction to mood 144 and by matricial analysis.

Q (complemented by P) is a partial and contingent cause of R;
P (complemented by S) is a partial cause of Q;
so, P (complemented by S) is a partial cause of R.

No. 148 = **pq/q/q** (similarly, through 144 and MA).

§26. Mood No. 174 = **p/pq/p**. **VALID** by reduction to mood 134 and by matricial analysis.

Q (complemented by P) is a partial cause of R;
P (complemented by S) is a partial and contingent cause of Q;
so, P (complemented by S) is a partial cause of R.

No. 184 = **q/pq/q** (similarly, through 124 and MA).

§27. Mood No. 155 = **m/m/m**. **VALID** by reduction to 111, 112 and by matricial analysis.

Q is a complete cause of R;
P is a complete cause of Q;
so, P is a complete cause of R.

No. 166 = **n/n/n** (similarly, through 111, 113 and MA).

§28. Mood No. 156 = **m/n**.

Q is a complete cause of R;
P is a necessary cause of Q;
does it follow that P is a cause of R? No!

INVALID by reduction to moods 111, 113, 121.

No. 165 = **n/m** (similarly, through 111, 112, 131).

§29. Mood No. 157 = **m/p**.

Q is a complete cause of R;
P (complemented by S) is a partial cause of Q;
does it follow that P is (complemented by S) a cause of R? No!

INVALID by reduction to moods 113, 114, 124.

No. 168 = **n/q** (similarly, through 112, 114, 134).

§30. Mood No. 158 = **m/q**.

Q is a complete cause of R;
P (complemented by S) is a contingent cause of Q;
does it follow that P is (complemented by S) a cause of R? No!

INVALID by reduction to moods 112, 114, 152.

No. 167 = **n/p** (similarly, through 113, 114, 163).

§31. Mood No. 175 = **p/m**.

Q (complemented by S) is a partial cause of R;
P is a complete cause of Q;
does it follow that P is (complemented by S) a cause of R? No!

INVALID by reduction to moods 131, 135.

No. 186 = **q/n** (similarly, through 121, 126).

§32. Mood No. 176 = **p/n**.

Q (complemented by S) is a partial cause of R;
P is a necessary cause of Q;
does it follow that P is (complemented by S) a cause of R? No!

INVALID by reduction to moods 131, 136, 141.

No. 185 = **q/m** (similarly, through 121, 125, 141).

§33. Mood No. 177 = **p/p**.

Q (complemented by P) is a partial cause of R;
P (complemented by S) is a partial cause of Q;
does it follow that P is (complemented by S) a cause of R? No!

INVALID by reduction to 133, 134 and by matricial analysis.

No. 188 = **q/q** (similarly, through 122, 124 and MA).

§34. Mood No. 178 = **p/q**.

Q (complemented by P) is a partial cause of R;
P (complemented by S) is a contingent cause of Q;
does it follow that P is (complemented by S) a cause of R? No!

INVALID by reduction to moods 134, 138.

No. 187 = **q/p** (similarly, through 124, 127).

3. Moods in Figure 2.

§1. Mood No. 211 = **mn/mn/mn**. **VALID** by reduction to mood 111.

R is a complete and necessary cause of Q;
P is a complete and necessary cause of Q;
so, P is a complete and necessary cause of R.

No mirror mood.

§2. Mood No. 212 = **mn/mq/mq**. **VALID** by reduction to mood 112.

R is a complete and necessary cause of Q;
P (complemented by S) is a complete and contingent cause of Q;
so, P (complemented by S) is a complete and contingent cause of R.

No. 213 = **mn/np/np** (similarly, through 113).

§3. Mood No. 221 = **mq/mn/n**. **VALID** by reduction to mood 256 and by matricial analysis.

R (complemented by S) is a complete and contingent cause of Q;
P is a complete and necessary cause of Q;
so, P is a necessary cause of R.

No. 231 = **np/mn/m** (similarly, through 265 and MA).

§4. Mood No. 222 = **mq/mq**. INVALID by matricial analysis.

R (complemented by P) is a complete and contingent cause of Q;
P (complemented by S) is a complete and contingent cause of Q;
does it follow that P is (complemented by S) a cause of R? No!

No. 233 = **np/np** (similarly, through MA).

§5. Mood No. 223 = **mq/np**. INVALID due to inconsistency of premises.

R (complemented by P) is a complete and contingent cause of Q;
P (complemented by S) is a partial and necessary cause of Q;
does it follow that P is (complemented by S) a cause of R? No!

No. 232 = **np/mq** (similarly).

§6. Mood No. 214 = **mn/pq/pq**. **VALID** by reduction to mood 114.

R is a complete and necessary cause of Q;
P (complemented by S) is a partial and contingent cause of Q;
so, P (complemented by S) is a partial and contingent cause of R.

No mirror mood.

§7. Mood No. 241 = **pq/mn**. INVALID by matricial analysis.

R (complemented by S) is a partial and contingent cause of Q;
P is a complete and necessary cause of Q;
does it follow that P is (complemented by S) a cause of R? No!

No mirror mood.

§8. Mood No. 224 = **mq/pq**.	INVALID
R (complemented by P) is a complete and contingent cause of Q; P (complemented by S) is a partial and contingent cause of Q; does it follow that P is (complemented by S) a cause of R? No!	by matricial analysis.

No. 234 = **np/pq** (similarly, through MA).

§9. Mood No. 242. **pq/mq**.	INVALID
R (complemented by P) is a partial and contingent cause of Q; P (complemented by S) is a complete and contingent cause of Q; does it follow that P is (complemented by S) a cause of R? No!	due to inconsistency of premises.

No. 243 = **pq/np** (similarly).

§10. Mood No. 244 = **pq/pq**.	INVALID
R (complemented by P) is a partial and contingent cause of Q; P (complemented by S) is a partial and contingent cause of Q; does it follow that P is (complemented by S) a cause of R? No!	by matricial analysis.

No mirror mood.

§11. Mood No. 215 = **mn/m/m**.	**VALID**
R is a complete and necessary cause of Q; P is a complete cause of Q; so, P is a complete cause of R.	by reduction to mood 115.

No. 216 = **mn/n/n** (similarly, through 116).

§12. Mood No. 251 = **m/mn/n**.	**VALID**
R is a complete cause of Q; P is a complete and necessary cause of Q; so, P is a necessary cause of R.	by reduction to mood 161.

No. 261 = **n/mn/m** (similarly, through 151).

§13. Mood No. 225 = **mq/m**.	INVALID
R (complemented by S) is a complete and contingent cause of Q; P is a complete cause of Q; does it follow that P is (complemented by S) a cause of R? No!	by reduction to moods 221, 222.

No. 236 = **np/n** (similarly, through 231, 233).

§14. Mood No. 226 = **mq/n/n**.	**VALID**
R (complemented by S) is a complete and contingent cause of Q; P is a necessary cause of Q; so, P is a necessary cause of R.	by reduction to moods 221, 256.

No. 235 = **np/m/m** (similarly, through 231, 265).

§15. Mood No. 252 = **m/mq**.	INVALID
R is a complete cause of Q; P (complemented by S) is a complete and contingent cause of Q; does it follow that P is (complemented by S) a cause of R? No!	by reduction to mood 162.

No. 263 = **n/np** (similarly, through 153).

§16. Mood No. 253 = **m/np/n**.	**VALID**
R is a complete cause of Q; P (complemented by S) is a partial and necessary cause of Q; so, P is a necessary cause of R.	by reduction to mood 163.

No. 262 = **n/mq/m** (similarly, through 152).

§17. Mood No. 217 = **mn/p/p**.	**VALID**
R is a complete and necessary cause of Q; P (complemented by S) is a partial cause of Q; so, P (complemented by S) is a partial cause of R.	by reduction to mood 117.

No. 218 = **mn/q/q** (similarly, through 118).

§18. Mood No. 271 = **p/mn**.	INVALID
R (complemented by S) is a partial cause of Q; P is a complete and necessary cause of Q; does it follow that P is (complemented by S) a cause of R? No!	by reduction to moods 231, 241.

No. 281 = **q/mn** (similarly, through 221, 241).

§19. Mood No. 227 = **mq/p**.	INVALID
R (complemented by P) is a complete and contingent cause of Q; P (complemented by S) is a partial cause of Q; does it follow that P is (complemented by S) a cause of R? No!	by reduction to mood 224.

No. 238 = **np/q** (similarly, through 234).

§20. Mood No. 228 = **mq/q**.	INVALID
R (complemented by P) is a complete and contingent cause of Q; P (complemented by S) is a contingent cause of Q; does it follow that P is (complemented by S) a cause of R? No!	by reduction to mood 222.

No. 237 = **np/p** (similarly, through 233).

§21. Mood No. 272 = **p/mq**.	INVALID
R (complemented by P) is a partial cause of Q; P (complemented by S) is a complete and contingent cause of Q; does it follow that P is (complemented by S) a cause of R? No!	due to inconsistency of premises.

No. 283 = **q/np** (similarly).

§22. Mood No. 273 = **p/np**. — INVALID by reduction to mood 233.

R (complemented by P) is a partial cause of Q;
P (complemented by S) is a partial and necessary cause of Q;
does it follow that P is (complemented by S) a cause of R? No!

No. 282 = **q/mq** (similarly, through 222).

§23. Mood No. 245 = **pq/m**. — INVALID by reduction to mood 241.

R (complemented by S) is a partial and contingent cause of Q;
P is a complete cause of Q;
does it follow that P is (complemented by S) a cause of R? No!

No. 246 = **pq/n** (similarly, through 241).

§24. Mood No. 254 = **m/pq**. — INVALID by reduction to mood 164.

R is a complete cause of Q;
P (complemented by S) is a partial and contingent cause of Q;
does it follow that P is (complemented by S) a cause of R? No!

No. 264 = **n/pq** (similarly, through 154).

§25. Mood No. 247 = **pq/p**. — INVALID by reduction to mood 241.

R (complemented by P) is a partial and contingent cause of Q;
P (complemented by S) is a partial cause of Q;
does it follow that P is (complemented by S) a cause of R? No!

No. 248 = **pq/q** (similarly, through 244).

§26. Mood No. 274 = **p/pq**. — INVALID by reduction to mood 234.

R (complemented by P) is a partial cause of Q;
P (complemented by S) is a partial and contingent cause of Q;
does it follow that P is (complemented by S) a cause of R? No!

No. 284 = **q/pq** (similarly, through 224).

§27. Mood No. 255 = **m/m**. — INVALID by reduction to mood 165.

R is a complete cause of Q;
P is a complete cause of Q;
does it follow that P is (complemented by S) a cause of R? No!

No. 266 = **n/n** (similarly, through 156).

§28. Mood No. 256 = **m/n/n**. — **VALID** by reduction to mood 166.

R is a complete cause of Q;
P is a necessary cause of Q;
so, P is a necessary cause of R.

No. 265 = **n/m/m** (similarly, through 155).

§29. Mood No. 257 = **m/p**. INVALID by reduction to mood 167.

R is a complete cause of Q;
P (complemented by S) is a partial cause of Q;
does it follow that P is (complemented by S) a cause of R? No!

No. 268 = **n/q** (similarly, through 158).

§30. Mood No. 258 = **m/q**. INVALID by reduction to mood 168.

R is a complete cause of Q;
P (complemented by S) is a contingent cause of Q;
does it follow that P is (complemented by S) a cause of R? No!

No. 267 = **n/p** (similarly, through 157).

§31. Mood No. 275 = **p/m**. INVALID by reduction to moods 231, 241.

R (complemented by S) is a partial cause of Q;
P is a complete cause of Q;
does it follow that P is (complemented by S) a cause of R? No!

No. 286 = **q/n** (similarly, through 221, 241).

§32. Mood No. 276 = **p/n**. INVALID by reduction to moods 231, 233.

R (complemented by S) is a partial cause of Q;
P is a necessary cause of Q;
does it follow that P is (complemented by S) a cause of R? No!

No. 285 = **q/m** (similarly, through 221, 222).

§33. Mood No. 277 = **p/p**. INVALID by reduction to mood 233.

R (complemented by P) is a partial cause of Q;
P (complemented by S) is a partial cause of Q;
does it follow that P is (complemented by S) a cause of R? No!

No. 288 = **q/q** (similarly, through 222).

§34. Mood No. 278 = **p/q**. INVALID by reduction to mood 234.

R (complemented by P) is a partial cause of Q;
P (complemented by S) is a contingent cause of Q;
does it follow that P is (complemented by S) a cause of R? No!

No. 287 = **q/p** (similarly, through 224).

4. Moods in Figure 3.

§1. Mood No. 311 = **mn/mn/mn**. **VALID** by reduction to mood 111.

Q is a complete and necessary cause of R;
Q is a complete and necessary cause of P;
so, P is a complete and necessary cause of R.

No mirror mood.

§2. Mood No. 312 = **mn/mq/n**.	**VALID**
Q is a complete and necessary cause of R; Q (complemented by S) is a complete and contingent cause of P; so, P is a necessary cause of R.	by reduction to mood 365 and matricial analysis.
No. 313 = **mn/np/m** (similarly, through 356 and MA).	
§3. Mood No. 321 = **mq/mn/mq**.	**VALID**
Q (complemented by S) is a complete and contingent cause of R; Q is a complete and necessary cause of P; so, P (complemented by S) is a complete and contingent cause of R.	by reduction to mood 121.
No. 331 = **np/mn/np** (similarly, through 131).	
§4. Mood No. 322 = **mq/mq**.	INVALID
Q (complemented by P) is a complete and contingent cause of R; Q (complemented by S) is a complete and contingent cause of P; does it follow that P is (complemented by S) a cause of R? No!	due to inconsistency of premises.
No. 333 = **np/np** (similarly).	
§5. Mood No. 323 = **mq/np**.	INVALID
Q (complemented by P) is a complete and contingent cause of R; Q (complemented by S) is a partial and necessary cause of P; does it follow that P is (complemented by S) a cause of R? No!	due to inconsistency of premises.
No. 332 = **np/mq** (similarly).	
§6. Mood No. 314 = **mn/pq**.	INVALID
Q is a complete and necessary cause of R; Q (complemented by S) is a partial and contingent cause of P; does it follow that P is (complemented by S) a cause of R? No!	by matricial analysis.
No mirror mood.	
§7. Mood No. 341 = **pq/mn/pq**.	**VALID**
Q (complemented by S) is a partial and contingent cause of R; Q is a complete and necessary cause of P; so, P (complemented by S) is a partial and contingent cause of R.	by reduction to mood 141.
No mirror mood.	
§8. Mood No. 324 = **mq/pq**.	INVALID
Q (complemented by P) is a complete and contingent cause of R; Q (complemented by S) is a partial and contingent cause of P; does it follow that P is (complemented by S) a cause of R? No!	by matricial analysis.
No. 334 = **np/pq** (similarly, through MA).	

§9. Mood No. 342. **pq/mq**.

Q (complemented by P) is a partial and contingent cause of R; Q (complemented by S) is a complete and contingent cause of P; does it follow that P is (complemented by S) a cause of R? No!	INVALID due to inconsistency of premises.

No. 343 = **pq/np** (similarly).

§10. Mood No. 344 = **pq/pq**.

Q (complemented by P) is a partial and contingent cause of R; Q (complemented by S) is a partial and contingent cause of P; does it follow that P is (complemented by S) a cause of R? No!	INVALID by matricial analysis.

No mirror mood.

§11. Mood No. 315 = **mn/m/n**.

Q is a complete and necessary cause of R; Q is a complete cause of P; so, P is a necessary cause of R.	**VALID** by reduction to mood 116.

No. 316 = **mn/n/m** (similarly, through 115).

§12. Mood No. 351 = **m/mn/m**.

Q is a complete cause of R; Q is a complete and necessary cause of P; so, P is a complete cause of R.	**VALID** by reduction to moods 151.

No. 361 = **n/mn/n** (similarly, through 161).

§13. Mood No. 325 = **mq/m**.

Q (complemented by S) is a complete and contingent cause of R; Q is a complete cause of P; does it follow that P is (complemented by S) a cause of R? No!	INVALID by reduction to mood 126.

No. 336 = **np/n** (similarly, through 135).

§14. Mood No. 326 = **mq/n/m**.

Q (complemented by S) is a complete and contingent cause of R; Q is a necessary cause of P; so, P is a complete cause of R.	**VALID** by reduction to mood 125.

No. 335 = **np/m/n** (similarly, through 136).

§15. Mood No. 352 = **m/mq**.

Q is a complete cause of R; Q (complemented by S) is a complete and contingent cause of P; does it follow that P is (complemented by S) a cause of R? No!	INVALID by reduction to mood 312 and by matricial analysis.

No. 363 = **n/np** (similarly, through 313 and MA).

§16. Mood No. 353 = **m/np/m**. Q is a complete cause of R; Q (complemented by S) is a partial and necessary cause of P; so, P is a complete cause of R. No. 362 = **n/mq/n** (similarly, through 312, 365).	**VALID** by reduction to moods 313, 356.
§17. Mood No. 317 = **mn/p**. Q is a complete and necessary cause of R; Q (complemented by S) is a partial cause of P; does it follow that P is (complemented by S) a cause of R? No! No. 318 = **mn/q** (similarly, through 312, 314).	INVALID by reduction to moods 313, 314.
§18. Mood No. 371 = **p/mn/p**. Q (complemented by S) is a partial cause of R; Q is a complete and necessary cause of P; so, P (complemented by S) is a partial cause of R. No. 381 = **q/mn/q** (similarly, through 181).	**VALID** by reduction to mood 171.
§19. Mood No. 327 = **mq/p**. Q (complemented by P) is a complete and contingent cause of R; Q (complemented by S) is a partial cause of P; does it follow that P is (complemented by S) a cause of R? No! No. 338 = **np/q** (similarly, through 334).	INVALID by reduction to mood 324.
§20. Mood No. 328 = **mq/q**. Q (complemented by P) is a complete and contingent cause of R; Q (complemented by S) is a contingent cause of P; does it follow that P is (complemented by S) a cause of R? No! No. 337 = **np/p** (similarly, through 334).	INVALID by reduction to mood 324.
§21. Mood No. 372 = **p/mq**. Q (complemented by P) is a partial cause of R; Q (complemented by S) is a complete and contingent cause of P; does it follow that P is (complemented by S) a cause of R? No! No. 383 = **q/np** (similarly).	INVALID due to inconsistency of premises.
§22. Mood No. 373 = **p/np**. Q (complemented by P) is a partial cause of R; Q (complemented by S) is a partial and necessary cause of P; does it follow that P is (complemented by S) a cause of R? No! No. 382 = **q/mq** (similarly).	INVALID due to inconsistency of premises.

§23. Mood No. 345 = **pq/m**.

Q (complemented by S) is a partial and contingent cause of R;
Q is a complete cause of P;
does it follow that P is (complemented by S) a cause of R? No!

No. 346 = **pq/n** (similarly, through 145).

INVALID by reduction to mood 146.

§24. Mood No. 354 = **m/pq**.

Q is a complete cause of R;
Q (complemented by S) is a partial and contingent cause of P;
does it follow that P is (complemented by S) a cause of R? No!

No. 364 = **n/pq** (similarly, through 314).

INVALID by reduction to mood 314.

§25. Mood No. 347 = **pq/p**.

Q (complemented by P) is a partial and contingent cause of R;
Q (complemented by S) is a partial cause of P;
does it follow that P is (complemented by S) a cause of R? No!

No. 348 = **pq/q** (similarly, through 344).

INVALID by reduction to mood 344.

§26. Mood No. 374 = **p/pq**.

Q (complemented by P) is a partial cause of R;
Q (complemented by S) is a partial and contingent cause of P;
does it follow that P is (complemented by S) a cause of R? No!

No. 384 = **q/pq** (similarly, through 324).

INVALID by reduction to mood 334.

§27. Mood No. 355 = **m/m**.

Q is a complete cause of R;
Q is a complete cause of P;
does it follow that P is a cause of R? No!

No. 366 = **n/n** (similarly, through 165).

INVALID by reduction to mood 156.

§28. Mood No. 356 = **m/n/m**.

Q is a complete cause of R;
Q is a necessary cause of P;
so, P is a complete cause of R.

No. 365 = **n/m/n** (similarly, through 166).

VALID by reduction to mood 155.

§29. Mood No. 357 = **m/p**.

Q is a complete cause of R;
Q (complemented by S) is a partial cause of P;
does it follow that P is (complemented by S) a cause of R? No!

No. 368 = **n/q** (similarly, through 312, 314).

INVALID by reduction to moods 313, 314.

§30. Mood No. 358 = **m/q**.	INVALID
Q is a complete cause of R; Q (complemented by S) is a contingent cause of P; does it follow that P is (complemented by S) a cause of R? No!	by reduction to moods 312, 314.
No. 367 = **n/p** (similarly, through 313, 314).	

§31. Mood No. 375 = **p/m**.	INVALID
Q (complemented by S) is a partial cause of R; Q is a complete cause of P; does it follow that P is (complemented by S) a cause of R? No!	by reduction to mood 176.
No. 386 = **q/n** (similarly, through 185).	

§32. Mood No. 376 = **p/n**.	INVALID
Q (complemented by S) is a partial cause of R; Q is a necessary cause of P; does it follow that P is (complemented by S) a cause of R? No!	by reduction to mood 175.
No. 385 = **q/m** (similarly, through 186).	

§33. Mood No. 377 = **p/p**.	INVALID
Q (complemented by P) is a partial cause of R; Q (complemented by S) is a partial cause of P; does it follow that P is (complemented by S) a cause of R? No!	by reduction to mood 334.
No. 388 = **q/q** (similarly, through 324).	

§34. Mood No. 378 = **p/q**.	INVALID
Q (complemented by P) is a partial cause of R; Q (complemented by S) is a contingent cause of P; does it follow that P is (complemented by S) a cause of R? No!	by reduction to mood 334.
No. 387 = **q/p** (similarly, through 324).	

Chapter 7. REDUCTION OF POSITIVE MOODS

1. Reduction.

The method of reduction was first theoretically identified by Aristotle, though of course it had been practically used by human beings long before. Reduction, in its broadest sense, consists in showing that an argument is valid or invalid because another argument is valid or invalid. Thus, reduction is not a primary process of evaluation but a method for transmitting validity or invalidity, and therefore presupposes that we have some other means for establishing certain fundamental validities or invalidities.

The 'other means', in our case, is matricial analysis; we shall use this method for a number of validations and invalidations, but before we do so we want to find out the minimum number of moods of causative syllogism which have to be so treated. For as already said, matricial analysis is a cumbersome, though essential and certain, method; and we wish to facilitate our task. Furthermore, while this method treats each mood as 'an island onto itself', reduction reveals the precise interrelations between moods, which we ought to be aware of.

Reduction is a short-cut. In the field of causative syllogism, reduction has many guises. The broad Aristotelian distinction between direct reduction and indirect (or *ad absurdum*) reduction is of course applicable here; but we may find fit to subdivide the concept of direct reduction.

Within the domain of positive moods of any figure, the validity of conclusions is transmitted by direct reduction and the invalidity of conclusions is transmitted by indirect reduction, within the same figure. The main implication which concerns us here is *subalternation* by joint determinations of generic determinations (thus, **mn** implies **m** and **n**, **mq** implies **m** and **q**, **np** implies **p** and **n**, and **pq** implies **p** and **q**). Since subalternation is one-way implication, different implications are used for validations and invalidations.

But there is also reduction from one figure to another, for which the eductive process of *conversion* is appropriate. Some second figure moods may be directly reduced to first figure moods, by conversion of the major premise; and some third figure moods may be directly reduced to first figure moods, by conversion of the minor premise. Since conversion works both ways, such reductions serve for both validation and invalidation.

We can thus distinguish between two sorts of direct reduction of positive moods, with reference to the precise sort of implication appealed to, i.e. subalternation (within the same figure) or conversion (across figures).[44]

It should be noted that the validity of any conclusion implies the invalidity of conflicting putative conclusions (thus if **m** is true, **p** cannot be, and vice versa; and if **n** is true, **q** cannot be, and vice versa); though note well that the invalidity of a putative conclusion does not imply the validity of its opposite, i.e. both **m** and **p** or both **n** and **q** may be invalid). We can

---

[44] Negative moods might be evaluated by indirect reductions to the positive moods, across figures. Imperfect moods might be evaluated by means of direct reductions consisting of eductive processes which result in negations of the complement. We'll check into that later.

on this basis save ourselves some work; this also might be viewed as a sort of indirect reduction.

Before going further, let us point out that some moods are composed of **incompatible premises**. Such moods may be declared invalid without further ado. This occurs specifically in subfigure (d) of each figure, where the premises have two items in common (namely, P and Q). Here, if the minor premise has a strong component, it may conflict with the weak component(s) of the major premise.

We shall now identify the **implications between moods** within any of the figures, due to inclusion within compound forms (joint determinations) of their constituent forms (generic determinations). The following table lists all implications between *premises* (note well) of moods; it is based on information given in Table 5.6, developed in the chapter on causative syllogism, listing the 64 moods conceivable in each figure.

Table 7.1. Implications between premises of moods, in all figures.															
The mood numbers on the left imply the adjacent mood numbers on the right.															
Along **rows** of table listing all moods:															
11	51	12	52	13	53	14	54	15	55	16	56	17	57	18	58
11	61	12	62	13	63	14	64	15	65	16	66	17	67	18	68
21	51	22	52	23	53	24	54	25	55	26	56	27	57	28	58
21	81	22	82	23	83	24	84	25	85	26	86	27	87	28	88
31	61	32	62	33	63	34	64	35	65	36	66	37	67	38	68
31	71	32	72	33	73	34	74	35	75	36	76	37	77	38	78
41	71	42	72	43	73	44	74	45	75	46	76	47	77	48	78
41	81	42	82	43	83	44	84	45	85	46	86	47	87	48	88
Down **columns** of table listing all moods:															
11	15	21	25	31	35	41	45	51	55	61	65	71	75	81	85
11	16	21	26	31	36	41	46	51	56	61	66	71	76	81	86
12	15	22	25	32	35	42	45	52	55	62	65	72	75	82	85
12	18	22	28	32	38	42	48	52	58	62	68	72	78	82	88
13	16	23	26	33	36	43	46	53	56	63	66	73	76	83	86
13	17	23	27	33	37	43	47	53	57	63	67	73	77	83	87
14	17	24	27	34	37	44	47	54	57	64	67	74	77	84	87
14	18	24	28	34	38	44	48	54	58	64	68	74	78	84	88

Note well that each implication may in turn imply others, i.e. one must follow up *implications of implications*. For instance, 11 implies 15 and 16, and 51 and 61; in turn, 15 implies 55 and 65, and 16 implies 56 and 66; also, 51 implies 55 and 56, and 61 implies 65 and 66. Similarly for other premises, as shown in the above table.

Also note, some of the implications shown in the above table may be useless in practice for a given figure: this occurs when a mood referred to has inconsistent premises.

The following are the principles for **inference of validity or invalidity**. Note well the condition that the validating or invalidating mood be internally *consistent*; as we explained, it can happen, in a given figure, that they are not so.

1. If the premises of one of the above moods, say Y, are consistent and *imply* those of another, say X, then any validated conclusion *of X*, say c1, is also a **valid** conclusion of Y. (But an invalidated conclusion *of X*, say c2, cannot be inferred to be an invalid conclusion of Y.)

Proof: Since Y implies X and X implies c1, it follows that Y implies c1.
(But that X does not imply c2, does not mean that Y does not imply c2.)

2. If the premises of one of the above moods, say Z, are consistent and *imply* those of another, say Y, then any invalidated conclusion *of Z*, say c2, is also an **invalid** conclusion of Y. (But a validated conclusion *of Z*, say c1, cannot be inferred to be a valid conclusion of Y.)

Proof: Since Z implies Y and Z does not imply c2, it follows that Y does not imply c2; for given that Z implies Y, if Y implied c2, Z would imply c2.
(But that Z implies c1, does not mean that Y implies c1.)

One should be careful not to confuse the premises of a mood with a mood as a whole. Referring to the above rules, in case (1), while Y implies X, the validity of X+c1 implies the validity of Y+c1 (this is a direct reduction). In case (2), while Z implies Y, the invalidity of Z+c2 implies the invalidity of Y+c2 (this is an indirect reduction).
Generally, then, to establish a mood Y+c1+notc2 by reduction, we must look for two moods X and Z, such that (1) Y implies X, which concludes c1, and (2) Y is implied by Z, which fails to conclude c2. The following diagram illustrates these principles:

Diagram 7.1. Pathways of Reduction, for Validation (right) and Invalidation (left).

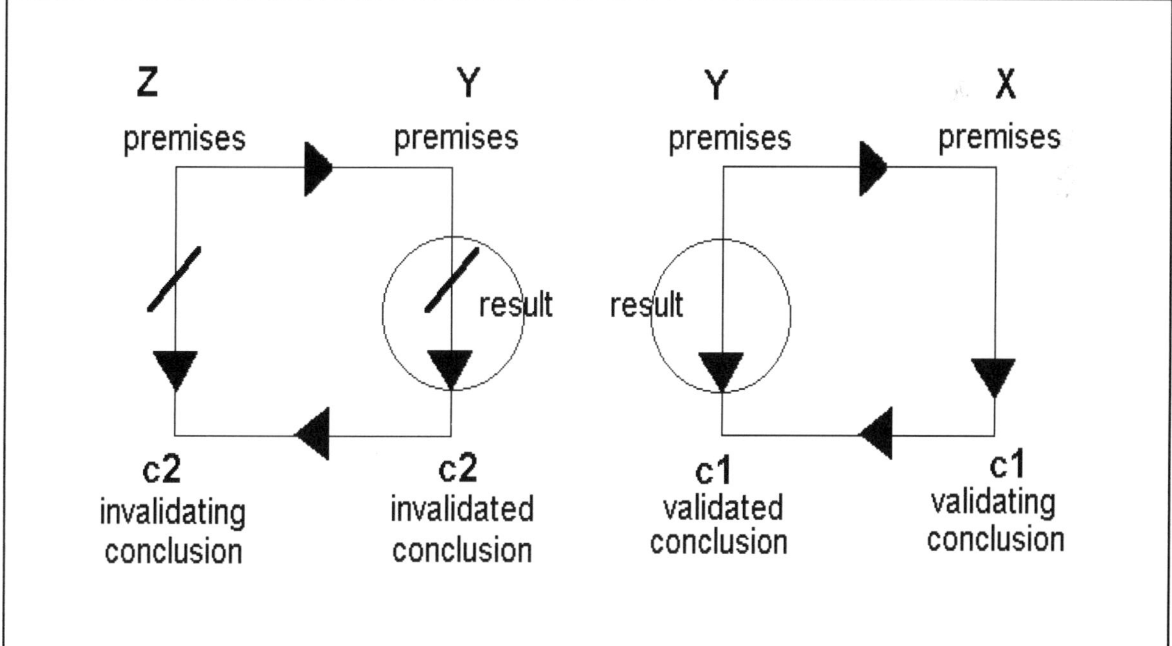

3. The above applies to reductions within a given figure, by subalternation. In the special case of direct reduction across figures, by conversion of the major premise (to derive figure 2) or the minor premise (to derive figure 3), the implications between the premises concerned are

two-way; it follows in such case that both validity and invalidity are transmitted by the same mood of figure 1.

The following table, based on the preceding one, shows more explicitly the possible sources of validity or invalidity by reduction, for each mood within any figure. It should be noted that we cannot (as far as I can see) predict from it, at the outset for all figures, which moods will require matricial analysis; such knowledge has to be acquired in each figure by judicious trial and error.

Table 7.2. For each mood (central col.), those which imply it and those it implies.

Moods implying central mood (if any of them is invalid, the central mood is also invalid)								**Mood**	Moods implied by central mood (if any of them is valid, the central mood is also valid)							
								11	15	16	51	55	56	61	65	66
								12	15	18	52	55	58	62	65	68
								13	16	17	53	56	57	63	66	67
								14	17	18	54	57	58	64	67	68
						11	12	**15**			55			65		
						11	13	**16**			56			66		
						13	14	**17**			57			67		
						12	14	**18**			58			68		
								21	25	26	51	55	56	81	85	86
								22	25	28	52	55	58	82	85	88
								23	26	27	53	56	57	83	86	87
								24	27	28	54	57	58	84	87	88
						21	22	**25**			55			85		
						21	23	**26**			56			86		
						23	24	**27**			57			87		
						22	24	**28**			58			88		
								31	35	36	61	65	66	71	75	76
								32	35	38	62	65	68	72	75	78
								33	36	37	63	66	67	73	76	77
								34	37	38	64	67	68	74	77	78
						31	32	**35**			65			75		
						31	33	**36**			66			76		
						33	34	**37**			67			77		
						32	34	**38**			68			78		
								41	45	46	71	75	76	81	85	86
								42	45	48	72	75	78	82	85	88
								43	46	47	73	76	77	83	86	87
								44	47	48	74	77	78	84	87	88
						41	42	**45**			75			85		

Table 7.2 continued.

						41	43	**46**			76			86		
						43	44	**47**			77			87		
						42	44	**48**			78			88		
		11			21			**51**	55	56						
		12			22			**52**	55	58						
		13			23			**53**	56	57						
		14			24			**54**	57	58						
11	12	15	21	22	25	51	52	**55**								
11	13	16	21	23	26	51	53	**56**								
13	14	17	23	24	27	53	54	**57**								
12	14	18	22	24	28	52	54	**58**								
		11			31			**61**	65	66						
		12			32			**62**	65	68						
		13			33			**63**	66	67						
		14			34			**64**	67	68						
11	12	15	31	32	35	61	62	**65**								
11	13	16	31	33	36	61	63	**66**								
13	14	17	33	34	37	63	64	**67**								
12	14	18	32	34	38	62	64	**68**								
		31			41			**71**	75	76						
		32			42			**72**	75	78						
		33			43			**73**	76	77						
		34			44			**74**	77	78						
31	32	35	41	42	45	71	72	**75**								
31	33	36	41	43	46	71	73	**76**								
33	34	37	43	44	47	73	74	**77**								
32	34	38	42	44	48	72	74	**78**								
		21			41			**81**	85	86						
		22			42			**82**	85	88						
		23			43			**83**	86	87						
		24			44			**84**	87	88						
21	22	25	41	42	45	81	82	**85**								
21	23	26	41	43	46	81	83	**86**								
23	24	27	43	44	47	83	84	**87**								
22	24	28	42	44	48	82	84	**88**								

Remember that breaks will occur in such implications, if any mood is invalid due to inconsistency between premises.

The following tables summarize the results obtained by such reductions, for each of the figures. Conclusions not validated or invalidated by such means must be evaluated through matricial analysis (which is done in the next chapter). The tables below may be read as follows:

yes = element of conclusion (**m**, **n**, **p** or **q**) are implied by the given premises.
no = element of conclusion (**m**, **n**, **p** or **q**) are not implied (which does not mean denied) by the given premises.
by = by any sort of reduction to (**number of mood used**) or **MA** (matricial analysis). Elements of conclusions for which matricial analysis is required are *shaded*.
since = for given premises, if an element of conclusion is valid (yes), then its contrary element is invalid (no).
** = incompatible premises.
nil = no valid conclusion.

2. Reductions in Figure 1.

First, note that ten moods in subfigure 1d have inconsistent premises. Specifically, if the minor premise (which has form P(S)Q) involves a strong determination, then it conflicts with the weak determination(s) of the major premise (which has form Q(P)R).
For if the minor concerns complete causation, clause (i) of which means that (P + notQ) is impossible – it is incompatible with the major, which implies (P + notQ) is possible, whether it concerns partial causation (see clause (ii) of that) or contingent causation (see clause (iii) of that). Similarly, if the minor concerns necessary causation, clause (i) of which means that (notP + Q) is impossible – it is incompatible with the major, which implies (notP + Q) is possible, whether it concerns partial causation (see clause (iii) of that) or contingent causation (see clause (ii) of that).

Table 7.3. Sources of validity or invalidity in figure 1.

Ref.	Mood #	Elements of conclusion implied?			
§1	**111**	m	n	p	q
major minor concl.	**mn** **mn** **mn**	**yes** by 155	**yes** by 166	no since m	no since n
§2	**112**	m	n	p	q
major minor concl.	**mn** **mq** **mq**	**yes** by 155	no since q	no since m	**yes** by 118
§2	**113**	m	n	p	q
major minor concl.	**mn** **np** **np**	no since p	**yes** by 166	**yes** by 117	no since n

Table 7.3 continued.

§6	**114**	m	n	p	q
major	**mn**	no	no	**yes**	**yes**
minor	**pq**	since	since	by	by
concl.	**pq**	p	q	117	118
§11	**115**	m	n	p	q
major	**mn**	**yes**	no	no	no
minor	**m**	by	by	since	by
concl.	**m**	155	112	m	111
§11	**116**	m	n	p	q
major	**mn**	no	**yes**	no	no
minor	**n**	by	by	by	since
concl.	**n**	113	166	111	n
§17	**117**	m	n	p	q
major	**mn**	no	no	**yes**	no
minor	**p**	since	by	by	by
concl.	**p**	p	114	MA	113
§17	**118**	m	n	p	q
major	**mn**	no	no	no	**yes**
minor	**q**	by	since	by	by
concl.	**q**	114	q	112	MA
§3	**121**	m	n	p	q
major	**mq**	**yes**	no	no	**yes**
minor	**mn**	by	since	since	by
concl.	**mq**	155	q	m	181
§4	**122**	m	n	p	q
major	**mq**	**q** of major premise and			
minor	**mq**	**m** of minor premise			
concl.	**	are incompatible			
§5	**123**	m	n	p	q
major	**mq**	**q** of major premise and			
minor	**np**	**n** of minor premise			
concl.	**	are incompatible			
§8	**124**	m	n	p	q
major	**mq**	no	no	no	**yes**
minor	**pq**	by	since	by	by
concl.	**q**	MA	q	MA	128,184
§13	**125**	m	n	p	q
major	**mq**	**yes**	no	no	no
minor	**m**	by	by	since	by
concl.	**m**	155	121	m	MA
§14	**126**	m	n	p	q
major	**mq**	no	no	no	no
minor	**n**	by	by	by	by
concl.	***nil***	MA	121	121	MA

Table 7.3 continued.

§19	**127**	m	n	p	q
major minor concl.	**mq** **p** ***nil***	no by 124	no by 124	no by 124	no by MA
§20	**128**	m	n	p	q
major minor concl.	**mq** **q** **q**	no by 124	no since q	no by 124	**yes** by MA
§3	**131**	m	n	p	q
major minor concl.	**np** **mn** **np**	no since p	**yes** by 166	**yes** by 171	no since n
§5	**132**	m	n	p	q
major minor concl.	**np** **mq** **	**p** of major premise and **m** of minor premise are incompatible			
§4	**133**	m	n	p	q
major minor concl.	**np** **np** **	**p** of major premise and **n** of minor premise are incompatible			
§8	**134**	m	n	p	q
major minor concl.	**np** **pq** **p**	no since p	no by MA	**yes** by 137,174	no by MA
§14	**135**	m	n	p	q
major minor concl.	**np** **m** ***nil***	no by 131	no by MA	no by MA	no by 131
§13	**136**	m	n	p	q
major minor concl.	**np** **n** **n**	no by 131	**yes** by 166	no by MA	no since n
§20	**137**	m	n	p	q
major minor concl.	**np** **p** **p**	no since p	no by 134	**yes** by MA	no by 134
§19	**138**	m	n	p	q
major minor concl.	**np** **q** ***nil***	no by 134	no by 134	no by MA	no by 134

Table 7.3 continued.

§7	**141**	m	n	p	q
major minor concl.	**pq** **mn** **pq**	no since p	no since q	**yes** by 171	**yes** by 181
§9	**142**	m	n	p	q
major minor concl.	**pq** **mq** **	**p**, **q** of major premise and **m** of minor premise are incompatible			
§9	**143**	m	n	p	q
major minor concl.	**pq** **np** **	**p**, **q** of major premise and **n** of minor premise are incompatible			
§10	**144**	m	n	p	q
major minor concl.	**pq** **pq** **pq**	no since p	no since q	**yes** by 147,174	**yes** by 148,184
§23	**145**	m	n	p	q
major minor concl.	**pq** **m** ***nil***	no by 141	no by 141	no by MA	no by MA
§23	**146**	m	n	p	q
major minor concl.	**pq** **n** ***nil***	no by 141	no by 141	no by MA	no by MA
§25	**147**	m	n	p	q
major minor concl.	**pq** **p** **p**	no since p	no by 144	**yes** by MA	no by MA
§25	**148**	m	n	p	q
major minor concl.	**pq** **q** **q**	no by 144	no since q	no by MA	**yes** by MA
§12	**151**	m	n	p	q
major minor concl.	**m** **mn** **m**	**yes** by 155	no by 121	no since m	no by 111
§15	**152**	m	n	p	q
major minor concl.	**m** **mq** **m**	**yes** by 155	no by 112	no since m	no by MA

Table 7.3 continued.

§16	**153**	m	n	p	q
major	**m**	no	no	no	no
minor	**np**	by	by	by	by
concl.	***nil***	113	MA	MA	113
§24	**154**	m	n	p	q
major	**m**	no	no	no	no
minor	**pq**	by	by	by	by
concl.	***nil***	114	114	124	MA
§27	**155**	m	n	p	q
major	**m**	**yes**	no	no	no
minor	**m**	by	by	since	by
concl.	**m**	MA	112	m	111
§28	**156**	m	n	p	q
major	**m**	no	no	no	no
minor	**n**	by	by	by	by
concl.	***nil***	113	121	111	111
§29	**157**	m	n	p	q
major	**m**	no	no	no	no
minor	**p**	by	by	by	by
concl.	***nil***	113	114	124	113
§30	**158**	m	n	p	q
major	**m**	no	no	no	no
minor	**q**	by	by	by	by
concl.	***nil***	114	112	112	152
§12	**161**	m	n	p	q
major	**n**	no	**yes**	no	no
minor	**mn**	by	by	by	since
concl.	**n**	131	166	111	n
§16	**162**	m	n	p	q
major	**n**	no	no	no	no
minor	**mq**	by	by	by	by
concl.	***nil***	MA	112	112	MA
§15	**163**	m	n	p	q
major	**n**	no	**yes**	no	no
minor	**np**	by	by	by	since
concl.	**n**	113	166	MA	n
§24	**164**	m	n	p	q
major	**n**	no	no	no	no
minor	**pq**	by	by	by	by
concl.	***nil***	114	114	MA	134

Table 7.3 continued.

§28	**165**	m	n	p	q
major minor concl.	**n** **m** ***nil***	no by 131	no by 112	no by 111	no by 111
§27	**166**	m	n	p	q
major minor concl.	**n** **n** **n**	no by 113	**yes** by MA	no by 111	no since n
§30	**167**	m	n	p	q
major minor concl.	**n** **p** ***nil***	no by 113	no by 114	no by 163	no by 113
§29	**168**	m	n	p	q
major minor concl.	**n** **q** ***nil***	no by 114	no by 112	no by 112	no by 134
§18	**171**	m	n	p	q
major minor concl.	**p** **mn** **p**	no since p	no by 141	**yes** by MA	no by 131
§21	**172**	m	n	p	q
major minor concl.	**p** **mq** **	**p** of major premise and **m** of minor premise are incompatible			
§22	**173**	m	n	p	q
major minor concl.	**p** **np** **	**p** of major premise and **n** of minor premise are incompatible			
§26	**174**	m	n	p	q
major minor concl.	**p** **pq** **p**	no since p	no by 134	**yes** by MA	no by 134
§31	**175**	m	n	p	q
major minor concl.	**p** **m** ***nil***	no by 131	no by 135	no by 135	no by 131
§32	**176**	m	n	p	q
major minor concl.	**p** **n** ***nil***	no by 131	no by 141	no by 136	no by 131

Table 7.3 continued.

§33	**177**	m	n	p	q
major	**p**	no	no	no	no
minor	**p**	by	by	by	by
concl.	***nil***	134	134	MA	134
§34	**178**	m	n	p	q
major	**p**	no	no	no	no
minor	**q**	by	by	by	by
concl.	***nil***	134	134	138	134
§18	**181**	m	n	p	q
major	**q**	no	no	no	**yes**
minor	**mn**	by	since	by	by
concl.	**q**	141	q	121	MA
§22	**182**	m	n	p	q
major	**q**	**q** of major premise and			
minor	**mq**	**m** of minor premise			
concl.	**	are incompatible			
§21	**183**	m	n	p	q
major	**q**	**q** of major premise and			
minor	**np**	**n** of minor premise			
concl.	**	are incompatible			
§26	**184**	m	n	p	q
major	**q**	no	no	no	**yes**
minor	**pq**	by	since	by	by
concl.	**q**	124	q	124	MA
§32	**185**	m	n	p	q
major	**q**	no	no	no	no
minor	**m**	by	by	by	by
concl.	***nil***	141	121	121	125
§31	**186**	m	n	p	q
major	**q**	no	no	no	no
minor	**n**	by	by	by	by
concl.	***nil***	126	121	121	126
§34	**187**	m	n	p	q
major	**q**	no	no	no	no
minor	**p**	by	by	by	by
concl.	***nil***	124	124	124	127
§33	**188**	m	n	p	q
major	**q**	no	no	no	no
minor	**q**	by	by	by	by
concl.	***nil***	124	124	124	MA

Summary of figure 1.

- 30 valid moods:
  111-118, 121, 124-125, 128, 131, 134, 136-137, 141, 144, 147-148, 151-152, 155, 161, 163, 166, 171, 174, 181, 184.
- 24 moods without conclusion (*nil*):
  126-127, 135, 138, 145-146, 153-154, 156-158, 162, 164-165, 167-168, 175-178, 185-188.
- 10 impossible moods (**):
  122-123, 132-133, 142-143, 172-173, 182-183.

Total of moods = 30 valid and 34 invalid = 64.

3. Reductions in Figure 2.

First, note that six moods in subfigure 2d have inconsistent premises. Specifically, if the minor premise (which has form P(S)Q) involves a strong determination, then it conflicts with any weak determination of same polarity in the major premise (which has form R(P)Q).
For if the minor concerns complete causation, clause (i) of which means that (P + notQ) is impossible – it is incompatible with the major, which implies (P + notQ) is possible when it concerns partial causation (see clause (ii) of that). Similarly, if the minor concerns necessary causation, clause (i) of which means that (notP + Q) is impossible – it is incompatible with the major, which implies (notP + Q) is possible when it concerns contingent causation (see clause (ii) of that).

Additionally, we may directly reduce a number of moods in figure 2 to figure 1, by converting the major premise. This is feasible when the major premise involves only strong causation; i.e. subfigures 2a and 2b are thus reducible respectively to subfigures 1a and 1b. This is not feasible when the major premise involves weak causation, since its conversion results in negation of the complement; which means that subfigures 2c and 2d have to be evaluated relatively independently (i.e. within the same figure, even if possibly through some moods reduced to figure 1).

Table 7.4. Sources of validity or invalidity in figure 2.					
Ref.	Mood #	Elements of conclusion implied?			
§1	**211**	m	n	p	q
major minor concl.	**mn** **mn** **mn**	**yes** by 111	**yes** by 111	no since m	no since n

Table 7.4 continued.

§2	**212**	m	n	p	q
major	**mn**	**yes**	no	no	**yes**
minor	**mq**	by	since	since	by
concl.	**mq**	112	q	m	112
§2	**213**	m	n	p	q
major	**mn**	no	**yes**	**yes**	no
minor	**np**	since	by	by	since
concl.	**np**	p	113	113	n
§6	**214**	m	n	p	q
major	**mn**	no	no	**yes**	**yes**
minor	**pq**	since	since	by	by
concl.	**pq**	p	q	114	114
§11	**215**	m	n	p	q
major	**mn**	**yes**	no	no	no
minor	**m**	by	by	since	by
concl.	**m**	115	115	m	115
§11	**216**	m	n	p	q
major	**mn**	no	**yes**	no	no
minor	**n**	by	by	by	since
concl.	**n**	116	116	116	n
§17	**217**	m	n	p	q
major	**mn**	no	no	**yes**	no
minor	**p**	since	by	by	by
concl.	**p**	p	117	117	117
§17	**218**	m	n	p	q
major	**mn**	no	no	no	**yes**
minor	**q**	by	since	by	by
concl.	**q**	118	q	118	118
§3	**221**	m	n	p	q
major	**mq**	no	**yes**	no	no
minor	**mn**	by	by	by	since
concl.	**n**	MA	256	MA	n
§4	**222**	m	n	p	q
major	**mq**	no	no	no	no
minor	**mq**	by	by	by	by
concl.	***nil***	MA	MA	MA	MA
§5	**223**	m	n	p	q
major	**mq**	**q** of major premise and			
minor	**np**	**n** of minor premise			
concl.	**	are incompatible			

Table 7.4 continued.

§8	**224**	m	n	p	q
major minor concl.	**mq** **pq** ***nil***	no by MA	no by MA	no by MA	no by MA
§13	**225**	m	n	p	q
major minor concl.	**mq** **m** ***nil***	no by 221	no by 222	no by 221	no by 221
§14	**226**	m	n	p	q
major minor concl.	**mq** **n** **n**	no by 221	**yes** by 256	no by 221	no since n
§19	**227**	m	n	p	q
major minor concl.	**mq** **p** ***nil***	no by 224	no by 224	no by 224	no by 224
§20	**228**	m	n	p	q
major minor concl.	**mq** **q** ***nil***	no by 222	no by 222	no by 222	no by 222
§3	**231**	m	n	p	q
major minor concl.	**np** **mn** **m**	**yes** by 265	no by MA	no since m	no by MA
§5	**232**	m	n	p	q
major minor concl.	**np** **mq** **	**p** of major premise and **m** of minor premise are incompatible			
§4	**233**	m	n	p	q
major minor concl.	**np** **np** ***nil***	no by MA	no by MA	no by MA	no by MA
§8	**234**	m	n	p	q
major minor concl.	**np** **pq** ***nil***	no by MA	no by MA	no by MA	no by MA
§14	**235**	m	n	p	q
major minor concl.	**np** **m** **m**	**yes** by 265	no by 231	no since m	no by 231

Table 7.4 continued.

§13	**236**	m	n	p	q
major	**np**	no	no	no	no
minor	**n**	by	by	by	by
concl.	***nil***	233	231	231	231
§20	**237**	m	n	p	q
major	**np**	no	no	no	no
minor	**p**	by	by	by	by
concl.	***nil***	233	233	233	233
§19	**238**	m	n	p	q
major	**np**	no	no	no	no
minor	**q**	by	by	by	by
concl.	***nil***	234	234	234	234
§7	**241**	m	n	p	q
major	**pq**	no	no	no	no
minor	**mn**	by	by	by	by
concl.	***nil***	MA	MA	MA	MA
§9	**242**	m	n	p	q
major	**pq**	**p** of major premise and			
minor	**mq**	**m** of minor premise			
concl.	**	are incompatible			
§9	**243**	m	n	p	q
major	**pq**	**q** of major premise and			
minor	**np**	**n** of minor premise			
concl.	**	are incompatible			
§10	**244**	m	n	p	q
major	**pq**	no	no	no	no
minor	**pq**	by	by	by	by
concl.	***nil***	MA	MA	MA	MA
§23	**245**	m	n	p	q
major	**pq**	no	no	no	no
minor	**m**	by	by	by	by
concl.	***nil***	241	241	241	241
§23	**246**	m	n	p	q
major	**pq**	no	no	no	no
minor	**n**	by	by	by	by
concl.	***nil***	241	241	241	241
§25	**247**	m	n	p	q
major	**pq**	no	no	no	no
minor	**p**	by	by	by	by
concl.	***nil***	244	244	244	244

Table 7.4 continued.

§25	**248**	m	n	p	q
major	**pq**	no	no	no	no
minor	**q**	by	by	by	by
concl.	***nil***	244	244	244	244
§12	**251**	m	n	p	q
major	**m**	no	**yes**	no	no
minor	**mn**	by	by	by	since
concl.	**n**	161	161	161	n
§15	**252**	m	n	p	q
major	**m**	no	no	no	no
minor	**mq**	by	by	by	by
concl.	***nil***	162	162	162	162
§16	**253**	m	n	p	q
major	**m**	no	**yes**	no	no
minor	**np**	by	by	by	since
concl.	**n**	163	163	163	n
§24	**254**	m	n	p	q
major	**m**	no	no	no	no
minor	**pq**	by	by	by	by
concl.	***nil***	164	164	164	164
§27	**255**	m	n	p	q
major	**m**	no	no	no	no
minor	**m**	by	by	by	by
concl.	***nil***	165	165	165	165
§28	**256**	m	n	p	q
major	**m**	no	**yes**	no	no
minor	**n**	by	by	by	since
concl.	**n**	166	166	166	n
§29	**257**	m	n	p	q
major	**m**	no	no	no	no
minor	**p**	by	by	by	by
concl.	***nil***	167	167	167	167
§30	**258**	m	n	p	q
major	**m**	no	no	no	no
minor	**q**	by	by	by	by
concl.	***nil***	168	168	168	168
§12	**261**	m	n	p	q
major	**n**	**yes**	no	no	no
minor	**mn**	by	by	since	by
concl.	**m**	151	151	m	151

Table 7.4 continued.

§16	**262**	m	n	p	q
major	**n**	**yes**	no	no	no
minor	**mq**	by	by	since	by
concl.	**m**	152	152	m	152
§15	**263**	m	n	p	q
major	**n**	no	no	no	no
minor	**np**	by	by	by	by
concl.	***nil***	153	153	153	153
§24	**264**	m	n	p	q
major	**n**	no	no	no	no
minor	**pq**	by	by	by	by
concl.	***nil***	154	154	154	154
§28	**265**	m	n	p	q
major	**n**	**yes**	no	no	no
minor	**m**	by	by	since	by
concl.	**m**	155	155	m	155
§27	**266**	m	n	p	q
major	**n**	no	no	no	no
minor	**n**	by	by	by	by
concl.	***nil***	156	156	156	156
§30	**267**	m	n	p	q
major	**n**	no	no	no	no
minor	**p**	by	by	by	by
concl.	***nil***	157	157	157	157
§29	**268**	m	n	p	q
major	**n**	no	no	no	no
minor	**q**	by	by	by	by
concl.	***nil***	158	158	158	158
§18	**271**	m	n	p	q
major	**p**	no	no	no	no
minor	**mn**	by	by	by	by
concl.	***nil***	241	231	231	231
§21	**272**	m	n	p	q
major	**p**	**p** of major premise and			
minor	**mq**	**m** of minor premise			
concl.	**	are incompatible			
§22	**273**	m	n	p	q
major	**p**	no	no	no	no
minor	**np**	by	by	by	by
concl.	***nil***	233	233	233	233

Table 7.4 continued.

§26	**274**	m	n	p	q
major minor concl.	**p** **pq** ***nil***	no by 234	no by 234	no by 234	no by 234
§31	**275**	m	n	p	q
major minor concl.	**p** **m** ***nil***	no by 241	no by 231	no by 231	no by 231
§32	**276**	m	n	p	q
major minor concl.	**p** **n** ***nil***	no by 233	no by 231	no by 231	no by 231
§33	**277**	m	n	p	q
major minor concl.	**p** **p** ***nil***	no by 233	no by 233	no by 233	no by 233
§34	**278**	m	n	p	q
major minor concl.	**p** **q** ***nil***	no by 234	no by 234	no by 234	no by 234
§18	**281**	m	n	p	q
major minor concl.	**q** **mn** ***nil***	no by 221	no by 241	no by 221	no by 221
§22	**282**	m	n	p	q
major minor concl.	**q** **mq** ***nil***	no by 222	no by 222	no by 222	no by 222
§21	**283**	m	n	p	q
major minor concl.	**q** **np** **	**q** of major premise and **n** of minor premise are incompatible			
§26	**284**	m	n	p	q
major minor concl.	**q** **pq** ***nil***	no by 224	no by 224	no by 224	no by 224
§32	**285**	m	n	p	q
major minor concl.	**q** **m** ***nil***	no by 221	no by 222	no by 221	no by 221

Table 7.4 continued.

§31	**286**	m	n	p	q
major minor concl.	**q** **n** ***nil***	no by 221	no by 241	no by 221	no by 221
§34	**287**	m	n	p	q
major minor concl.	**q** **p** ***nil***	no by 224	no by 224	no by 224	no by 224
§33	**288**	m	n	p	q
major minor concl.	**q** **q** ***nil***	no by 222	no by 222	no by 222	no by 222

Summary of figure 2.

- 18 valid moods:
  211-218, 221, 226, 231, 235, 251, 253, 256, 261-262, 265.
- 40 moods without conclusion (*nil*):
  222, 224-225, 227-228, 233-234, 236-238, 241, 244-248, 252, 254-255, 257-258, 263-264, 266-268, 271, 273-278, 281-282, 284-288.
- 6 impossible moods (**):
  223, 232, 242-243, 272, 283.

Total of moods = 18 valid and 46 invalid = 64.

4. Reductions in Figure 3.

First, note that ten moods in subfigure 3d have inconsistent premises. Specifically, if the minor premise (which has form Q(S)P) involves a strong determination, then it conflicts with the weak determination(s) of the major premise (which has form Q(P)R).

For if the minor concerns complete causation, clause (i) of which means that (notP + Q) is impossible – it is incompatible with the major, which implies (notP + Q) is possible, whether it concerns partial causation (see clause (iii) of that) or contingent causation (see clause (ii) of that). Similarly, if the minor concerns necessary causation, clause (i) of which means that (P + notQ) is impossible – it is incompatible with the major, which implies (P + notQ) is possible, whether it concerns partial causation (see clause (ii) of that) or contingent causation (see clause (iii) of that).

Additionally, we may directly reduce a number of moods in figure 3 to figure 1, by converting the minor premise. This is feasible when the minor premise involves only strong causation; i.e. subfigures 3a and 3c are thus reducible respectively to subfigures 1a and 1c. This is not feasible when the minor premise involves weak causation, since its conversion results in negation of the complement; which means that subfigures 3b and 3d have to be evaluated relatively independently (i.e. within the same figure, even if possibly through some moods reduced to figure 1).

Table 7.5. Sources of validity or invalidity in figure 3.

Ref.	Mood #	Elements of conclusion implied?			
§1	**311**	m	n	p	q
major minor concl.	**mn** **mn** **mn**	**yes** by 111	**yes** by 111	no since m	no since n
§2	**312**	m	n	p	q
major minor concl.	**mn** **mq** **n**	no by MA	**yes** by 365	no by MA	no since n
§2	**313**	m	n	p	q
major minor concl.	**mn** **np** **m**	**yes** by 356	no by MA	no since m	no by MA
§6	**314**	m	n	p	q
major minor concl.	**mn** **pq** ***nil***	no by MA	no by MA	no by MA	no by MA
§11	**315**	m	n	p	q
major minor concl.	**mn** **m** **n**	no by 116	**yes** by 116	no by 116	no since n
§11	**316**	m	n	p	q
major minor concl.	**mn** **n** **m**	**yes** by 115	no by 115	no since m	no by 115
§17	**317**	m	n	p	q
major minor concl.	**mn** **p** ***nil***	no by 314	no by 313	no by 313	no by 313
§17	**318**	m	n	p	q
major minor concl.	**mn** **q** ***nil***	no by 312	no by 314	no by 312	no by 312
§3	**321**	m	n	p	q
major minor concl.	**mq** **mn** **mq**	**yes** by 121	no since q	no since m	**yes** by 121
§4	**322**	m	n	p	q
major minor concl.	**mq** **mq** **	**q** of major premise and **m** of minor premise are incompatible			

Table 7.5 continued.

§5	**323**	m	n	p	q
major minor concl.	**mq** **np** **	**q** of major premise and **n** of minor premise are incompatible			
§8	**324**	m	n	p	q
major minor concl.	**mq** **pq** ***nil***	no by MA	no by MA	no by MA	no by MA
§13	**325**	m	n	p	q
major minor concl.	**mq** **m** ***nil***	no by 126	no by 126	no by 126	no by 126
§14	**326**	m	n	p	q
major minor concl.	**mq** **n** **m**	**yes** by 125	no by 125	no since m	no by 125
§19	**327**	m	n	p	q
major minor concl.	**mq** **p** ***nil***	no by 324	no by 324	no by 324	no by 324
§20	**328**	m	n	p	q
major minor concl.	**mq** **q** ***nil***	no by 324	no by 324	no by 324	no by 324
§3	**331**	m	n	p	q
major minor concl.	**np** **mn** **np**	no since p	**yes** by 131	**yes** by 131	no since n
§5	**332**	m	n	p	q
major minor concl.	**np** **mq** **	**p** of major premise and **m** of minor premise are incompatible			
§4	**333**	m	n	p	q
major minor concl.	**np** **np** **	**p** of major premise and **n** of minor premise are incompatible			
§8	**334**	m	n	p	q
major minor concl.	**np** **pq** ***nil***	no by MA	no by MA	no by MA	no by MA

Table 7.5 continued.

§14	**335**	m	n	p	q
major	**np**	no	**yes**	no	no
minor	**m**	by	by	by	since
concl.	**n**	136	136	136	n
§13	**336**	m	n	p	q
major	**np**	no	no	no	no
minor	**n**	by	by	by	by
concl.	***nil***	135	135	135	135
§20	**337**	m	n	p	q
major	**np**	no	no	no	no
minor	**p**	by	by	by	by
concl.	***nil***	334	334	334	334
§19	**338**	m	n	p	q
major	**np**	no	no	no	no
minor	**q**	by	by	by	by
concl.	***nil***	334	334	334	334
§7	**341**	m	n	p	q
major	**pq**	no	no	**yes**	**yes**
minor	**mn**	since	since	by	by
concl.	**pq**	p	q	141	141
§9	**342**	m	n	p	q
major	**pq**	**p** and **q** of major premise and			
minor	**mq**	**m** of minor premise			
concl.	**	are incompatible			
§9	**343**	m	n	p	q
major	**pq**	**p** and **q** of major premise and			
minor	**np**	**n** of minor premise			
concl.	**	are incompatible			
§10	**344**	m	n	p	q
major	**pq**	no	no	no	no
minor	**pq**	by	by	by	by
concl.	***nil***	MA	MA	MA	MA
§23	**345**	m	n	p	q
major	**pq**	no	no	no	no
minor	**m**	by	by	by	by
concl.	***nil***	146	146	146	146
§23	**346**	m	n	p	q
major	**pq**	no	no	no	no
minor	**n**	by	by	by	by
concl.	***nil***	145	145	145	145

Table 7.5 continued.

§25	**347**	m	n	p	q
major minor concl.	**pq** **p** ***nil***	no by 344	no by 344	no by 344	no by 344
§25	**348**	m	n	p	q
major minor concl.	**pq** **q** ***nil***	no by 344	no by 344	no by 344	no by 344
§12	**351**	m	n	p	q
major minor concl.	**m** **mn** **m**	**yes** by 151	no by 151	no since m	no by 151
§15	**352**	m	n	p	q
major minor concl.	**m** **mq** ***nil***	no by 312	no by MA	no by 312	no by 312
§16	**353**	m	n	p	q
major minor concl.	**m** **np** **m**	**yes** by 356	no by 313	no since m	no by 313
§24	**354**	m	n	p	q
major minor concl.	**m** **pq** ***nil***	no by 314	no by 314	no by 314	no by 314
§27	**355**	m	n	p	q
major minor concl.	**m** **m** ***nil***	no by 156	no by 156	no by 156	no by 156
§28	**356**	m	n	p	q
major minor concl.	**m** **n** **m**	**yes** by 155	no by 155	no since m	no by 155
§29	**357**	m	n	p	q
major minor concl.	**m** **p** ***nil***	no by 314	no by 313	no by 313	no by 313
§30	**358**	m	n	p	q
major minor concl.	**m** **q** ***nil***	no by 312	no by 314	no by 312	no by 312

Table 7.5 continued.

§12	**361**	m	n	p	q
major	**n**	no	**yes**	no	no
minor	**mn**	by	by	by	since
concl.	**n**	161	161	161	n
§16	**362**	m	n	p	q
major	**n**	no	**yes**	no	no
minor	**mq**	by	by	by	since
concl.	**n**	312	365	312	n
§15	**363**	m	n	p	q
major	**n**	no	no	no	no
minor	**np**	by	by	by	by
concl.	***nil***	MA	313	313	313
§24	**364**	m	n	p	q
major	**n**	no	no	no	no
minor	**pq**	by	by	by	by
concl.	***nil***	314	314	314	314
§28	**365**	m	n	p	q
major	**n**	no	**yes**	no	no
minor	**m**	by	by	by	since
concl.	**n**	166	166	166	n
§27	**366**	m	n	p	q
major	**n**	no	no	no	no
minor	**n**	by	by	by	by
concl.	***nil***	165	165	165	165
§30	**367**	m	n	p	q
major	**n**	no	no	no	no
minor	**p**	by	by	by	by
concl.	***nil***	314	313	313	313
§29	**368**	m	n	p	q
major	**n**	no	no	no	no
minor	**q**	by	by	by	by
concl.	***nil***	312	314	312	312
§18	**371**	m	n	p	q
major	**p**	no	no	**yes**	no
minor	**mn**	since	by	by	by
concl.	**p**	p	171	171	171
§21	**372**	m	n	p	q
major	**p**	**p** of major premise and			
minor	**mq**	**m** of minor premise			
concl.	**	are incompatible			

Table 7.5 continued.

§22	**373**	m	n	p	q
major minor concl.	**p** **np** **	**p** of major premise and **n** of minor premise are incompatible			
§26	**374**	m	n	p	q
major minor concl.	**p** **pq** ***nil***	no by 334	no by 334	no by 334	no by 334
§31	**375**	m	n	p	q
major minor concl.	**p** **m** ***nil***	no by 176	no by 176	no by 176	no by 176
§32	**376**	m	n	p	q
major minor concl.	**p** **n** ***nil***	no by 175	no by 175	no by 175	no by 175
§33	**377**	m	n	p	q
major minor concl.	**p** **p** ***nil***	no by 334	no by 334	no by 334	no by 334
§34	**378**	m	n	p	q
major minor concl.	**p** **q** ***nil***	no by 334	no by 334	no by 334	no by 334
§18	**381**	m	n	p	q
major minor concl.	**q** **mn** **q**	no by 181	no since q	no by 181	**yes** by 181
§22	**382**	m	n	p	q
major minor concl.	**q** **mq** **	**q** of major premise and **m** of minor premise are incompatible			
§21	**383**	m	n	p	q
major minor concl.	**q** **np** **	**q** of major premise and **n** of minor premise are incompatible			
§26	**384**	m	n	p	q
major minor concl.	**q** **pq** ***nil***	no by 324	no by 324	no by 324	no by 324

Table 7.5 continued.

§32	**385**	m	n	p	q
major minor concl.	**q** **m** ***nil***	no by 186	no by 186	no by 186	no by 186
§31	**386**	m	n	p	q
major minor concl.	**q** **n** ***nil***	no by 185	no by 185	no by 185	no by 185
§34	**387**	m	n	p	q
major minor concl.	**q** **p** ***nil***	no by 324	no by 324	no by 324	no by 324
§33	**388**	m	n	p	q
major minor concl.	**q** **q** ***nil***	no by 324	no by 324	no by 324	no by 324

Summary of figure 3.

- 18 valid moods:
  311-313, 315-316, 321, 326, 331, 335, 341, 351, 353, 356, 361-362, 365, 371, 381.
- 36 moods without conclusion (*nil*):
  314, 317-318, 324-325, 327-328, 334, 336-338, 344-348, 352, 354-355, 357-358, 363-364, 366-368, 374-378, 384-388.
- 10 impossible moods (**):
  322-323, 332-333, 342-343, 372-373, 382-383.

Total of moods = 18 valid and 46 invalid = 64.

Chapter 8. MATRICIAL ANALYSES

1. Matricial Analysis.

We will in this chapter show the matricial analyses on the basis of which moods were declared valid or invalid in previous chapters.

Now, the *matrix* underlying a syllogism may be defined as a table with a listing of all conceivable conjunctions of all the items involved in its premises and conclusion and/or the negations of these items. Thus, for instance, the matrix of three items P, Q, R, will look like this:

P	Q	R
P	Q	notR
P	notQ	R
P	notQ	notR
notP	Q	R
notP	Q	notR
notP	notQ	R
notP	notQ	notR

(With four items, the table would be twice as long; with five items, four times as long; and so on.)

Briefly put, matricial analysis is a process which seeks to answer, for each of the conceivable conjunctions in the matrix, the question as to whether it is implied *impossible* or *possible* or neither; in the latter case, if the conjunction is neither implied impossible nor implied possible, it is declared *open*. In other words, matricial analysis is a pursuit of the 'modus' of the matrix.

The answers to this question for each row must be derived from the premises, singly or together, by established means; the conclusion is valid only if it may be *entirely* (with all its implicit clauses) constructed from these answers. As we shall see, this process relies heavily on paradoxical logic, or more simply put, on dilemmatic argument.

The process is, as you will presently discover, long and difficult. I have unfortunately found no better short-cut, maybe other logicians have or eventually do. However, its advantage over reduction is that it provides us with sure results; for with reduction we cannot always be sure to have applied all possible means, whereas with a matricial analysis we know we have exhausted the available information. We are free to choose the appropriate method in each case: certain crucial syllogisms are best subjected to matricial analysis, and then we can use reduction to derive others.

To begin with, here is a step by step description of this method of evaluation. The reader is requested to follow the procedure concretely by referring to one of the examples given in the following sections. It is much less complicated than it sounds.

a. Write down the premises constituting the mood and the putative conclusion(s) to be evaluated. Translate all these causative propositions into conditional or conjunctive propositions, i.e. make their constituent clauses (as elucidated in the chapter on the determinations of causation) explicit. Number the clauses involved for purposes of reference (Roman numerals are used for this, here).

b. Construct a table with a matrix involving all the items and negations of items concerned, in orderly sequences. If there are three items (P, Q, R), the table will have $2^3 = 8$ rows; if there are four items (P, Q, R, S), it will have $2^4 = 16$ rows.

c. Consider first *all* the positive conditional propositions found in the premises. Every causative proposition contains at least one positive conditional clause; therefore, there will be at least two such clauses per mood. These tell us which conjunctions in the matrix (rows) are implied *impossible*.

In a three-item matrix, each such statement (e.g. 'if P, then Q', which means 'P+notQ is impossible') will imply two rows to be impossible (namely, 'P+notQ+R' and 'P+notQ+notR'). Similarly, in a four-item matrix, each if/then statement involving two items will imply four rows impossible; while each if/then statement involving three items will imply two rows impossible.

d. Only thereafter, deal with the remaining clauses found in the premises (because their impact will depend on the results of the preceding step), which imply the possibility of certain conjunctions of two or three items in the matrix.

These include negative conditional propositions (e.g. 'if P, not-then Q', which means 'P+notQ is possible'); as well as bare statements of possibility of an item or of a conjunction of items. The latter are to be enlarged with reference to the corresponding positive conditional proposition (e.g. 'P is possible' and 'if P, then Q', together imply 'P+Q is possible').

In a three-item matrix, a possibility of conjunction of two items will imply that *at least one* of two rows is possible. In a four-item matrix, a possibility of conjunction of two items will imply that *at least one* of four rows is possible; while a possibility of conjunction of three items will imply that *at least one* of two rows is possible.

Note this well: whereas the impossibility of a conjunction entails the impossibility of all its expressions in the matrix, the possibility of a conjunction is satisfied by only one expression. Thus, the knowledge that two or more rows are collectively possible does not settle the question of the possibility of each of these rows individually.

Only if *all but one* of these rows are declared impossible by other means (i.e. the preceding step of the procedure), can we declare the remaining one possible. Otherwise, if two or more rows are left unsettled, they must each be considered 'open' (i.e. 'possible or impossible'). That is, even though we know that at least one of them must be possible, we cannot specify which one.

e. When all the information implicit in the premises has been thus systematically included in the table, we can evaluate the putative conclusion(s). Taking one of the clauses at a time, check out whether it can be inferred from the table.

If the clause in question is a positive conditional, *every* row corresponding to it in the matrix must have been declared impossible to allow us to accept the clause as implied. If the clause in question is a negative conditional or bare statement of possibility, it suffices that *one* row in the matrix has been declared possible, even if the other(s) was/were declared impossible or left open. (Often, the last clause of the putative conclusion can be inferred directly from a premise, note.)

f. If, and only if, *all* the clauses of the putative conclusion are thus found to be implied by the data in our table, we may admit that conclusion to be drawable from the premises. If any clause(s) of the putative conclusion is/are left open or worse still denied, by the table, that conclusion must be declared a non-sequitur or antinomy, respectively.

A computer could be programmed to carry out this evaluation process. Once it is understood, it requires no great intelligence to perform. It is tedious detail work, no more.[45]

What is matricial analysis, essentially? A causative proposition is a complex of simpler statements, which affirm the impossibility or possibility of certain conjunctions of items or individual items. But causative propositions differ in their forms and implications, so that comparisons between them are difficult. By recapitulating or recoding all the information in a table, we are better able to judge their mutual impact. The matrix is the common denominator of these disparate forms. The annotations down the comments column of the table comprising it, record the answers to the question we must settle for each row (or conjunction of all the items involved): is this possible, impossible or open (i.e. unsettled)?

The premises collectively structure this table, filling in all or only some of the answers to the question. The mutual impact of the determinations of the premises produces the result. This in turn allows us to judge, *with absolute certainty*, the logical impact on any of the four putative generic conclusions, and thence evaluate the validity or invalidity of that conclusion.

The main form of reasoning used in matricial analysis is dilemmatic argument – that is, we use paradoxical logic (the branch of logic concerned with paradoxes[46]). This is clear in the above account.

First, with regard to *expansion*: for example, knowing that P+Q is impossible, we can infer that P+Q+R and P+Q+notR are both impossible, or again knowing that P+Q+R is impossible, we can infer that P+Q+R+S and P+Q+R+notS are both impossible. And conversely, regarding *contraction*: we can infer a two-item impossibility from two three-item ones or a three-item impossibility from two four-item ones.

Likewise, by contraposition, when we argue from a two-item or three-item possibility to the possibility of *at least one* of two contrary conjunctions (e.g. from P+Q is possible, to P+Q+R and/or P+Q+notR is/are possible; or from P+Q+R is possible, to P+Q+R+S and/or P+Q+R+notS is/are possible), we rely on dilemma.

Such reasoning is especially productive when, from one clause we know some row(s) of the matrix to be impossible (say, P+Q+R), while from another clause we know that a set of rows including the preceding row(s) is possible (say, P+Q+R and P+Q+notR); for then, if the latter set exceeds the former by only one row, we can infer that remaining row (viz. P+Q+notR, in this example) to be possible.

Matricial analysis can be used to evaluate, and validate or invalidate, *any* putative conclusion of *any* mood in *any* figure of the syllogism (as well as in immediate inferences). It is a universal method.

---

45 Of course, it was not immediately evident; I had to develop the method gradually. The work took me a month or so. I was motivated to do it by the thought that once done by one human being, the formal research would never have to be repeated.

46 See *Future Logic*, ch. 31.

It could in principle be replaced by use of logical compositions[47]; but we are more likely to be confused in practice by the techniques of symbolic logic; and they are in any event themselves ultimately based on matricial analysis. Similarly, reduction of causative syllogisms to conditional syllogisms[48] is a conceivable method; but in practice we are much more likely to make mistakes with it, so that we will at the end remain uncertain about the reliability of our results; and anyway, matricial analysis is the ultimate basis of conditional syllogism, too.

As we have seen, to avoid having to use matricial analysis everywhere, since it is time-consuming, it is wise to first identify the minimum number of conclusions, required to be validated or invalidated that way, from which all other conclusions can be derived by direct or indirect reduction. This is the approach adopted here.

Another short-cut resorted to, here, is to identify moods which are 'mirror images' of each other, i.e. whose forms are identical in every respect, except that each item in the one is replaced by the contradictory item in the other. In such cases, the matrices of both moods are bound to be identical, except that the polarity of every *symbol* will be reversed, i.e. notP will replace P wherever it occurs, P will replace notP, and so forth.

Since the great majority of moods have a mirror image (all but four in each figure, viz. **mn/mn**, **mn/pq**, **pq/mn**, and **pq/pq**), this diminishes the work required by almost half (at most thirty moods per figure, instead of sixty).

We shall in the next three sections evaluate by matricial analysis the positive moods we identified in the preceding chapter as needing evaluation, with respect to some or all (as the case may be) of their conceivable conclusions.

2. Crucial Matricial Analyses in Figure 1.

Evaluation of mood # 117. (Similarly, *mutatis mutandis*, for **mood # 118.**)

Major premise: Q is a complete and necessary cause of R:

(i) If Q, then R;
(ii) if notQ, not-then R;
(iii) where: Q is possible.
(iv) If notQ, then notR;
(v) if Q, not-then notR;
(vi) where: notQ is possible.

Minor premise: P (complemented by S) is a partial cause of Q:

(vii) If (P + S), then Q;
(viii) if (notP + S), not-then Q;
(ix) if (P + notS), not-then Q;
(x) where: (P + S) is possible.

47 See *Future Logic*, ch. 28.
48 See *Future Logic*, ch. 29.

Putative conclusion: is P (complemented by S) a partial cause of R?

YES! P is a partial cause of R:

If (P + S), then R	implied by (i) + (vii);
if (notP + S), not-then R	implied by (iv) + (viii);
if (P + notS), not-then R	implied by (iv) + (ix);
where: (P + S) is possible	same as (x).

Table 8.1. Evaluation of mood # 117.

P	Q	R	S	implied possible by (i) + (vii) + (x)
P	Q	R	notS	see (i) + (iii), or (v)
P	Q	notR	S	implied impossible by (i)
P	Q	notR	notS	implied impossible by (i)
P	notQ	R	S	implied impossible by (iv) or (vii)
P	notQ	R	notS	implied impossible by (iv)
P	notQ	notR	S	implied impossible by (vii)
P	notQ	notR	notS	implied possible by (iv) + (ix)
notP	Q	R	S	see (i) + (iii), or (v)
notP	Q	R	notS	see (i) + (iii), or (v)
notP	Q	notR	S	implied impossible by (i)
notP	Q	notR	notS	implied impossible by (i)
notP	notQ	R	S	implied impossible by (iv)
notP	notQ	R	notS	implied impossible by (iv)
notP	notQ	notR	S	implied possible by (iv) + (viii)
notP	notQ	notR	notS	see (ii), or (iv) + (vi)

Evaluation of mood # 124. (Similarly, *mutatis mutandis*, for **mood # 134.**)

Major premise: Q is a complete and (complemented by P) a contingent cause of R:

(i) If Q, then R;
(ii) if notQ, not-then R;
(iii) where: Q is possible.
(iv) If (notQ + notP), then notR;
(v) if (Q + notP), not-then notR;
(vi) if (notQ + P), not-then notR;
(vii) where: (notQ + notP) is possible.

Minor premise: P (complemented by S) is a partial and contingent cause of Q:

(viii) If (P + S), then Q;
(ix) if (notP + S), not-then Q;
(x) if (P + notS), not-then Q;
(xi) where: (P + S) is possible.
(xii) If (notP + notS), then notQ;
(xiii) if (P + notS), not-then notQ;
(xiv) if (notP + S), not-then notQ;
(xv) where: (notP + notS) is possible.

Putative conclusion: is P a complete or (complemented by S) a partial cause of R?

NO! P is not implied to be a complete cause of R:

If P, then R?	open;
if notP, not-then R	implied by (iv) + (ix), or (iv) + (xii) + (xv);
where: P is possible	implied by (vi) or (x) or (xi) or (xiii).

Nor (complemented by S) a partial cause of R:

If (P + S), then R	implied by (i) + (viii);
if (notP + S), not-then R	implied by (iv) + (ix);
if (P + notS), not-then R?	open;
where: (P + S) is possible	same as (xi).

Table 8.2. Evaluation of mood # 124.

P	Q	R	S	implied possible by (i) + (viii) + (xi)
P	Q	R	notS	implied possible by (i) + (xiii)
P	Q	notR	S	implied impossible by (i)
P	Q	notR	notS	implied impossible by (i)
P	notQ	R	S	implied impossible by (viii)
P	notQ	R	notS	implied possible by (ii) + (iv) + (viii), or (vi) + (viii)
P	notQ	notR	S	implied impossible by (viii)
P	notQ	notR	notS	see (ii) or (x)
notP	Q	R	S	implied possible by (v) + (xii), or (i) + (xiv)
notP	Q	R	notS	implied impossible by (xii)
notP	Q	notR	S	implied impossible by (i)
notP	Q	notR	notS	implied impossible by (i) or (xii)
notP	notQ	R	S	implied impossible by (iv)
notP	notQ	R	notS	implied impossible by (iv)
notP	notQ	notR	S	implied possible by (iv) + (ix)
notP	notQ	notR	notS	implied possible by (iv) + (xii) + (xv)

Evaluation of mood # 125. (Similarly, *mutatis mutandis*, for **mood # 136.**)

Major premise: Q is a complete and (complemented by S) a contingent cause of R:

(i) If Q, then R;
(ii) if notQ, not-then R;
(iii) where: Q is possible.
(iv) If (notQ + notS), then notR;
(v) if (Q + notS), not-then notR;
(vi) if (notQ + S), not-then notR;
(vii) where: (notQ + notS) is possible.

Minor premise: P is a complete cause of Q:

(viii) If P, then Q;
(ix) if notP, not-then Q;
(x) where: P is possible.

Putative conclusion: is P (complemented by S) a contingent cause of R?

NO! P (complemented by S) is not implied to be a contingent cause of R:

If (notP + notS), then notR?	open;
if (P + notS), not-then notR?	open;
if (notP + S), not-then notR	implied by (vi) + (viii);
where: (notP + notS) is possible	implied by (iv) + (vii) + (viii).

Table 8.3. Evaluation of mood # 125.

P	Q	R	S	see (i) + (iii), or (viii) + (x)
P	Q	R	notS	see (i) + (iii), or (v), or (viii) + (x)
P	Q	notR	S	implied impossible by (i)
P	Q	notR	notS	implied impossible by (i)
P	notQ	R	S	implied impossible by (viii)
P	notQ	R	notS	implied impossible by (iv) or (viii)
P	notQ	notR	S	implied impossible by (viii)
P	notQ	notR	notS	implied impossible by (viii)
notP	Q	R	S	see (i) + (iii)
notP	Q	R	notS	see (i) + (iii), or (v)
notP	Q	notR	S	implied impossible by (i)
notP	Q	notR	notS	implied impossible by (i)
notP	notQ	R	S	implied possible by (vi) + (viii)
notP	notQ	R	notS	implied impossible by (iv)
notP	notQ	notR	S	see (ii) or (ix)
notP	notQ	notR	notS	implied possible by (iv) + (vii) + (viii)

Evaluation of mood # 126. (Similarly, *mutatis mutandis*, for **mood # 135.**)

Major premise: Q is a complete and (complemented by S) a contingent cause of R:

(i) If Q, then R;
(ii) if notQ, not-then R;
(iii) where: Q is possible.
(iv) If (notQ + notS), then notR;
(v) if (Q + notS), not-then notR;
(vi) if (notQ + S), not-then notR;
(vii) where: (notQ + notS) is possible.

Minor premise: P is a necessary cause of Q:

(viii) If notP, then notQ;
(ix) if P, not-then notQ;
(x) where: notP is possible.

Putative conclusion: is P a complete or (complemented by S) a contingent cause of R?

NO! P is not implied to be a complete cause of R:

If P, then R?	open;
if notP, not-then R?	open;
where: P is possible	implied by (v) + (viii), or (ix).

Nor (complemented by S) a contingent cause of R:

If (notP + notS), then notR — implied by (iv) + (viii);
if (P + notS), not-then notR — implied by (v) + (viii);
if (notP + S), not-then notR? — open;
where: (notP + notS) is possible? — open.

Table 8.4. Evaluation of mood # 126.				
P	Q	R	S	see (i) + (iii), or (ix)
P	Q	R	notS	implied possible by (v) + (viii)
P	Q	notR	S	implied impossible by (i)
P	Q	notR	notS	implied impossible by (i)
P	notQ	R	S	see (vi)
P	notQ	R	notS	implied impossible by (iv)
P	notQ	notR	S	see (ii)
P	notQ	notR	notS	see (ii), or (iv) + (vii)
notP	Q	R	S	implied impossible by (viii)
notP	Q	R	notS	implied impossible by (viii)
notP	Q	notR	S	implied impossible by (i) or (viii)
notP	Q	notR	notS	implied impossible by (i) or (viii)
notP	notQ	R	S	see (vi), or (viii) + (x)
notP	notQ	R	notS	implied impossible by (iv)
notP	notQ	notR	S	see (ii), or (viii) + (x)
notP	notQ	notR	notS	see (ii), or (iv) + (vii), or (viii) + (x)

Evaluation of mood # 127. (Similarly, *mutatis mutandis*, for **mood # 138.**)

Major premise: Q is a complete and (complemented by P) a contingent cause of R:
(i) If Q, then R;
(ii) if notQ, not-then R;
(iii) where: Q is possible.
(iv) If (notQ + notP), then notR;
(v) if (Q + notP), not-then notR;
(vi) if (notQ + P), not-then notR;
(vii) where: (notQ + notP) is possible.

Minor premise: P (complemented by S) is a partial cause of Q:
(viii) If (P + S), then Q;
(ix) if (notP + S), not-then Q;
(x) if (P + notS), not-then Q;
(xi) where: (P + S) is possible.

Putative conclusion: is P (complemented by S) a contingent cause of R?
NO! P (complemented by S) is not implied to be a contingent cause of R:
If (notP + notS), then notR? — open;
if (P + notS), not-then notR — implied by (vi) + (viii);
if (notP + S), not-then notR? — open;
where: (notP + notS) is possible? — open.

Table 8.5.	Evaluation of mood # 127.			
P	Q	R	S	implied possible by (i) +(viii) + (xi)
P	Q	R	notS	see (i) + (iii)
P	Q	notR	S	implied impossible by (i)
P	Q	notR	notS	implied impossible by (i)
P	notQ	R	S	implied impossible by (viii)
P	notQ	R	notS	implied possible by (vi) + (viii)
P	notQ	notR	S	implied impossible by (viii)
P	notQ	notR	notS	see (ii), or (x)
notP	Q	R	S	see (i) + (iii), or (v)
notP	Q	R	notS	see (i) + (iii), or (v)
notP	Q	notR	S	implied impossible by (i)
notP	Q	notR	notS	implied impossible by (i)
notP	notQ	R	S	implied impossible by (iv)
notP	notQ	R	notS	implied impossible by (iv)
notP	notQ	notR	S	implied possible by (iv) + (ix)
notP	notQ	notR	notS	see (ii), or (iv) + (vii)

Evaluation of mood # 128. (Similarly, *mutatis mutandis*, for **mood # 137.**)

Major premise: Q is a complete and (complemented by P) a contingent cause of R:

(i) If Q, then R;
(ii) if notQ, not-then R;
(iii) where: Q is possible.
(iv) If (notQ + notP), then notR;
(v) if (Q + notP), not-then notR;
(vi) if (notQ + P), not-then notR;
(vii) where: (notQ + notP) is possible.

Minor premise: P (complemented by S) is a contingent cause of Q:

(viii) If (notP + notS), then notQ;
(ix) if (P + notS), not-then notQ;
(x) if (notP + S), not-then notQ;
(xi) where: (notP + notS) is possible.

Putative conclusion: is P (complemented by S) a contingent cause of R?

YES! P is a contingent cause of R:

If (notP + notS), then notR	implied by (iv) + (viii);
if (P + notS), not-then notR	implied by (i) + (ix);
if (notP + S), not-then notR	implied by (v) + (viii), or (i) + (x);
where: (notP + notS) is possible	same as (xi).

Table 8.6. Evaluation of mood # 128.				
P	Q	R	S	see (i) + (iii)
P	Q	R	notS	implied possible by (i) + (ix)
P	Q	notR	S	implied impossible by (i)
P	Q	notR	notS	implied impossible by (i)
P	notQ	R	S	see (vi)
P	notQ	R	notS	see (vi)
P	notQ	notR	S	see (ii)
P	notQ	notR	notS	see (ii)
notP	Q	R	S	implied possible by (v) + (viii), or (i) + (x)
notP	Q	R	notS	implied impossible by (viii)
notP	Q	notR	S	implied impossible by (i)
notP	Q	notR	notS	implied impossible by (i) or (viii)
notP	notQ	R	S	implied impossible by (iv)
notP	notQ	R	notS	implied impossible by (iv)
notP	notQ	notR	S	see (ii), or (iv) + (vii)
notP	notQ	notR	notS	implied possible by (iv) + (viii) + (xi)

Evaluation of mood # 145. (Similarly, *mutatis mutandis*, for **mood # 146.**)

Major premise: Q (complemented by S) is a partial and contingent cause of R:
- **(i) If (Q + S), then R;**
- (ii) if (notQ + S), not-then R;
- (iii) if (Q + notS), not-then R;
- (iv) where: (Q + S) is possible.
- **(v) If (notQ + notS), then notR;**
- (vi) if (Q + notS), not-then notR;
- (vii) if (notQ + S), not-then notR;
- (viii) where: (notQ + notS) is possible.

Minor premise: P is a complete cause of Q:
- **(ix) If P, then Q;**
- (x) if notP, not-then Q;
- (xi) where: P is possible.

Putative conclusion: is P (complemented by S) a partial or contingent cause of R?

NO! P (complemented by S) is not implied to be a partial cause of R:

If (P + S), then R	implied by (i) + (ix);
if (notP + S), not-then R	implied by (ii) + (ix);
if (P + notS), not-then R?	open;
where: (P + S) is possible?	open.

Nor (complemented by S) a contingent cause of R:

If (notP + notS), then notR?	open;
if (P + notS), not-then notR?	open;
if (notP + S), not-then notR	implied by (vii) + (ix);
where: (notP + notS) is possible	implied by (v) + (viii) + (ix).

Table 8.7. Evaluation of mood # 145.				
P	Q	R	S	see (i) + (iv), or (ix) + (xi)
P	Q	R	notS	see (vi), or (ix) + (xi)
P	Q	notR	S	implied impossible by (i)
P	Q	notR	notS	see (iii), or (ix) + (xi)
P	notQ	R	S	implied impossible by (ix)
P	notQ	R	notS	implied impossible by (v) or (ix)
P	notQ	notR	S	implied impossible by (ix)
P	notQ	notR	notS	implied impossible by (ix)
notP	Q	R	S	see (i) + (iv)
notP	Q	R	notS	see (vi)
notP	Q	notR	S	implied impossible by (i)
notP	Q	notR	notS	see (iii)
notP	notQ	R	S	implied possible by (vii) + (ix)
notP	notQ	R	notS	implied impossible by (v)
notP	notQ	notR	S	implied possible by (ii) + (ix)
notP	notQ	notR	notS	implied possible by (v) + (viii) + (ix)

Evaluation of mood # 147. (Similarly, *mutatis mutandis*, for **mood # 148.**)

Major premise: Q (complemented by P) is a partial and contingent cause of R:

- **(i) If (Q + P), then R;**
- (ii) if (notQ + P), not-then R;
- (iii) if (Q + notP), not-then R;
- (iv) where: (Q + P) is possible.
- **(v) If (notQ + notP), then notR;**
- (vi) if (Q + notP), not-then notR;
- (vii) if (notQ + P), not-then notR;
- (viii) where: (notQ + notP) is possible.

Minor premise: P (complemented by S) is a partial cause of Q:

- **(ix) If (P + S), then Q;**
- (x) if (notP + S), not-then Q;
- (xi) if (P + notS), not-then Q;
- (xii) where: (P + S) is possible.

Putative conclusion: is P (complemented by S) a partial or contingent cause of R?

YES! P is a partial cause of R:

If (P + S), then R	implied by (i) + (ix);
if (notP + S), not-then R	implied by (v) + (x);
if (P + notS), not-then R	implied by (ii) + (ix);
where: (P + S) is possible	same as (xii).

NO! P (complemented by S) is not implied to be a contingent cause of R:

If (notP + notS), then notR?	open;
if (P + notS), not-then notR	implied by (vii) + (ix);

if (notP + S), not-then notR? open;
where: (notP + notS) is possible? open.

Table 8.8. Evaluation of mood # 147.

P	Q	R	S	implied possible by (i) + (ix) + (xii)
P	Q	R	notS	see (i) + (iv)
P	Q	notR	S	implied impossible by (i)
P	Q	notR	notS	implied impossible by (i)
P	notQ	R	S	implied impossible by (ix)
P	notQ	R	notS	implied possible by (vii) + (ix)
P	notQ	notR	S	implied impossible by (ix)
P	notQ	notR	notS	implied possible by (ii) + (ix)
notP	Q	R	S	see (vi)
notP	Q	R	notS	see (vi)
notP	Q	notR	S	see (iii)
notP	Q	notR	notS	see (iii)
notP	notQ	R	S	implied impossible by (v)
notP	notQ	R	notS	implied impossible by (v)
notP	notQ	notR	S	implied possible by (v) + (x)
notP	notQ	notR	notS	see (v) + (viii)

Evaluation of mood # 152. (Similarly, *mutatis mutandis*, for **mood # 163.**)

Major premise: Q is a complete cause of R:

(i) If Q, then R;
(ii) if notQ, not-then R;
(iii) where: Q is possible.

Minor premise: P is a complete and (complemented by S) a contingent cause of Q:

(iv) If P, then Q;
(v) if notP, not-then Q;
(vi) where: P is possible.
(vii) If (notP + notS), then notQ;
(viii) if (P + notS), not-then notQ;
(ix) if (notP + S), not-then notQ;
(x) where: (notP + notS) is possible.

Putative conclusion: is P (complemented by S) a contingent cause of R?

NO! P (complemented by S) is not implied to be a contingent cause of R:

If (notP + notS), then notR? open;
if (P + notS), not-then notR implied by (i) + (viii);
if (notP + S), not-then notR implied by (i) + (ix);
where: (notP + notS) is possible same as (x).

Table 8.9. Evaluation of mood # 152.				
P	Q	R	S	see (i) + (iii), or (iv) + (vi)
P	Q	R	notS	implied possible by (i) + (viii)
P	Q	notR	S	implied impossible by (i)
P	Q	notR	notS	implied impossible by (i)
P	notQ	R	S	implied impossible by (iv)
P	notQ	R	notS	implied impossible by (iv)
P	notQ	notR	S	implied impossible by (iv)
P	notQ	notR	notS	implied impossible by (iv)
notP	Q	R	S	implied possible by (i) + (ix)
notP	Q	R	notS	implied impossible by (vii)
notP	Q	notR	S	implied impossible by (i)
notP	Q	notR	notS	implied impossible by (i) or (vii)
notP	notQ	R	S	see (v)
notP	notQ	R	notS	see (v), or (vii) + (x)
notP	notQ	notR	S	see (ii) or (v)
notP	notQ	notR	notS	see (ii), or (v), or (vii) + (x)

Note that although the above matrix does not show it, the conclusion **m** of mood 152 (which we uncovered through reduction) is valid. Judging from the last two lines of this matrix, one would think that (notP + notR) is open. However, it is in fact possible. This can be seen as follows.

Clause (ii) above tells us that (notQ + notR) is possible; but two of its possible expressions are implied impossible by (iv); therefore *at least one* of the remaining two possible expressions has to be possible. This implicit disjunctive result suffices to prove that (notP + notR) is possible, considering that its other two possible expressions are implied impossible by (i).

Compare the matrix of mood 155, whose last line corresponds to the last two lines of the matrix of mood 152. Thus, the above matrix fails to make something significant explicit for us. But this is a mere difficulty of notation, when something about more than one line has to be specified.

The same can be said for the conclusion **n** of mirror mood 163 and other cases. However, **a similar problem does not arise with regard to any of the conclusions tested by matricial analysis in this chapter** (as can be verified by reexamining all clauses of tested conclusions which were left open where a possibility was required). So it does not seem worthwhile our trying to remedy this difficulty with more elaborate notational artifices (better than "see (ii)").

The lesson taught us by this special case is the wisdom of using matricial analysis only for crucial questions and using reduction for all others, as we did. Without awareness of the relation between moods 152 and 155 (or similarly 163 and 166), we might not have spotted the positive conclusion's validity, unless we had developed a straddling notation.

Evaluation of mood # 153. (Similarly, *mutatis mutandis*, for **mood # 162.**)

Major premise: Q is a complete cause of R:

(i) If Q, then R;
(ii) if notQ, not-then R;
(iii) where: Q is possible.

Minor premise: P is a necessary and (complemented by S) a partial cause of Q:

(iv) If (P + S), then Q;
(v) if (notP + S), not-then Q;
(vi) if (P + notS), not-then Q;
(vii) where: (P + S) is possible.
(viii) If notP, then notQ;
(ix) if P, not-then notQ;
(x) where: notP is possible.

Putative conclusion: is P a necessary or (complemented by S) a partial cause of R?

NO! P (complemented by S) is not implied to be a partial cause of R:

If (P + S), then R	implied by (i) + (iv);
if (notP + S), not-then R?	open;
if (P + notS), not-then R?	open;
where: (P + S) is possible	same as (vii).

Nor a necessary cause of R:

If notP, then notR?	open;
if P, not-then notR	implied by (i) + (iv) + (vii);
where: notP is possible	same as (x).

Table 8.10. Evaluation of mood # 153.

P	Q	R	S	implied possible by (i) + (iv) + (vii)
P	Q	R	notS	see (i) + (iii), or (ix)
P	Q	notR	S	implied impossible by (i)
P	Q	notR	notS	implied impossible by (i)
P	notQ	R	S	implied impossible by (iv)
P	notQ	R	notS	see (vi)
P	notQ	notR	S	implied impossible by (iv)
P	notQ	notR	notS	see (ii) or (vi)
notP	Q	R	S	implied impossible by (viii)
notP	Q	R	notS	implied impossible by (viii)
notP	Q	notR	S	implied impossible by (i) or (viii)
notP	Q	notR	notS	implied impossible by (i) or (viii)
notP	notQ	R	S	see (v), or (viii) + (x)
notP	notQ	R	notS	see (viii) + (x)
notP	notQ	notR	S	see (ii), or (v), or (viii) + (x)
notP	notQ	notR	notS	see (ii), or (viii) + (x)

Evaluation of mood # 154. (Similarly, *mutatis mutandis*, for **mood # 164.**)

Major premise: Q is a complete cause of R:

- **(i) If Q, then R;**
- (ii) if notQ, not-then R;
- (iii) where: Q is possible.

Minor premise: P (complemented by S) is a partial and contingent cause of Q:

- **(iv) If (P + S), then Q;**
- (v) if (notP + S), not-then Q;
- (vi) if (P + notS), not-then Q;
- (vii) where: (P + S) is possible.
- **(viii) If (notP + notS), then notQ;**
- (ix) if (P + notS), not-then notQ;
- (x) if (notP + S), not-then notQ;
- (xi) where: (notP + notS) is possible.

Putative conclusion: is P (complemented by S) a contingent cause of R?

NO! P (complemented by S) is not implied to be a contingent cause of R:

If (notP + notS), then notR?	open;
if (P + notS), not-then notR	implied by (i) + (ix);
if (notP + S), not-then notR	implied by (i) + (x);
where: (notP + notS) is possible	same as (xi).

Table 8.11. Evaluation of mood # 154.

P	Q	R	S	implied possible by (i) + (iv) + (vii)
P	Q	R	notS	implied possible by (i) + (ix)
P	Q	notR	S	implied impossible by (i)
P	Q	notR	notS	implied impossible by (i)
P	notQ	R	S	implied impossible by (iv)
P	notQ	R	notS	see (vi)
P	notQ	notR	S	implied impossible by (iv)
P	notQ	notR	notS	see (ii) or (vi)
notP	Q	R	S	implied possible by (i) + (x)
notP	Q	R	notS	implied impossible by (viii)
notP	Q	notR	S	implied impossible by (i)
notP	Q	notR	notS	implied impossible by (i) or (viii)
notP	notQ	R	S	see (v)
notP	notQ	R	notS	see (viii) + (xi)
notP	notQ	notR	S	see (ii) or (v)
notP	notQ	notR	notS	see (ii), or (viii) + (xi)

Evaluation of mood # 155. (Similarly, *mutatis mutandis*, for **mood # 166.**)

Major premise: Q is a complete cause of R:
(i) If Q, then R;
(ii) if notQ, not-then R;
(iii) where: Q is possible.
Minor premise: P is a complete cause of Q:
(iv) If P, then Q;
(v) if notP, not-then Q;
(vi) where: P is possible.
Putative conclusion: P is a complete cause of R?
YES! P is a complete cause of R:
If P, then R — implied by (i) + (iv);
if notP, not-then R — implied by (ii) + (iv);
where: P is possible — same as (vi).

Table 8.12. Evaluation of mood # 155.			
P	Q	R	implied possible by (i) + (iv) + (vi)
P	Q	notR	implied impossible by (i)
P	notQ	R	implied impossible by (iv)
P	notQ	notR	implied impossible by (iv)
notP	Q	R	see (i) + (iii)
notP	Q	notR	implied impossible by (i)
notP	notQ	R	see (v)
notP	notQ	notR	implied possible by (ii) + (iv)

Evaluation of mood # 171. (Similarly, *mutatis mutandis*, for **mood # 181.**)

Major premise: Q (complemented by S) is a partial cause of R:
(i) If (Q + S), then R;
(ii) if (notQ + S), not-then R;
(iii) if (Q + notS), not-then R;
(iv) where: (Q + S) is possible.
Minor premise: P is a complete and necessary cause of Q:
(v) If P, then Q;
(vi) if notP, not-then Q;
(vii) where: P is possible.
(viii) If notP, then notQ;
(ix) if P, not-then notQ;

(x) where: notP is possible.

Putative conclusion: is P (complemented by S) a partial cause of R?

YES! P is a partial cause of R:

If (P + S), then R	implied by (i) + (v);
if (notP + S), not-then R	implied by (ii) + (v);
if (P + notS), not-then R	implied by (iii) + (viii);
where: (P + S) is possible	implied by (i) + (iv) + (viii).

Table 8.13. Evaluation of mood # 171.

P	Q	R	S	
P	Q	R	S	implied possible by (i) + (iv) + (viii)
P	Q	R	notS	see (v) + (vii), or (ix)
P	Q	notR	S	implied impossible by (i)
P	Q	notR	notS	implied possible by (iii) + (viii)
P	notQ	R	S	implied impossible by (v)
P	notQ	R	notS	implied impossible by (v)
P	notQ	notR	S	implied impossible by (v)
P	notQ	notR	notS	implied impossible by (v)
notP	Q	R	S	implied impossible by (viii)
notP	Q	R	notS	implied impossible by (viii)
notP	Q	notR	S	implied impossible by (i) or (viii)
notP	Q	notR	notS	implied impossible by (viii)
notP	notQ	R	S	see (vi), or (viii) + (x)
notP	notQ	R	notS	see (vi), or (viii) + (x)
notP	notQ	notR	S	implied possible by (ii) + (v)
notP	notQ	notR	notS	see (vi), or (viii) + (x)

Evaluation of mood # 174. (Similarly, *mutatis mutandis*, for **mood # 184.**)

Major premise: Q (complemented by P) is a partial cause of R:

(i) If (Q + P), then R;
(ii) if (notQ + P), not-then R;
(iii) if (Q + notP), not-then R;
(iv) where: (Q + P) is possible.

Minor premise: P (complemented by S) is a partial and contingent cause of Q:

(v) If (P + S), then Q;
(vi) if (notP + S), not-then Q;
(vii) if (P + notS), not-then Q;
(viii) where: (P + S) is possible.
(ix) If (notP + notS), then notQ;
(x) if (P + notS), not-then notQ;
(xi) if (notP + S), not-then notQ;
(xii) where: (notP + notS) is possible.

Putative conclusion: is P (complemented by S) a partial cause of R?

YES! P is a partial cause of R:

If (P + S), then R	implied by (i) + (v);
if (notP + S), not-then R	implied by (iii) + (ix);
if (P + notS), not-then R	implied by (ii) + (v);
where: (P + S) is possible	same as (viii).

Table 8.14. Evaluation of mood # 174.

P	Q	R	S	implied possible by (i) + (v) + (viii)
P	Q	R	notS	implied possible by (i) + (x)
P	Q	notR	S	implied impossible by (i)
P	Q	notR	notS	implied impossible by (i)
P	notQ	R	S	implied impossible by (v)
P	notQ	R	notS	see (vii)
P	notQ	notR	S	implied impossible by (v)
P	notQ	notR	notS	implied possible by (ii) + (v)
notP	Q	R	S	see (xi)
notP	Q	R	notS	implied impossible by (ix)
notP	Q	notR	S	implied possible by (iii) + (ix)
notP	Q	notR	notS	implied impossible by (ix)
notP	notQ	R	S	see (vi)
notP	notQ	R	notS	see (ix) + (xii)
notP	notQ	notR	S	see (vi)
notP	notQ	notR	notS	see (ix) + (xii)

Evaluation of mood # 177. (Similarly, *mutatis mutandis*, for **mood # 188.**)

Major premise: Q (complemented by P) is a partial cause of R:

(i) If (Q + P), then R;
(ii) if (notQ + P), not-then R;
(iii) if (Q + notP), not-then R;
(iv) where: (Q + P) is possible.

Minor premise: P (complemented by S) is a partial cause of Q:

(v) If (P + S), then Q;
(vi) if (notP + S), not-then Q;
(vii) if (P + notS), not-then Q;
(viii) where: (P + S) is possible.

Putative conclusion: is P (complemented by S) a partial cause of R?

NO! P (complemented by S) is not implied to be a partial cause of R:

If (P + S), then R	implied by (i)+ (v);
if (notP + S), not-then R?	open;
if (P + notS), not-then R	implied by (ii) + (v);
where: (P + S) is possible	same as (viii).

Table 8.15.	Evaluation of mood # 177.			
P	Q	R	S	implied possible by (i) + (v) + (viii)
P	Q	R	notS	see (i) + (iv)
P	Q	notR	S	implied impossible by (i)
P	Q	notR	notS	implied impossible by (i)
P	notQ	R	S	implied impossible by (v)
P	notQ	R	notS	see (vii)
P	notQ	notR	S	implied impossible by (v)
P	notQ	notR	notS	implied possible by (ii) + (v)
notP	Q	R	S	
notP	Q	R	notS	
notP	Q	notR	S	see (iii)
notP	Q	notR	notS	see (iii)
notP	notQ	R	S	see (vi)
notP	notQ	R	notS	
notP	notQ	notR	S	see (vi)
notP	notQ	notR	notS	

3. Crucial Matricial Analyses in Figure 2.

Evaluation of mood # 221. (Similarly, *mutatis mutandis*, for **mood # 231.**)

Major premise: R is a complete and (complemented by S) a contingent cause of Q:
- **(i) If R, then Q;**
- (ii) if notR, not-then Q;
- (iii) where: R is possible.
- **(iv) If (notR + notS), then notQ;**
- (v) if (R + notS), not-then notQ;
- (vi) if (notR + S), not-then notQ;
- (vii) where: (notR + notS) is possible.

Minor premise: P is a complete and necessary cause of Q:
- **(viii) If P, then Q;**
- (ix) if notP, not-then Q;
- (x) where: P is possible.
- **(xi) If notP, then notQ;**
- (xii) if P, not-then notQ;
- (xiii) where: notP is possible.

Putative conclusion is P a complete or (complemented by S) a partial cause of R?

NO! P is not implied to be a complete cause of R:

If P, then R?	denied by (vi) + (xi);
if notP, not-then R	implied by (iv) + (vii) + (viii);
where: P is possible	same as (x).

Nor (complemented by S) a partial cause of R:

If (P + S), then R? denied by (vi) + (xi);
if (notP + S), not-then R? open;
if (P + notS), not-then R? denied by (iv) + (viii);
where: (P + S) is possible implied by (vi) + (xi).

Table 8.16. Evaluation of mood 221.

P	Q	R	S	see (i) + (iii), or (viii) + (x), or (xii)
P	Q	R	notS	implied possible by (v) + (xi)
P	Q	notR	S	implied possible by (vi) + (xi)
P	Q	notR	notS	implied impossible by (iv)
P	notQ	R	S	implied impossible by (i) or (viii)
P	notQ	R	notS	implied impossible by (i) or (viii)
P	notQ	notR	S	implied impossible by (viii)
P	notQ	notR	notS	implied impossible by (viii)
notP	Q	R	S	implied impossible by (xi)
notP	Q	R	notS	implied impossible by (xi)
notP	Q	notR	S	implied impossible by (xi)
notP	Q	notR	notS	implied impossible by (iv) or (xi)
notP	notQ	R	S	implied impossible by (i)
notP	notQ	R	notS	implied impossible by (i)
notP	notQ	notR	S	see (ii), or (ix), or (xi) + (xiii)
notP	notQ	notR	notS	implied possible by (iv) + (vii) + (viii)

Evaluation of mood # 222. (Similarly, *mutatis mutandis*, for **mood # 233.**)

Major premise: R is a complete and (complemented by P) a contingent cause of Q:
(i) If R, then Q;
(ii) if notR, not-then Q;
(iii) where: R is possible.
(iv) If (notR + notP), then notQ;
(v) if (R + notP), not-then notQ;
(vi) if (notR + P), not-then notQ;
(vii) where: (notR + notP) is possible.

Minor premise: P is a complete and (complemented by S) a contingent cause of Q:
(viii) If P, then Q;
(ix) if notP, not-then Q;
(x) where: P is possible.
(xi) If (notP + notS), then notQ;
(xii) if (P + notS), not-then notQ;
(xiii) if (notP + S), not-then notQ;
(xiv) where: (notP + notS) is possible.

Putative conclusion is P (complemented by S) a cause of R?
NO! P is not implied to be a complete cause of R:
If P, then R? open;
if notP, not-then R implied by (i) + (xi) + (xiv);

where: P is possible	same as (x).
Nor (complemented by S) a partial cause of R:	
If (P + S), then R?	open;
if (notP + S), not-then R?	open;
if (P + notS), not-then R?	open;
where: (P + S) is possible?	open.
Nor a necessary cause of R:	
If notP, then notR?	denied by (v) + (xi), or (iv) + (xiii);
if P, not-then notR?	open;
where: notP is possible	implied by (v) or (vii) or (ix) or (xiii) or (xiv).
Nor (complemented by S) a contingent cause of R:	
If (notP + notS), then notR	implied by (i) + (xi);
if (P + notS), not-then notR?	open;
if (notP + S), not-then notR	implied by (v) + (xi), or (iv) + (xiii);
where: (notP + notS) is possible	same as (xiv).

Table 8.17. Evaluation of mood 222.

P	Q	R	S	see (i) + (iii), or (viii) + (x)
P	Q	R	notS	see (i) + (iii), or (viii) + (x), or (xii)
P	Q	notR	S	see (vi), or (viii) + (x)
P	Q	notR	notS	see (vi), or (viii) + (x), or (xii)
P	notQ	R	S	implied impossible by (i) or (viii)
P	notQ	R	notS	implied impossible by (i) or (viii)
P	notQ	notR	S	implied impossible by (viii)
P	notQ	notR	notS	implied impossible by (viii)
notP	Q	R	S	implied possible by (v) + (xi), or (iv) + (xiii)
notP	Q	R	notS	implied impossible by (xi)
notP	Q	notR	S	implied impossible by (iv)
notP	Q	notR	notS	implied impossible by (iv) or (xi)
notP	notQ	R	S	implied impossible by (i)
notP	notQ	R	notS	implied impossible by (i)
notP	notQ	notR	S	see (ii), or (iv) + (vii), or (ix)
notP	notQ	notR	notS	implied possible by (i) + (xi) + (xiv)

Evaluation of mood # 224. (Similarly, *mutatis mutandis*, for **mood # 234.**)

Major premise: R is a complete and (complemented by P) a contingent cause of Q:

(i) If R, then Q;
(ii) if notR, not-then Q;
(iii) where: R is possible.
(iv) If (notR + notP), then notQ;
(v) if (R + notP), not-then notQ;
(vi) if (notR + P), not-then notQ;

(vii) where: (notR + notP) is possible.

Minor premise: P (complemented by S) is a partial and contingent cause of Q:

(viii) If (P + S), then Q;
(ix) if (notP + S), not-then Q;
(x) if (P + notS), not-then Q;
(xi) where: (P + S) is possible.
(xii) If (notP + notS), then notQ;
(xiii) if (P + notS), not-then notQ;
(xiv) if (notP + S), not-then notQ;
(xv) where: (notP + notS) is possible.

Putative conclusion is P (complemented by S) a cause of R?

NO! P is not implied to be a complete cause of R:

If P, then R?	denied by (i) + (x);
if notP, not-then R	implied by (i) + (ix), or (i) + (xii) + (xv);
where: P is possible	implied by (vi) or (x) or (xi) or (xiii).

Nor (complemented by S) a partial cause of R:

If (P + S), then R?	open;
if (notP + S), not-then R	implied by (i) + (ix);
if (P + notS), not-then R	implied by (i) + (x);
where: (P + S) is possible	same as (xi).

Nor a necessary cause of R:

If notP, then notR?	denied by (iv) + (xiv), or (v) + (xii);
if P, not-then notR?	open;
where: notP is possible	implied by (v) or (vii) or (ix) or (xiv) or (xv).

Nor (complemented by S) a contingent cause of R:

If (notP + notS), then notR	implied by (i) + (xii);
if (P + notS), not-then notR?	open;
if (notP + S), not-then notR	implied by (iv) + (xiv), or (v) + (xii);
where: (notP + notS) is possible	same as (xv).

Table 8.18. Evaluation of mood 224.				
P	Q	R	S	see (i) + (iii), or (viii) + (xi)
P	Q	R	notS	see (i) + (iii), or (xiii)
P	Q	notR	S	see (vi), or (viii) + (xi)
P	Q	notR	notS	see (vi), or (xiii)
P	notQ	R	S	implied impossible by (i) or (viii)
P	notQ	R	notS	implied impossible by (i)
P	notQ	notR	S	implied impossible by (viii)
P	notQ	notR	notS	implied possible by (i) + (x)
notP	Q	R	S	implied possible by (iv) + (xiv), or (v) + (xii)
notP	Q	R	notS	implied impossible by (xii)
notP	Q	notR	S	implied impossible by (iv)
notP	Q	notR	notS	implied impossible by (iv) or (xii)
notP	notQ	R	S	implied impossible by (i)
notP	notQ	R	notS	implied impossible by (i)
notP	notQ	notR	S	implied possible by (i) + (ix)
notP	notQ	notR	notS	implied possible by (i) + (xii) + (xv)

Evaluation of mood # 241.

Major premise: R (complemented by S) is a partial and contingent cause of Q:

(i) If (R + S), then Q;
(ii) if (notR + S), not-then Q;
(iii) if (R + notS), not-then Q;
(iv) where: (R + S) is possible.
(v) If (notR + notS), then notQ;
(vi) if (R + notS), not-then notQ;
(vii) if (notR + S), not-then notQ;
(viii) where: (notR + notS) is possible.

Minor premise: P is a complete and necessary cause of Q:

(ix) If P, then Q;
(x) if notP, not-then Q;
(xi) where: P is possible.
(xii) If notP, then notQ;
(xiii) if P, not-then notQ;
(xiv) where: notP is possible.

Putative conclusion is P (complemented by S) a cause of R?

NO! P is not implied to be a complete cause of R:

If P, then R?	denied by (vii) + (xii);
if notP, not-then R	implied by (ii) + (ix), or (v) + (viii) + (ix);
where: P is possible	same as (xi).

Nor (complemented by S) a partial cause of R:

If (P + S), then R?	denied by (vii) + (xii);

if (notP + S), not-then R	implied by (ii) + (ix);
if (P + notS), not-then R?	denied by (v) + (ix);
where: (P + S) is possible	implied by (i) + (iv) + (xii), or (vii) + (xii).

Nor a necessary cause of R:

If notP, then notR?	denied by (iii) + (ix);
if P, not-then notR	implied by (i) + (iv) + (xii), or (vi) + (xii);
where: notP is possible	same as (xiv).

Nor (complemented by S) a contingent cause of R:

If (notP + notS), then notR?	denied by (iii) + (ix);
if (P + notS), not-then notR	implied by (vi) + (xii);
if (notP + S), not-then notR?	denied by (i) + (xii);
where: (notP + notS) is possible	implied by (iii) + (ix), or (v) + (viii) + (ix).

Table 8.19. Evaluation of mood 241.

P	Q	R	S	implied possible by (i) + (iv) + (xii)
P	Q	R	notS	implied possible by (vi) + (xii)
P	Q	notR	S	implied possible by (vii) + (xii)
P	Q	notR	notS	implied impossible by (v)
P	notQ	R	S	implied impossible by (i) or (ix)
P	notQ	R	notS	implied impossible by (ix)
P	notQ	notR	S	implied impossible by (ix)
P	notQ	notR	notS	implied impossible by (ix)
notP	Q	R	S	implied impossible by (xii)
notP	Q	R	notS	implied impossible by (xii)
notP	Q	notR	S	implied impossible by (xii)
notP	Q	notR	notS	implied impossible by (v) or (xii)
notP	notQ	R	S	implied impossible by (i)
notP	notQ	R	notS	implied possible by (iii) + (ix)
notP	notQ	notR	S	implied possible by (ii) + (ix)
notP	notQ	notR	notS	implied possible by (v) + (viii) + (ix)

Evaluation of mood # 244.

Major premise: R (complemented by P) is a partial and contingent cause of Q:

(i) If (R + P), then Q;
(ii) if (notR + P), not-then Q;
(iii) if (R + notP), not-then Q;
(iv) where: (R + P) is possible.
(v) If (notR + notP), then notQ;
(vi) if (R + notP), not-then notQ;
(vii) if (notR + P), not-then notQ;

(viii) where: (notR + notP) is possible.

Minor premise: P (complemented by S) is a partial and contingent cause of Q:

(ix) If (P + S), then Q;
(x) if (notP + S), not-then Q;
(xi) if (P + notS), not-then Q;
(xii) where: (P + S) is possible.
(xiii) If (notP + notS), then notQ;
(xiv) if (P + notS), not-then notQ;
(xv) if (notP + S), not-then notQ;
(xvi) where: (notP + notS) is possible.

Putative conclusion is P (complemented by S) a cause of R?

NO! P is not implied to be a complete cause of R:

If P, then R?	denied by (i) + (xi), or (ii) + (ix);
if notP, not-then R?	open;
where: P is possible	implied by (ii) or (iv) or (vii) or (xi) or (xii) or (xiv).

Nor (complemented by S) a partial cause of R:

If (P + S), then R?	open;
if (notP + S), not-then R?	open;
if (P + notS), not-then R	implied by (i) + (xi), or (ii) + (ix);
where: (P + S) is possible	same as (xii).

Nor a necessary cause of R:

If notP, then notR?	denied by (v) + (xv), or (vi) + (xiii);
if P, not-then notR?	open;
where: notP is possible	implied by (iii) or (vi) or (viii) or (x) or (xv) or (xvi).

Nor (complemented by S) a contingent cause of R:

If (notP + notS), then notR?	open;
if (P + notS), not-then notR?	open;
if (notP + S), not-then notR	implied by (v) + (xv), or (vi) + (xiii);
where: (notP + notS) is possible	same as (xvi).

Table 8.20.	Evaluation of mood 244.			
P	Q	R	S	see (i) + (iv), or (ix) + (xii)
P	Q	R	notS	see (i) + (iv), or (xiv)
P	Q	notR	S	see (vii), or (ix) + (xii)
P	Q	notR	notS	see (vii), or (xiv)
P	notQ	R	S	implied impossible by (i) or (ix)
P	notQ	R	notS	implied impossible by (i)
P	notQ	notR	S	implied impossible by (ix)
P	notQ	notR	notS	implied possible by (i) + (xi), or (ii) + (ix)
notP	Q	R	S	implied possible by (v) + (xv), or (vi) + (xiii)
notP	Q	R	notS	implied impossible by (xiii)
notP	Q	notR	S	implied impossible by (v)
notP	Q	notR	notS	implied impossible by (v) or (xiii)
notP	notQ	R	S	see (iii) or (x)
notP	notQ	R	notS	see (iii), or (xiii) + (xvi)
notP	notQ	notR	S	see (v) + (viii), or (x)
notP	notQ	notR	notS	see (v) + (viii), or (xiii) + (xvi)

4. Crucial Matricial Analyses in Figure 3.

Evaluation of mood # 312. (Similarly, *mutatis mutandis*, for **mood # 313.**)

Major premise: Q is a complete and necessary cause of R:

- **(i) If Q, then R;**
- (ii) if notQ, not-then R;
- (iii) where: Q is possible.
- **(iv) If notQ, then notR;**
- (v) if Q, not-then notR;
- (vi) where: notQ is possible.

Minor premise: Q is a complete and (complemented by S) a contingent cause of P:

- **(vii) If Q, then P;**
- (viii) if notQ, not-then P;
- (ix) where: Q is possible.
- **(x) If (notQ + notS), then notP;**
- (xi) if (Q + notS), not-then notP;
- (xii) if (notQ + S), not-then notP;
- (xiii) where: (notQ + notS) is possible.

Putative conclusion: is P a complete or (complemented by S) a partial cause of R?

NO! P is not implied to be a complete cause of R:

If P, then R?	denied by (iv) + (xii);
if notP, not-then R	implied by (iv) + (x) + (xiii);
where: P is possible	implied by (vii) + (ix), or (xi), or (xii).

Nor (complemented by S) a partial cause of R:

If (P + S), then R?	denied by (iv) + (xii);
if (notP + S), not-then R?	open;
if (P + notS), not-then R?	denied by (i) + (x);
where: (P + S) is possible	implied by (xii).

Table 8.21 Evaluation of mood # 312.

P	Q	R	S	see (i) + (iii), or (v), or (vii) + (ix)
P	Q	R	notS	implied possible by (i) + (xi)
P	Q	notR	S	implied impossible by (i)
P	Q	notR	notS	implied impossible by (i)
P	notQ	R	S	implied impossible by (iv)
P	notQ	R	notS	implied impossible by (iv) or (x)
P	notQ	notR	S	implied possible by (iv) + (xii)
P	notQ	notR	notS	implied impossible by (x)
notP	Q	R	S	implied impossible by (vii)
notP	Q	R	notS	implied impossible by (vii)
notP	Q	notR	S	implied impossible by (i) or (vii)
notP	Q	notR	notS	implied impossible by (i) or (vii)
notP	notQ	R	S	implied impossible by (iv)
notP	notQ	R	notS	implied impossible by (iv)
notP	notQ	notR	S	see (ii), or (iv) + (vi), or (viii)
notP	notQ	notR	notS	implied possible by (iv) + (x) + (xiii)

Evaluation of mood # 314.

Major premise: Q is a complete and necessary cause of R:

(i) If Q, then R;
(ii) if notQ, not-then R;
(iii) where: Q is possible.
(iv) If notQ, then notR;
(v) if Q, not-then notR;
(vi) where: notQ is possible.

Minor premise: Q (complemented by S) is a partial and contingent cause of P:

(vii) If (Q + S), then P;
(viii) if (notQ + S), not-then P;
(ix) if (Q + notS), not-then P;
(x) where: (Q + S) is possible.
(xi) If (notQ + notS), then notP;
(xii) if (Q + notS), not-then notP;
(xiii) if (notQ + S), not-then notP;
(xiv) where: (notQ + notS) is possible.

<u>Putative conclusion</u>: is P (complemented by S) a cause of R?

NO! P is not implied to be a complete cause of R:

If P, then R?	denied by (iv) + (xiii);
if notP, not-then R	implied by (iv) + (viii), or (iv) + (xi) + (xiv);
where: P is possible	implied by (vii) + (x), or (xii), or (xiii).

Nor (complemented by S) a partial cause of R:

If (P + S), then R?	denied by (iv) + (xiii);
if (notP + S), not-then R	implied by (iv) + (viii);
if (P + notS), not-then R?	denied by (i) + (xi);
where: (P + S) is possible	implied by (vii) + (x), or (xiii).

Nor a necessary cause of R:

If notP, then notR?	denied by (i) + (ix);
if P, not-then notR	implied by (i) + (vii) + (x), or (i) + (xii);
where: notP is possible	implied by (viii), or (ix), or (xi) + (xiv).

Nor (complemented by S) a contingent cause of R:

If (notP + notS), then notR?	denied by (i) + (ix);
if (P + notS), not-then notR	implied by (i) + (xii);
if (notP + S), not-then notR?	denied by (iv) + (vii);
where: (notP + notS) is possible	implied by (ix), or (xi) + (xiv).

Table 8.22 Evaluation of mood # 314.

P	Q	R	S	implied possible by (i) + (vii) + (x)
P	Q	R	notS	implied possible by (i) + (xii)
P	Q	notR	S	implied impossible by (i)
P	Q	notR	notS	implied impossible by (i)
P	notQ	R	S	implied impossible by (iv)
P	notQ	R	notS	implied impossible by (iv) or (xi)
P	notQ	notR	S	implied possible by (iv) + (xiii)
P	notQ	notR	notS	implied impossible by (xi)
notP	Q	R	S	implied impossible by (vii)
notP	Q	R	notS	implied possible by (i) + (ix)
notP	Q	notR	S	implied impossible by (i) or (vii)
notP	Q	notR	notS	implied impossible by (i)
notP	notQ	R	S	implied impossible by (iv)
notP	notQ	R	notS	implied impossible by (iv)
notP	notQ	notR	S	implied possible by (iv) + (viii)
notP	notQ	notR	notS	implied possible by (iv) + (xi) + (xiv)

Evaluation of mood # 324. (Similarly, *mutatis mutandis*, for **mood # 334.**)

Major premise: Q is a complete and (complemented by P) a contingent cause of R:

(i) If Q, then R;
(ii) if notQ, not-then R;
(iii) where: Q is possible.
(iv) If (notQ + notP), then notR;
(v) if (Q + notP), not-then notR;
(vi) if (notQ + P), not-then notR;
(vii) where: (notQ + notP) is possible.

Minor premise: Q (complemented by S) is a partial and contingent cause of P:

(viii) If (Q + S), then P;
(ix) if (notQ + S), not-then P;
(x) if (Q + notS), not-then P;
(xi) where: (Q + S) is possible.
(xii) If (notQ + notS), then notP;
(xiii) if (Q + notS), not-then notP;
(xiv) if (notQ + S), not-then notP;
(xv) where: (notQ + notS) is possible.

Putative conclusion: is P (complemented by S) a cause of R?

NO! P is not implied to be a complete cause of R:

If P, then R?	open;
if notP, not-then R	implied by (iv) + (ix), or (iv) + (xii) + (xv);
where: P is possible	implied by (vi), or (viii) + (xi), or (xiii), or (xiv).

Nor (complemented by S) a partial cause of R:

If (P + S), then R?	open;
if (notP + S), not-then R	implied by (iv) + (ix);
if (P + notS), not-then R?	denied by (i) + (xii);
where: (P + S) is possible	implied by (vi) + (xii), or (viii) + (xi), or (xiv).

Nor a necessary cause of R:

If notP, then notR?	denied by (v) + (viii), or (i) + (x);
if P, not-then notR	implied by (i) + (viii) + (xi), or (i) + (xiii), or (vi) + (xii);
where: notP is possible	implied by (v), or (ix), or (x), or (xii) + (xv).

Nor (complemented by S) a contingent cause of R:

If (notP + notS), then notR?	denied by (v) + (viii), or (i) + (x);
if (P + notS), not-then notR	implied by (i) + (xiii);
if (notP + S), not-then notR?	denied by (iv) + (viii);
where: (notP + notS) is possible	implied by (v) + (viii), or (x), or (xii) + (xv).

Table 8.23	Evaluation of mood # 324.			
P	Q	R	S	implied possible by (i) + (viii) + (xi)
P	Q	R	notS	implied possible by (i) + (xiii)
P	Q	notR	S	implied impossible by (i)
P	Q	notR	notS	implied impossible by (i)
P	notQ	R	S	implied possible by (vi) + (xii)
P	notQ	R	notS	implied impossible by (xii)
P	notQ	notR	S	see (ii) or (xiv)
P	notQ	notR	notS	implied impossible by (xii)
notP	Q	R	S	implied impossible by (viii)
notP	Q	R	notS	implied possible by (v) + (viii), or (i) + (x)
notP	Q	notR	S	implied impossible by (i) or (viii)
notP	Q	notR	notS	implied impossible by (i)
notP	notQ	R	S	implied impossible by (iv)
notP	notQ	R	notS	implied impossible by (iv)
notP	notQ	notR	S	implied possible by (iv) + (ix)
notP	notQ	notR	notS	implied possible by (iv) + (xii) + (xv)

Evaluation of mood # 344.

Major premise: Q (complemented by P) is a partial and contingent cause of R:

(i) If (Q + P), then R;
(ii) if (notQ + P), not-then R;
(iii) if (Q + notP), not-then R;
(iv) where: (Q + P) is possible.
(v) If (notQ + notP), then notR;
(vi) if (Q + notP), not-then notR;
(vii) if (notQ + P), not-then notR;
(viii) where: (notQ + notP) is possible.

Minor premise: Q (complemented by S) is a partial and contingent cause of P:

(ix) If (Q + S), then P;
(x) if (notQ + S), not-then P;
(xi) if (Q + notS), not-then P;
(xii) where: (Q + S) is possible.
(xiii) If (notQ + notS), then notP;
(xiv) if (Q + notS), not-then notP;
(xv) if (notQ + S), not-then notP;
(xvi) where: (notQ + notS) is possible.

Putative conclusion: is P (complemented by S) a cause of R?

NO! P is not implied to be a complete cause of R:

If P, then R?	denied by (ii) + (xiii);
if notP, not-then R	implied by (iii) + (ix), or (v) + (x), or (v) + (xiii) + (xvi);
where: P is possible	implied by (ii), or (iv), or (vii) + (xiii), or (ix) + (xii), or (xiv) or

	(xv).
Nor (complemented by S) a partial cause of R:	
If (P + S), then R?	denied by (ii) + (xiii);
if (notP + S), not-then R	implied by (v) + (x);
if (P + notS), not-then R?	denied by (i) + (xiii);
where: (P + S) is possible	implied by (ii) + (xiii), or (vii) + (xiii), or (ix) + (xii), or (xv).
Nor a necessary cause of R:	
If notP, then notR?	denied by (vi) + (ix);
if P, not-then notR	implied by (i) + (ix) + (xii), or (i) + (xiv), or (vii) + (xiii);
where: notP is possible	implied by (iii), or (vi), or (viii), or (x), or (xi), or (xiii) + (xvi).
Nor (complemented by S) a contingent cause of R:	
If (notP + notS), then notR?	denied by (vi) + (ix);
if (P + notS), not-then notR	implied by (i) + (xiv);
if (notP + S), not-then notR?	denied by (v) + (ix);
where: (notP + notS) is possible	implied by (iii) + (ix), or (vi) + (ix), or (xi), or (xiii) + (xvi).

Table 8.24 Evaluation of mood # 344.

P	Q	R	S	implied possible by (i) + (ix) + (xii)
P	Q	R	notS	implied possible by (i) + (xiv)
P	Q	notR	S	implied impossible by (i)
P	Q	notR	notS	implied impossible by (i)
P	notQ	R	S	implied possible by (vii) + (xiii)
P	notQ	R	notS	implied impossible by (xiii)
P	notQ	notR	S	implied possible by (ii) + (xiii)
P	notQ	notR	notS	implied impossible by (xiii)
notP	Q	R	S	implied impossible by (ix)
notP	Q	R	notS	implied possible by (vi) + (ix)
notP	Q	notR	S	implied impossible by (ix)
notP	Q	notR	notS	implied possible by (iii) + (ix)
notP	notQ	R	S	implied impossible by (v)
notP	notQ	R	notS	implied impossible by (v)
notP	notQ	notR	S	implied possible by (v) + (x)
notP	notQ	notR	notS	implied possible by (v) + (xiii) + (xvi)

Evaluation of mood # 352. (Similarly, *mutatis mutandis*, for **mood # 363**.)

Major premise: Q is a complete cause of R:
(i) If Q, then R;
(ii) if notQ, not-then R;
(iii) where: Q is possible.
Minor premise: Q is a complete and (complemented by S) a contingent cause of P:
(iv) If Q, then P;
(v) if notQ, not-then P;
(vi) where: Q is possible.
(vii) If (notQ + notS), then notP;
(viii) if (Q + notS), not-then notP;
(ix) if (notQ + S), not-then notP;
(x) where: (notQ + notS) is possible.
Putative conclusion: is P a necessary cause of R?
NO! P is not implied to be a necessary cause of R:
If notP, then notR? open;
if P, not-then notR implied by (i) + (viii);
where: notP is possible? implied by (v), or (vii) + (x).

Table 8.25 Evaluation of mood # 352.

P	Q	R	S	see (i) + (iii), or (iv) + (vi)
P	Q	R	notS	implied possible by (i) + (viii)
P	Q	notR	S	implied impossible by (i)
P	Q	notR	notS	implied impossible by (i)
P	notQ	R	S	see (ix)
P	notQ	R	notS	implied impossible by (vii)
P	notQ	notR	S	see (ii) or (ix)
P	notQ	notR	notS	implied impossible by (vii)
notP	Q	R	S	implied impossible by (iv)
notP	Q	R	notS	implied impossible by (iv)
notP	Q	notR	S	implied impossible by (i) or (iv)
notP	Q	notR	notS	implied impossible by (i) or (iv)
notP	notQ	R	S	see (v)
notP	notQ	R	notS	see (v), or (vii) + (x)
notP	notQ	notR	S	see (ii) or (v)
notP	notQ	notR	notS	see (ii) or (v), or (vii) + (x)

Chapter 9. SQUEEZING OUT MORE INFORMATION

1. The Interactions of Determinations.

Before considering the possibility of other inferences from causative propositions, let us summarize and extend the results obtained thus far, and especially try and understand them in a global perspective. We have in the preceding chapters identified, in the three figures, 66 valid positive conclusions obtainable from positive premises, out of 192 (3*8*8) possible combinations of generic and joint premises. We thus found a *validity rate* of 34.4% - meaning that reasoning with causative propositions cannot be left to chance, since we would likely be wrong two times out of three! The table shows the distribution of valid and invalid moods in the three figures:

Table 9.1. Valid and Invalid Moods		
Figure	Valid Moods (positive)	Invalid Moods (impossible or nil)
1	30	34
2	18	46
3	18	46
Total	66	126

Moreover, not all of the valid moods have equal significance. As the table below shows, some moods (20, shaded) are conceptually basic, while others (46) are mere derivatives of these, in the sense of *compounds* (16) or *subalterns* (30) of them. We shall call the former '**primary**' moods, and the latter '**secondary**' moods. Note that these terms are not intended as references to validation processes, but to comparisons of results. By which I mean that some of the moods here classed as 'primary' (such as #217, to cite one case) were validated by reduction to others; whereas some of the moods here classed as 'secondary' (such as #117, for example) were among those that had to be validated by matricial analysis.

A primary mood teaches us a lesson in reasoning. For instance, mood 1/**m**/**m**/**m** (#155) teaches us that in Figure 1, the premises **m** and **m** *yield the conclusion* **m**. A secondary (subaltern or compound) mood has premises that teach us nothing new (compared to the corresponding primary), *except to tell us that no additional information is implied.* For instances, 1/**m**/**mq**/**m** (#152) is equivalent (subaltern) to 1/**m**/**m**/**m**; and 1/**mn**/**mn**/**mn** (#111) is equivalent to (a compound of) 1/**m**/**m**/**m** plus 1/**n**/**n**/**n**.

Such equivalencies are due to the fact that *the premises of the secondary mood imply those of the primary mood(s), while the conclusion(s) of the latter imply that of the former.* We can thus 'reconstruct' the derivative mood from its conceptual source(s). Effectively, primary moods represent general truths, of which secondary moods are specific expressions. This ordering of the valid moods signifies that we do not have to memorize them all, but only 20 out of 66.

In the following table, the valid positive moods of causative syllogism are listed for each figure in order of the strength of their conclusions (joint determinations before generics). Within each group of moods yielding a given conclusion, moods are ordered in the reverse order with reference to their premises (the weakest premises capable of yielding a certain conclusion being listed first, so far as possible - some are of course incomparable). Explanations will be given further on.

Primary moods (shaded) are distinguished from compounds and subalterns, and the primary sources of the secondaries are specified. *Notice that all moods with a joint determination as conclusion are compounds.*

Table 9.2. Valid Positive Moods, Primaries and Secondaries.

No.	Major	Minor	Conclusion	Relation	to mood
Figure 1 (12 primaries, 8 compounds and 10 subalterns)					
111	**mn**	**mn**	**mn**	compound	155 + 166
121	**mq**	**mn**	**mq**	compound	155 + 181
112	**mn**	**mq**	**mq**	compound	155 + 118
131	**np**	**mn**	**np**	compound	166 + 171
113	**mn**	**np**	**np**	compound	166 + 117
141	**pq**	**mn**	**pq**	compound	171 + 181
114	**mn**	**pq**	**pq**	compound	117 + 118
144	**pq**	**pq**	**pq**	compound compound	147 + 148, 174 + 184
155	**m**	**m**	**m**	**primary**	
152	**m**	**mq**	**m**	subaltern	155
125	**mq**	**m**	**m**	subaltern	155
151	**m**	**mn**	**m**	subaltern	155
115	**mn**	**m**	**m**	subaltern	155
166	**n**	**n**	**n**	**primary**	
163	**n**	**np**	**n**	subaltern	166
136	**np**	**n**	**n**	subaltern	166
161	**n**	**mn**	**n**	subaltern	166
116	**mn**	**n**	**n**	subaltern	166
147	**pq**	**p**	**p**	**primary**	
174	**p**	**pq**	**p**	**primary**	
137	**np**	**p**	**p**	**primary**	
134	**np**	**pq**	**p**	subaltern	137 or 174
171	**p**	**mn**	**p**	**primary**	
117	**mn**	**p**	**p**	**primary**	
148	**pq**	**q**	**q**	**primary**	
184	**q**	**pq**	**q**	**primary**	
128	**mq**	**q**	**q**	**primary**	
124	**mq**	**pq**	**q**	subaltern	128 or 184
181	**q**	**mn**	**q**	**primary**	
118	**mn**	**q**	**q**	**primary**	

Table 9.2 continued.

Figure 2 (4 primaries, 4 compounds and 10 subalterns)					
211	**mn**	**mn**	**mn**	compound	256 + 265
212	**mn**	**mq**	**mq**	compound	265 + 218
213	**mn**	**np**	**np**	compound	256 + 217
214	**mn**	**pq**	**pq**	compound	217 + 218
265	**n**	**m**	**m**	**primary**	
262	**n**	**mq**	**m**	subaltern	265
235	**np**	**m**	**m**	subaltern	265
261	**n**	**mn**	**m**	subaltern	265
215	**mn**	**m**	**m**	subaltern	265
231	**np**	**mn**	**m**	subaltern	265
256	**m**	**n**	**n**	**primary**	
253	**m**	**np**	**n**	subaltern	256
226	**mq**	**n**	**n**	subaltern	256
251	**m**	**mn**	**n**	subaltern	256
216	**mn**	**n**	**n**	subaltern	256
221	**mq**	**mn**	**n**	subaltern	256
217	**mn**	**p**	**p**	**primary**	
218	**mn**	**q**	**q**	**primary**	
Figure 3 (4 primaries, 4 compounds and 10 subalterns)					
311	**mn**	**mn**	**mn**	compound	356 + 365
321	**mq**	**mn**	**mq**	compound	356 + 381
331	**np**	**mn**	**np**	compound	365 + 371
341	**pq**	**mn**	**pq**	compound	371 + 381
356	**m**	**n**	**m**	**primary**	
353	**m**	**np**	**m**	subaltern	356
326	**mq**	**n**	**m**	subaltern	356
351	**m**	**mn**	**m**	subaltern	356
316	**mn**	**n**	**m**	subaltern	356
313	**mn**	**np**	**m**	subaltern	356
365	**n**	**m**	**n**	**primary**	
362	**n**	**mq**	**n**	subaltern	365
335	**np**	**m**	**n**	subaltern	365
361	**n**	**mn**	**n**	subaltern	365
315	**mn**	**m**	**n**	subaltern	365
312	**mn**	**mq**	**n**	subaltern	365
371	**p**	**mn**	**p**	**primary**	
381	**q**	**mn**	**q**	**primary**	

As already stated, we need only keep in mind the 20 primaries, the remaining 46 secondaries being obvious corollaries. It is implicitly understood that, had any of the latter been primary (e.g. if 1/**m**/**mq** had concluded **mq**, say, instead of just **m**), it would have been classified as such among the former.

We can further cut down the burden on memory by taking stock of 'mirror' moods. As we can see on the table above, among the primaries (shaded): in Figure 1, mood 166 is a mirror of mood 155, 148 of 147, 184 of 174, 128 of 137, 118 of 117, and 181 of 171. In Figure 2, mood 256 is a mirror of mood 265, and 218 of 217. In Figure 3, mood 365 is a mirror of mood 356, and 381 of 371. In this way, we need only remember 10 primary moods (6 in the first figure, 2 in the second and 2 in the third), and the 10 others follow by mirroring.

To better understand the results obtained, we ought to notice the phenomenon of *transposition of determinations* in the premises. Moods can be paired-off if they have the same premises in reverse order. Note that, for each pair, the figure number (hundreds) is the same, while the numbers of the major and minor premises (tens and units, respectively) are transposed.

- Thus, among primaries, we should mentally pair off the following: 147 and 174, 117 and 171, 148 and 184, 118 and 181. In these paired cases, the combination of the determinations involved has the same conclusion, however ordered in the premises. Take, for instance, moods 147 and 174, i.e. 1/**pq**/**p**/**p** and 1/**p**/**pq**/**p**; the conclusion has the same determination **p**, whether the determinations of the premises are **pq**/**p** or **p**/**pq**. *This allows us to regard, in such cases, the determination of the conclusion as a product of the determinations of the premises, irrespective of their ordering.* (We can similarly pair off many secondary moods: for instances, 125 and 152, 115 and 151, etc.)
- In the case of the following transposed pairs, 265 and 256, 356 and 365, the conclusions are of similar strength, but not identical determination. Thus, for instance, 1/**n**/**m**/**m** (#265) and 1/**m**/**n**/**n** (#256) are comparable although only by way of mirroring. (We can similarly pair off some secondaries, like 226 and 262, 235 and 253, etc.)
- Some individual moods have the same determination in both premises, and thus cannot be paired with others. These, we might say, pair off with themselves. Thus with Nos. 155, 166 among primaries; and some likewise among secondaries.
- But, note well, some moods are not similarly paired; specifically, the primary moods 128, 137, 217, 218, 371, 381 are not; similarly some of the secondaries. For instance, mood 1/**np**/**p**/**p** (#137) is valid, but mood 1/**p**/**np**/**p** (#173) is invalid. *This teaches us not to indiscriminately look upon the order of the determinations in the premises as irrelevant.*

Moreover, transposition of the determinations of the premises should not be confused with *transposition of the premises* themselves. For if the premises are transposed, the conclusion obtained from them is converted. Additionally, in the case of the first figure, transposition of the premises would take us out of the first figure (into the so-called fourth figure[49]), since the middle item changes position in them. As for the second and third figures, though transposition of premises does not entail a change of figure (the middle item remains in the same position either way), it entails a change of determination in the conclusion (since the items in the premises change place); see for instances moods 256 and 265, or 356 and 365.

Nevertheless, awareness of the phenomenon of transposition of determinations is valuable, because it allows us to make an analogy with composition of forces in mechanics. Syllogism in general may be viewed as a doctrine concerning the interactions of different propositional forms. With regard to the determinations of causation, we learn from the cases mentioned above something about *the interactions of determinations*, i.e. how their 'forces' combine.

---

[49]Aristotle regarded the fourth figure (PQ/QR/PR) as an impractical way of thinking, and so ignored it. My own position is more mitigated (see discussion in *FL*, p. 38). I have nevertheless disregarded it in the present treatise, to avoid excessive detail.

We can push this insight further, with reference to the hierarchies between the significant moods (primaries) and their respective derivatives (secondaries). Consider, for instance, the primary mood 1/**m**/**m**, which has conclusion **m**; if we gradually increase the strength of the major premise (to **mq** or **mn**), while keeping the minor premise the same (**m**), or vice versa, the determination of the conclusion remains unaffected (**m**). In contrast, if we increase the strength of both premises at once to **mn**, the conclusion increases in strength to **mn**. Similarly in many other cases. Thus, *some increases in strength in the premises produce no additional strength in the conclusion; but at some threshold, the intensification may get sufficient to produce an upward shift in determination.*

We can in like manner view changes in conclusion from **p** to **m** or from **q** to **n** (and likewise to the joint determinations **mq** or **np**). For instance, compare moods 1/**mn**/**p**/**p** and 1/**mn**/**m**/**m**; here, keeping the major premise constant (**mn**), as we upgrade the minor premise from **p** to **m**, we find the conclusion upgraded from **p** to **m**. Similarly with moods 1/**p**/**mn**/**p** and 1/**m**/**mn**/**m**, keeping the minor constant while varying the major. Let us not forget that the determinations of causation were conceived essentially as modalities: **p** and **m**, though defined as mutually exclusive, are meant as different *degrees* of positive causation; similarly for the negative aspects of causation, **q** and **n**. Thus, some such transitions were to be expected.

We can in this way interpret our list of valid moods as ***a map of the changing topography in the field of determination***. This gives us an interesting overview of the whole domain of causation. This is the intent of Table 9.2, above.

2. Negative Moods.

Thus far, we have only validated causative syllogisms with positive premises and positive conclusions. We will now look into the possibility of obtaining, at least by derivation from the foregoing, additional *valid moods involving a negative premise* and, consequently, a negative conclusion.

This is made possible by using Aristotle's **method of indirect reduction, or reduction *ad absurdum***. To begin with, let us describe the various reduction processes involved. Note the changed positions of items P, Q, R, in each situation. The mood to be validated (left) involves a positive premise (indicated by a + sign) and a negative premise (-) yielding a negative putative conclusion. The reduction process keeps one of the original premises (the positive one), and shows that contradicting the putative conclusion would result, through an already validated positive mood (right), in contradiction of the other premise (the negative one). Notice the figure used for validation purposes depends on which original premise is the positive one, staying constant in the process.

	Figure 1	Reduction process:	Figure 2
major premise	+QR	*keeping the same major,*	+QR
minor premise	-PQ	*if we deny the conclusion,*	+PR
Conclusion	-PR	*then we deny the minor.*	+PQ

	Figure 1	Reduction process:	Figure 3
major premise	-QR	*if we deny the conclusion,*	+PR
minor premise	+PQ	*keeping the same minor,*	+PQ
Conclusion	-PR	*then we deny the major.*	+QR

	Figure 2	Reduction process:	Figure 1
major premise	+RQ	*keeping the same major,*	+RQ
minor premise	-PQ	*if we deny the conclusion,*	+PR
Conclusion	-PR	*then we deny the minor.*	+PQ

	Figure 2	Reduction process:	Figure 3
major premise	-RQ	*keeping the minor as a major,*	+PQ
minor premise	+PQ	*if we deny the conclusion,*	+PR
Conclusion	-PR	*then we deny the major.*	+RQ

	Figure 3	Reduction process:	Figure 1
major premise	-QR	*if we deny the conclusion,*	+PR
minor premise	+QP	*keeping the same minor,*	+QP
Conclusion	-PR	*then we deny the major.*	+QR

	Figure 3	Reduction process:	Figure 2
major premise	+QR	*keeping the major as a minor,*	+PR
minor premise	-QP	*if we deny the conclusion,*	+QR
Conclusion	-PR	*then we deny the minor.*	+QP

Consider, for instance, a first figure syllogism QR/PQ/PR, which we wish to reduce *ad absurdum* to a second figure syllogism of established validity. Knowing that the given major premise (QR), and the negation of the putative conclusion (PR), together imply (in Figure 2) the negation of the given minor premise (PQ) - we are logically forced to admit the putative conclusion from the given premises (in Figure 1). Similar arguments apply to the other three cases, as indicated above.

Using these reduction arguments, we can validate the following moods, in the three figures. In the following table, all I have done is apply indirect reduction to the primary moods listed in Table 9.2. I ignored all subaltern and compound moods in it, since they would only give rise to other derivatives.

Table 9.3 Valid Negative Moods, Primaries only.			
Major	**Minor**	**Conclusion**	**Source**
Figure 1 from Figure 2 - keep same major			
n	**not-m**	**not-m**	265
m	**not-n**	**not-n**	256
mn	**not-p**	**not-p**	217
mn	**not-q**	**not-q**	218
Figure 1 from Figure 3 - keep same minor			
not-m	**n**	**not-m**	356
not-n	**m**	**not-n**	365
not-p	**mn**	**not-p**	371
not-q	**mn**	**not-q**	381

Table 9.3 continued.

Figure 2 from Figure 1 - keep same major			
m	**not-m**	**not-m**	155
n	**not-n**	**not-n**	166
mn	**not-p**	**not-p**	117
np	**not-p**	**not-p**	137
pq	**not-p**	**not-p**	147
mn	**not-q**	**not-q**	118
mq	**not-q**	**not-q**	128
pq	**not-q**	**not-q**	148
p	**not-p**	**not(mn)**	171
q	**not-q**	**not(mn)**	181
p	**not-p**	**not(pq)**	174
q	**not-q**	**not(pq)**	184
Figure 2 from Figure 3 - keep the minor as a major			
not-n	**n**	**not-m**	365
not-m	**m**	**not-n**	356
not-p	**p**	**not(mn)**	371
not-q	**q**	**not(mn)**	381
Figure 3 from Figure 2 - keep the major as a minor			
n	**not-n**	**not-m**	256
m	**not-m**	**not-n**	265
p	**not-p**	**not(mn)**	217
q	**not-q**	**not(mn)**	218
Figure 3 from Figure 1 - keep same minor			
not-m	**m**	**not-m**	155
not-n	**n**	**not-n**	166
not-p	**mn**	**not-p**	171
not-p	**pq**	**not-p**	174
not-q	**mn**	**not-q**	181
not-q	**pq**	**not-q**	184
not-p	**p**	**not(mn)**	117
not-q	**q**	**not(mn)**	118
not-q	**q**	**not(mq)**	128
not-p	**p**	**not(np)**	137
not-p	**p**	**not(pq)**	147
not-q	**q**	**not(pq)**	148

Obviously, the significance of **not-p** or **not-q** in a premise or conclusion must be carefully assessed in each case. This is best done by writing it out in full.

Take for example 1/**mn/not-p/not-p**, which we derived ad absurdum from mood 217, i.e. 2/**mn/p/p**. The major premise in both cases has form QR. The Figure 1 mood has minor premise of form P(S)R and conclusion of form P(S)Q. The Figure 2 minor premise and conclusion have form P(S)Q and P(S)R, respectively. We thus indirectly reduce subfigure 1b to subfigure 2b. The complement is S everywhere and the negative propositions **not-p** can be read as **not-p**$_S$. We may also generalize this argument to all complements, since whatever the complement happen to be it will return in the conclusion. It follows that if the minor premise is absolute, so is the conclusion.

In some other cases, however, the transition is not so simple. For example, when 2/**p/not-p/not(mn)** is reduced ad absurdum to 1/**p/mn/p**, we apparently have subfigure 2d (say) derived from subfigure 1c. But the number of complements does not match, so this case is rather artificial in construction. But I will not delve further into such issues here, not wanting to complicate matters unnecessarily. The conscientious reader will find personal investigation of these details a rewarding exercise.

Nevertheless, many of the above results are not without practical interest and value. For a start, they allow us to squeeze a bit more information out of causative propositions, and thus tell us a little more about the topography of the field of determination mentioned earlier. Most importantly, all the moods listed in this table *involve a negative generic premise*. Until now, we have only managed to validate moods with positive premises, i.e. positive moods. These are the first **negative moods** we manage to validate, by indirect reduction to (primary) positive moods.

This supplementary class of valid moods yields negative conclusions, whether the negation of a generic determination or that of a joint determination. Remember that the conclusions **not(mn)**, **not(mq)**, **not(np)**, or **not(pq)** can be interpreted as disjunctive propositions involving all remaining (i.e. not negated) formal possibilities. Thus, for instance, **not(mn)** means "either **mq** or **np** or **pq** or **non-causation**".

Summarizing, we have a total of 20 valid moods with a negative major premise, and 20 with a negative minor premise, making a total of 40 new moods. In Figure 1, the statistics are 4 + 4 = 8; in Figure2, they are 4 + 12 = 16; and in Figure 3, they are 12+ 4 = 16. We could similarly derive additional negative moods, by indirect reduction to compound and subaltern moods: this exercise is left to the reader.

3. Negative Conclusions from Positive Moods.

We have in the preceding chapters evaluated all conceivable *positive* conclusions from positive moods, i.e. from moods both of whose premises are positive (generic and/or joint) causative propositions. But we have virtually ignored *negative conclusions* from these (positive) moods, effectively lumping them with 'non-conclusions' (labeled *nil*), which they are not. We shall consider the significance of negative conclusions now[50].

In this context, it is important to keep in mind the distinction between a mood **not implying** a certain conclusion (which is therefore a ***non-sequitur***, an 'it does not follow', which is invalid, but whose contradictory may yet be a valid or invalid conclusion), and a mood **implying the negation of** (i.e. denying) a certain conclusion (which is therefore more specifically an ***antinomy***, so that not only is it invalid, but moreover it is so because its contradictory is a valid conclusion).

a) For a start, we have to note that wherever a positive mood yields a valid positive conclusion, it also incidentally yields a valid negative conclusion, namely one denying the

[50]That is to say, more precisely, conclusions that deny *generic* determinations. As we shall see further on, there are additionally (and derivatively from the present investigation) positive moods yielding negations of *joint* determinations, such as mood numbers 126, 135, 326, 335 (see Table 9.5, below).

contrary determination(s). Thus, for example, mood 111 (**mn/mn**) yields the positive conclusions "P is a complete and necessary cause of Q" (**mn**); it therefore also yields as negative conclusions "P is *not* a partial and *not* a contingent cause of Q" (**not-p** and **not-q**). We thus have *at least as many* valid negative conclusions as we have valid positive ones. Such syllogisms with negative conclusions are, of course, mere *subalterns* of those with positive conclusions they are derived from.

b) Moreover, we may notice that some of the crucial matricial analyses developed in the previous chapter invalidated certain conclusions, not merely by leaving one or more of their constituent clauses open, but more radically by ***denying***, i.e. implying the negation of, some clause(s). Specifically, this occurred in the 14 cases listed in the following table (where '+' means implied, '-' means *denied*, and '?' means neither implied nor denied).
Notice that this table concerns negations of **p** or **q** *relative to the complement S* (whence my use here of the notation $\mathbf{p}_S$ or $\mathbf{q}_S$), which is not the same as absolute negation. It is very important to specify the complement, otherwise contradictions might wrongly be thought to appear at later stages. In the case of negations of **m** or **n**, they are absolute anyway since there are no complements for them. Also note that:

- Where **m** or **n** is affirmed (as in moods 221, 231, 312, 313), then **p** or **q** (respectively) may be denied absolutely, i.e. whatever complement (S, notS or any other) be considered for **p** or **q**. That is, **m** implies **not-p** and **n** implies **not-q**. This can also be stated as **m** = **mn** or **mq** and **n** = **mn** or **np**, wherein the complement is unspecified (possibly but not necessarily S, or notS, or any other).
- Although **not-m** by itself does not imply **p**, **not-m** + **n** = **np** (moods 221, 312). Likewise, although **not-n** by itself does not imply **q**, **not-n** + **m** = **mq** (moods 231, 313). This is evident from the fact that absolute lone determinations are impossible. Here again, note well, the complement concerned is not specified (i.e. it may be, but need not be, S, or notS, or any other, say T).
- Furthermore, where **m** and/or **n** is/are denied (as occurs in all 14 cases to some extent), the additional denial *if any* of **p** and/or **q** (as in 221, 231, 241, and the six moods of Figure 3) has to initially be understood as a restricted negation, i.e. as **not-**$\mathbf{p}_S$ or **not-**$\mathbf{q}_S$. Additional work is required to prove radical negation of the weak determinations.
- Since causation is by joint determination or not at all, **not-m** + **not-n** = **pq** or **no-causation**. But, **not-m** + **not-n** + **not-**$\mathbf{p}_S$ + **not-**$\mathbf{q}_S$ may not offhand be interpreted as **no-causation**, since **pq** remains conceivable as $\mathbf{p}_{notS}\mathbf{q}_{notS}$ or *relative to some other complement T*. Note well that **p**+ **not-**$\mathbf{p}_S$ does *not* imply $\mathbf{p}_{notS}$ and likewise **q**+ **not-**$\mathbf{q}_S$ does *not* imply $\mathbf{q}_{notS}$.

Table 9.4. Positive (generic and/or joint) premises whose conclusion includes additional negative elements.

No.	Premises		m	n	p_S	q_S	Full conclusion	Comments
Figure 1								
None								
Figure 2								
231	**np**	**mn**	+	-	-	-	**mq** and **not-q_S**	Since **m** + **not-n** = **mq**
221	**mq**	**mn**	-	+	-	-	**np** and **not-p_S**	Since **n** + **not-m** = **np**
233	**np**	**np**	-	?	?	?	**not-m**	Many outcomes possible
222	**mq**	**mq**	?	-	?	?	**not-n**	Many outcomes possible
224	**mq**	**pq**	-	-	?	?	**pq** or **no-causation**	Since **not-m** + **not-n**
234	**np**	**pq**	-	-	?	?	**pq** or **no-causation**	Since **not-m** + **not-n**
244	**pq**	**pq**	-	-	?	?	**pq** or **no-causation**	Since **not-m** + **not-n**
241	**pq**	**mn**	-	-	-	-	**pq** but **not-p_S** + **not-q_S** or **no-causation**	Since *if* causation, then **not-m** + **not-n** = **pq**
Figure 3								
313	**mn**	**np**	+	-	-	-	**mq** and **not-q_S**	Since **m** + **not-n** = **mq**
312	**mn**	**mq**	-	+	-	-	**np** and **not-p_S**	Since **n** + **not-m** = **np**
324	**mq**	**pq**	?	-	-	-	**not-n** and **not-p_S** + **not-q_S**	Various outcomes possible
334	**np**	**pq**	-	?	-	-	**not-m** and **not-p_S** + **not-q_S**	Various outcomes possible
314	**mn**	**pq**	-	-	-	-	**pq** but **not-p_S** + **not-q_S** or **no-causation**	Since *if* causation, then **not-m** + **not-n** = **pq**
344	**pq**	**pq**	-	-	-	-	**pq** but **not-p_S** + **not-q_S** or **no-causation**	Since *if* causation, then **not-m** + **not-n** = **pq**

There are thus 8 moods in the second figure and 6 in the third figure with additional negative conclusions (as revealed by matricial analysis in the preceding chapter). The differences between these two figures are simply due to moods 322 and 342 being self-contradictory, as already seen.

Note in passing that the conclusions of moods 231, 313 and 221, 312 may be read as the ***relative to S*** "lone determinations" **m-alone**$_{rel}$ and **n-alone**$_{rel}$, respectively; but it of course does not follow from this that absolute lone determinations exist – indeed we see here that in *absolute* terms the respective conclusion is **mq** or **np**. The latter imply that relative to some item other than S, be it notS or some other item T, **q** or **p** (as applicable) is true. That is of course not much information, but better than nothing.

It should be noted that none of these moods is implied by others, so that the negative conclusions implied by them are not repeated in such putative other moods. (See Diagram 7.1 and Table 7.2, in chapter 7, on reduction.) An issue nevertheless arises, as to whether the moods mentioned, above under (a) and (b), *exhaust* negative conclusions drawable from positive moods. The answer seems to be *yes*, we have covered all negative conclusions. This may be demonstrated as follows.

Suppose a mood (i.e. premises) labeled 'A' is found by matricial analysis to *not-imply* some positive conclusion 'C'. Consider another mood 'B', such that A implies B. It follows that B does not imply C, since if B implied C, then A would imply C - in contradiction to what was given. But our question is: may B still formally *imply notC*? Well, suppose B indeed implied notC, then A would imply notC, in conflict with the subalternative result of our matricial analysis that A does not imply C. *Granting that matricial analysis yields the maximum result*, such conflict is unacceptable. Therefore, it is not logically conceivable that B imply notC as a rule.

We can thus remain confident that the negative conclusions of positive moods mentioned above make up an exhaustive list, provided of course that we remain conscious of the complement under discussion at all times.

In any case, we have in this way succeeded in squeezing some more information out of causative propositions occurring in syllogistic conjunctions. No moods of this sort were found in Figure 1. In Figure 2, two moods (221, 231) were already valid in the sense of yielding positive conclusions; their validity has now been reinforced with additional information; six other moods in this figure (222, 224, 233, 234, 241, 244) were previously classed as 'invalid,' *in the sense of* yielding no positive conclusions; but here they have been declared 'valid' *with regard to* certain negative conclusions. Similarly, in Figure 3, two moods (312, 313) have increased in validity, while another four (314, 324, 334, 344) have acquired some validity. So, in sum, we have four moods with reinforced validity and ten with newly acquired validity.[51]

We can derive additional valid moods from these, as we did before, by use of indirect reduction, or reduction *ad absurdum*. If we focus, for the purpose of illustration, on the negative conclusions **not-m** and/or **not-n** in Table 9.4, we obtain the following:

Table 9.5 Positive moods with a negative conclusion

Major	Minor	Conclusion	Source
Figure 1 from Figure 2 - keep same major			
pq	**n** and/or **m**	**not(mn)**	241
mq	**m**	**not(mn)**	221
np	**n**	**not(mn)**	231
mq	**n**	**not(mq)**	222
np	**m**	**not(np)**	233
mq or **np** or **pq**	**n** and/or **m**	**not(pq)**	224, 234, 244
Figure 1 from Figure 3 - keep same minor			
n and/or **m**	**pq**	**not(mn)** and **not(pq)**	314, 344
n	**mq**	**not(mn)**	312
m	**np**	**not(mn)**	313
n	**pq**	**not(mq)**	324

51 Note also that some of these are pairs of mirror moods (viz. 221-231, 222-233, 224-234, 312-313, 324-334), others (241, 244, 314, 344) have no mirrors.

Table 9.5 continued.

Figure 2 from Figure 3 - keep the minor as a major			
m	**mn**	**not(mq)**	312
n	**mn**	**not(np)**	313
n and/or **m**	**mn** or **pq**	**not(pq)**	314, 344
n	**mq**	**not(pq)**	324
m	**np**	**not(pq)**	334
Figure 3 from Figure 2 - keep the major as a minor			
pq	**n** and/or **m**	**not(mq)** and **not(np)**	224, 234
mn	**m**	**not(mq)**	221
mq	**n**	**not(mq)**	222
mn	**n**	**not(np)**	231
np	**m**	**not(np)**	233
mn or pq	**n** and/or **m**	**not(pq)**	241, 244

We can analyze these results as follows, for examples.

With regard to the Figure 1 moods in the above table derived *ad absurdum* from Nos. 222 and 233, namely **mq/n/not(mq)** and **np/m/not(np)**, they correspond respectively to moods 126 and 135. Until here, these moods were invalid, because we had no positive conclusions from them. But here we have found some very vague conclusions, which negate joint determinations (a relatively indefinite result, since it signifies a disjunction of possible conclusions: i.e. either the remaining joint determinations or no-causation).

The same moods in Figure 3, correspond to the moods 326 and 335. In their case, however, we had positive conclusions from them, namely **m** from **mq/n** and **n** from **np/m**. The additional negative conclusions obtained from them here, namely **not(mq)** and **not(np)**, respectively, constitute further information extraction, since they are not formally implied by the previous conclusions.

Note well that **p** and **q** in these four cases mean $\mathbf{p}_S$ and $\mathbf{q}_S$, respectively, since we are in subfigure (c). Therefore, in Figure 3, we should *not* go on to infer that **m + not(mq) = mn**, or that **n + not(np) = mn**, i.e. that both moods 326 and 335 yield the full conclusion **mn**! They only in fact yield **m** and **n** in absolute terms, the rest of the conclusions being only relative to S. It would not be reasonable to expect more determination than that, because it would mean we are getting more out of our syllogism than we put in to it, contrary to the rules of inference.

4. Imperfect Moods.

Imperfect moods[52] of causative syllogism are those involving negative items as terms. That is, instead of directly concerning P, Q, R, S, they might relate to notP, notQ, notR and/or notS. We would not expect the investigation of such negative terms to enrich us with any new formal information, but rather to unnecessarily burden us with useless repetition. All the logic of such propositions can be derived quite easily from that of propositions with positive terms. We certainly will not engage in that exercise here (although some logician may be tempted to

52 The expression is Aristotelian in origin.

develop this field once and for all for the record). But we need to point out a couple of interesting facets of this issue.

a) As pointed out in a footnote in the chapter on immediate inferences, we commonly use positive forms with a negative intent, i.e. whose terms are positive on the surface but negative under it. Thus, the expression "P **prevents** Q" may be explicated as "P causes *not*Q". Rather than work out all the logical properties of this new copula called "prevention," we can simply reduce it to that of causation, by changing all occurrences of Q in causative logic to notQ. We could thus speak of complete or partial prevention, necessary or contingent prevention; and we could correlate such various forms with each other, in oppositions, eductions and syllogisms. However, we could additionally *correlate the forms of prevention in every which way with the forms of causation*. It is in the event that we wish to do this, that the need to develop a logic of imperfect moods would arise. Such an enlarged logic would concern not only forms like "P causes Q" (causation) and "P causes notQ" (prevention), but also forms like "notP causes Q" and "notP causes notQ." The latter two may be called *inverse* forms of causation and prevention, respectively.

b) A particularly interesting negative term is when a partial or contingent causative proposition involves a **negative complement**. For example, the proposition "P (with complement notR) is a partial cause of Q," involving the negative complement notR, needs to be investigated to fully comprehend the proposition "P (with complement R) is a partial cause of Q," involving the positive complement R. Some of this work has been done in the chapter on immediate inferences.

We saw there that the 'absolute' proposition $\mathbf{p}_{abs}$ "P is a partial cause of Q" (irrespective of complement) is implied by *either* of those 'relative' propositions $\mathbf{p}_{R}$ or $\mathbf{p}_{notR}$ (that specify the complement). It follows of course that the negation of the absolute implies the negation of *both* the relatives. Also, $\mathbf{p}_{abs}$ may be true while only one of $\mathbf{p}_{R}$ or $\mathbf{p}_{notR}$ is true and the other is false. That is, the conjunctions '$\mathbf{p}_{abs}$ + **not-**$\mathbf{p}_{R}$' or '$\mathbf{p}_{abs}$ + **not-**$\mathbf{p}_{notR}$' are logically possible. Similarly with regard to contingent causation, **q**.

Now, what shall arouse our interest in syllogistic theory are occurrences of a negative minor or subsidiary item. As the reader may recall, in Table 5.2 we identified four 'subfigures' (labeled a, b, c, d) for each of the three figures of causative syllogism, according to the presence and position of a positive complement in either premise or in the conclusion. We can here identify five more subfigures (to be labeled e, f, g, h, i) for each of the three figures. These 'imperfect' subfigures are clarified in the table below:

Table 9.6. Imperfect subfigures of each figure.

Subfigures	**e**	**f**	**g**	**h**	**i**
Figure 1	QR P(S)Q P(notS)R	Q(S)R PQ P(notS)R	Q(P)R P(S)Q P(notS)R	Q(notP)R P(S)Q P(S)R	Q(notP)R P(S)Q P(notS)R
Figure 2	RQ P(S)Q P(notS)R	R(S)Q PQ P(notS)R	R(P)Q P(S)Q P(notS)R	R(notP)Q P(S)Q P(S)R	R(notP)Q P(S)Q P(notS)R
Figure 3	QR Q(S)P P(notS)R	Q(S)R QP P(notS)R	Q(P)R Q(S)P P(notS)R	Q(notP)R Q(S)P P(S)R	Q(notP)R Q(S)P P(notS)R

Subfigures 'e' and 'f' are the most interesting. In both, the complement in the conclusion is negative compared to its origin in one of the premises; the subsidiary term has thus changed polarity. In subfigure 'e', the original complement is in the minor premise; in 'f', it is in the major premise. Subfigures 'g,' 'h,' 'i' are more complicated, since they involve the minor item or its negation as complement in the major premise. This is a conceivable situation, though one we are not likely to encounter often.

The layouts described by 'e' and 'f' are relatively common in our causative reasoning, inasmuch as we often have to distinguish between absolute and relative partial or contingent causation, or their negations. To make such distinctions, and decide just how much can be inferred from given premises, we have to refer to these subfigures. Logicians are therefore called upon to develop this particular field further, although the information is already tacit in the results of the subfigures we have already dealt with.

This work will not be pursued further here, except for the following general contribution. The table below predicts how subfigures may be derived from others by *direct reduction* (i.e. conversion of major or minor premise), i.e. it shows the logical interrelationships between the various subfigures in the different figures. Included in this table are indications for the reduction of perfect as well as imperfect subfigures of Figures 2 and 3 to subfigures of Figure 1. In one case, we reduce a subfigure of Figure 1 to subfigures of Figures 2, 3. This table, obtained by reflection on Tables 5.2 and 9.4, can be viewed as a guide to action for a future logician who may volunteer to finish this job.

Table 9.7. Reductions of Moods between Figures.

Stages of development of study	If mood is evaluated in subfigure	Then mood is derivable in subfigure
Firstly, perfect moods	2a 2b 2c 2d	1a 1b 1f 1h
	3a 3b 3c 3d	1a 1e 1c 1g
Secondly, main imperfect moods	2e 2f 2h 2i	1e 1c 1d 1g
	3e 3f 3g 3i	1b 1f 1d 1h
Thirdly, remaining imperfect moods	1i	2g, 3h

Chapter 10. WRAPPING UP PHASE ONE

1. Highlights of Findings.

I will stop the first phase of my research on the logic of causation at this point. Not just because I do not think it is worth going further into minutiae. I in fact do not consider that all the important formal issues have been covered. However, I do regard the logical techniques applied so far to have come close to the limits of their utility. That is why I have been developing more precise techniques, which I will publish eventually as Phase Two. Let us meanwhile review some of our main findings thus far in Phase One, and what information we are still missing.

We have succeeded in **defining** the various determinations of causation, by means of propositional forms already known to logic. These forms involve conjunctions ('and'), conditionings ('if-then'), modalities ('possibly', 'actually'), and of course negations of all those ('not').

The mechanics of these various source forms are thoroughly treated in my work *Future Logic*, and need not be reviewed here. Since we already know the **deductive** properties of these underlying forms (how they logically interact) and how they can ultimately be **induced** from experience (abstraction, adduction, generalization and particularization, factorial analysis, factor selection and formula revision), these formal problems are *in principle already solved* for causative propositions. It is only a question of finding ways and means to extract the implicit information systematically and reliably.

> I have tried to perform just this job in the preceding pages. The difficulties encountered are never such as to put the whole enterprise in doubt, note well. They are only due to the *complexity of forms involved*, since each positive causative is a conjunctive compound of several simpler forms, and all the more so in the case of negative propositions, which are disjunctive compounds of such simpler forms. The main problem is thus one of *volume of information to be treated*; there is so much data to sort out, order and organize, that we can easily get lost, forget things, make minor errors with numerous hidden repercussions.
>
> I am only human, and may well have made some mistakes in this process. A major annoyance for me is that I am often forced to interrupt my research work due to the need to earn my living by other means. In such circumstances, my attention is diverted for long periods; my mind loses its thorough concentration on the subject matter, and I have to later re-learn it all. Hopefully, I have nevertheless succeeded in spotting and removing all eventual inconsistencies. Certainly, I have tried: always making consistency checks, painstakingly reviewing large bodies of data and long chains of reasoning, doing what I call "quality control".

The best way to do this is to arrive at the same results using different means. That is one reason why, although the above Phase One work apparently stands up well on its own, I will not be entirely satisfied until Phase Two is complete and I arrive there at consistent results. But to return for now to our findings thus far…

It must be understood that this research has not been idle reshuffling of information and symbols. It had *both practical and theoretical* purposes in mind.

The practical questions relate to everyday reasoning about causes and effects. One of the principal questions we posed, you will recall, was *whether the cause of the cause of something is itself a cause of that thing or not, and if it is, to whether it is so to the same degree or a lesser degree*. This issue of causal (or effectual) chains is what the investigation of causal syllogism is all about. What our dispassionate research has shown is that it is absurd to expect ordinary reasoning, unaided by such patient formal reflections, to arrive at accurate results. The answer to the question about chains is resounding and crucial: **the cause of a cause is not necessarily itself a cause, and if it is a cause it need not be one to the same degree**. Once the scientific impact of this is understood, the importance of such research becomes evident.

But this syllogistic issue has not been the only one dealt with. We have in the process engaged in many other investigations of practical value. The definitions of the determinations causation by means of **matrixes** can help both laypeople and scientists to classify particular causative relations, simply by observing conjunctions of *presences and absences of various items*. Generalizations may occur thereafter, but they should always be checked by further empirical observation (at least, a readiness to notice; eventually, active experiment) and adjusted as new data appears (or is uncovered).

Another interesting finding has been the clarification of the relationships between positive and negative, absolute and relative causative propositions: for instance, that **we may affirm partial or contingent causation, while denying it of a particular complement**. One very important principle – that we have assumed in this volume, but not proved, because the proof is only possible in the later phase of research – is that *(absolute) "lone determinations" are logically impossible*. This means that we may in practice consider that **if there is causation at all, it must be in one or the other of the four "joint" determinations**.

Another finding worth highlighting is that **non-causation is denial of the four genera (or four species) of causation**, and before these can be definitely denied we have to go through a long process of empirical verification, observing presences and absences of items or their negations in all logically possible conjunctions. It is thus in practice as difficult to prove non-causation as to prove causation! Indeed, to be concluded the former requires a lot more careful analysis of data than the latter. Of course, in practice (as with all induction) we assume causation absent, except where it is proved present. But if we want to check the matter out closely, a more sustained effort is required.

With regard to the theoretical significance of our findings, now. By theoretical, here, I mean: relevant to philosophical discussions and debates about causality. Obviously, so far we have only treated causation, and said nothing about volition and allied cause-effect relations, so we cannot talk about causality in its broadest sense.

What our perspective makes clear is that **the existence of "causation" is indubitable**, once we apprehend it as a set of experiential yes or no answers to simple questions, leaving aside references to some underlying "force" or "connection" (which might be discussed as a later explanatory hypothesis). If we look upon causation in a positivistic manner, and avoid metaphysical discussions that tend to mystify, it is a simple matter. *Causation is an abstraction, in response to phenomenologically evident data*. It is a summary of data.

> It is not purely empirical, in the sense of a concept only summarizing *presences* of phenomena. It involves a rational element, in that it also summarizes *absences* of phenomena. Affirmation may only be acknowledgment of the empirically apparent. But

negation, as I have stressed in my work *Phenomenology*[53], is a partly rational act (a question is asked: is the thing I remember or imagine now present to my senses?), as well as a partly empirical act (the answer is no: I see or hear or otherwise sense nothing equivalent to that image!). Absence does not exist independently like presence, but signifies an empirically disappointed mental expectation.

Reading debates between philosophers (for example, David Hume's discussions), one might get the impression that non-causation is an obvious concept, while causation needs to be defined and justified. But, as we have seen here, *non-causation can only be understood and proven with reference to causation*. Before we can project a world without causation, we have to first understand what we mean by causation, its different determinations, their interactions, and so forth. But the moment we do that, the existence of causation is already obvious. However, this does not mean that non-causation does not exist. Quite the contrary. Since, as we have seen, some formal processes like syllogism with premises of causation are inconclusive, we may say that the existence of causation implies that of non-causation! This finding has two aspects:

(a) The more immediate aspect is inferred from the fact that the cause of a cause of something is not necessarily itself a cause of it: **taking any two things at random, they may or not be causatively related**. This implication is valuable to contradict the Buddhist notion that "everything is caused by everything". But the possibility of independence from *some* things does not exclude dependence on *other* things. Each of the two things taken at random may well have other causes and effects than each other.

(b) A more radical aspect is the issue of spontaneity, or no causation *by anything at all*. We can only touch upon this issue here, since we have only dealt with causation so far. But what our formal study of causation has made clear is that we cannot say offhand whether or not spontaneity in this sense is possible. **There is no "law of causation" that spontaneity is impossible**, i.e. that "everything has a cause", as far as I can see. Nothing we have come across so far implies such a universal law; it can only be affirmed by generalization. Spontaneity (chance, the haphazard) remains conceivable.

I think the point is made: that formal research such as the present one has both practical and theoretical value. Let us now explain why the research undertaken so far is insufficient.

2. The Modes of Causation.

The observant reader will have noticed that throughout the present study we have concentrated on **logical causation**, i.e. on causative propositions based on logical conditioning. But of course, this is but one aspect of human aetiological reasoning. To be thorough, we need to consider not only such "*de dicta*" forms, but also the "*de re*" modes of causation, i.e. **natural, temporal, extensional and spatial causation**. In many ways, the latter are more interesting than the former. We have focused our attention on logical causation because it is the most widely known theoretically, although not necessarily the most widely used in practice.

Each of these modes of causation is derived on one of the modes of conditioning. A thorough study of the underlying forms of conditioning may be found in my work *Future Logic* (Part IV, Chapters 33-42)[54]. What is evident from that study is that natural, temporal, extensional

---

53 This final chapter of Phase One was written in 2003, after publication of *Phenomenology*.

54 I do not there treat *spatial* modality, but it is easy enough to do eventually.

and spatial conditioning, are in most respects similar to logical conditioning, but in significant respects different. The difference is essentially due to the fact that logical conditional propositions (like "if P then Q") distinctively cannot be made to universally imply the "bases" (i.e. "P is possible, Q is possible") – *because if they were made to, we would not be able to express paradoxes*[55]. From this structural difference, various differences in behavior (during inference) emerge.

However, this distinction dissolves in the context of causation, because here logical causation like all other types implies the bases. We have specified this fact as the last clause of each of the definitions of the determinations. Complete or partial causation implied the cause, or the conjunction of causes, and therefore the effect, to be possible; necessary and contingent causation implied them to be unnecessary. It follows that all the logical properties of the different modes of causation will be comparable. The subdivision of each mode of causation into different determinations will be the same, as will the underlying interplay of presences and absences, possibilities and impossibilities, in every conceivable combination and permutation. All the matrixes of their forms will be identical and all arguments will have the same conclusions.

The only difference between these different logics is simply that the "possibility" and "impossibility" referred to in the definitions and matrices have a different sense in each case. In logical causation, they refer to logical modalities; in natural causation, to natural modality; in extensional causation, to extensional modality; and so forth. The only task left to logicians, therefore, is to more closely examine the interrelationships between these different modes of causation. That is, for instance, how any two natural and extensional causative propositions are opposed to each other, and how they behave in combination (i.e. within arguments). This complex work will not be attempted here.

> Nevertheless, I have already in *Future Logic* clarified the following essential relationships. Logical necessity implies but is not implied by the *de re* necessities. Logical possibility is implied by but does not imply the *de re* possibilities. Similarly on the negative side, for impossibility and unnecessity. Thus, the logical mode lies on the outer edges of rectangles of oppositions including the *de re* modes.

For now, let us only clarify in what context each mode is used. Logical (or *de dicta*) causation is concerned with causes in the literal sense of "reasons"; that is to say, it helps us to order our discourse and eventual knowledge with reference to logical implications, presuppositions, disconnections, contradictions, or consistencies, between hypotheses and/or apparent evidences. In contrast, the *de re* modes of causation are more directly object-oriented.

- The paradigm of *natural* causation is:

> **When the individual X *actually* is, has or does C (the cause),**
> **then it (or some other individual Y) must (i.e. in all circumstances) be, have or do E (the effect);**
> **and when C is not actual, neither is E.**

In this context, C and E are qualities, properties or activities of any sort, relative to some *individual* entity X (or pair of individuals X, Y, respectively). Presence, here, is called "actuality" to refer us to the underlying natural modality. Necessity, here, means *in all circumstances* relative to this X in the antecedent. The implied basis of such propositions is that "this X can both C and E" (or "X+C and Y+E is potential for the individual(s)

---

55 In paradox, either P or Q is implied impossible. See *Future Logic*, chapter 31.

concerned", as appropriate) – no need of additional clauses in that respect. The antecedent and consequent may be static or dynamic, and may or may not be temporally separated.

- The paradigm of *temporal* causation is very similar, save that "must" becomes "always" (all units of time) in the body of time concerned. The form is "When… at some time, then… at all times".
- The paradigm of *extensional* causation is a bit different:

  In such cases as class X *in some instance* is, has or does C (the cause), then it (or another instance of class X or an instance of some other class Y) must (i.e. in all instances) be, have or do E (the effect); and in such cases as C does not have an instance, neither does E.

In this context, C and E are qualities, properties or activities of any sort, relative to some *class* of entities X (or pair of classes X, Y, respectively). Presence, here, is called "instancing" to refer us to the underlying extensional modality. Necessity, here, means *in all instances* of X in the antecedent. The implied basis of such propositions is that "some X are both C and E" (or "X+C and Y+E is extensionally possible for the class(es) concerned", as appropriate) – no need of additional clauses in that respect. The antecedent and consequent may be static or dynamic, and may or may not be temporally separated. *They distinctively need not be actualities, but may be potentialities or necessities*, note well, since extensional conditioning refers only to quantity.

The paradigm of *spatial* causation is very similar, except that "must" becomes "everywhere" (all units of space) in the body of space concerned. The form is "Where… at some place, there… at all places".

What I want to make sure here is that the reader understands that there are different modes of causation, and that the differences between them are significant to ordinary and scientific thought or discourse.

> For example, the theory of Evolution is based partly on observation or experiment on *individual* biological specimens (spatial, temporal and natural causation) and partly on putting together the jigsaw puzzle of scattered findings relating to *a class* of individuals in different times and places (extensional causation), as well as partly on theoretical insights about consistency and implications between postulates and experiences (logical causation). All these involve induction and deduction, hypothetical reasoning and generalizations, but their focal center changes.
>
> When, for instance, we take note of the structural or even genetic similarities of all vertebrates, and presume them to have a common ancestor, we are engaged in *extensional* causative reasoning. We would be engaged in *natural* causative reasoning, only if we could trace the ascendancy from individual child to individual parent all the way back to the first vertebrate specimen. In the extensional mode, the different individuals (e.g. paleontological findings) are regarded as expressions of a single class (genus, species, variation, whatever). In the natural mode, our focus is on the life of individuals as such (irrespective of their class appurtenance).

People, and even scientists, often confuse these different ways of thinking, and remain unaware that **they may lead to different conclusions, or at least nuance our conclusions considerably**. For this reason, the study of the modes of causation needs to be carried out in appropriate detail.

3. Gaps and Loose Ends.

The main characteristics and limits of Phase One of our research into the logic of (logical) causation are two:

(a) The methodology of matricial analysis used for validation of inferences is cumbersome, bulky, manual, and therefore susceptible to human error.

(b) We are only able to deal systematically and exhaustively with positive causative propositions; negative causative propositions can only be treated incidentally, not directly.

For all the achievements of our research so far, these two defects leave us with an aftertaste of dissatisfaction. We have not till here succeeded in completely automating validation: human attention and intelligence are required at every step to ensure consistency and exhaustiveness. This does not prevent us from a thorough and reliable treatment of positive propositions, provided we have the requisite patience and carefulness. But the task becomes too daunting when dealing with negatives, in view of their disjunctive nature and of the sheer volume of data involved.

To overcome these handicaps, we have to greatly simplify matricial analysis, make it so *digital* that a computer program could operate it. This is what we shall endeavor to do in Phase Two. There we shall refer to the method of matricial analysis used in the present Phase One as **macroanalysis**, in contrast to the more pointed methodology of **microanalysis** used in Phase Two. In the latter case, we shall be able to develop a versatile logical mechanics, wherein any conjunctive, conditional or causative proposition, positive or negative, individually or in combination with any other(s), can be fully interpreted or evaluated in a matricial analysis and in ordinary language. This is no promise or vain boast: it is already largely done, needing only to be completed.

One important practical consequence of this new approach is our ability to freely handle negative causative propositions, and draw inferences from them (if they imply anything) in any arguments wherein they appear. Another is the crucial finding that absolute "lone" determinations are logically impossible; this refers, the reader will recall, to propositions involving only one positive generic determination, all three others being denied. But most importantly, it allows us to demonstrate everything demonstrable in causative logic without a drop of lingering doubt, since human error is eliminated.

APPENDIX 1: J. S. MILL'S METHODS: A CRITICAL ANALYSIS.

Revised version*. The present essay was originally written, or at least published, in 1999; but I decided to rewrite almost all of it in March 2005, when I found the time to engage in more detailed hermeneutics. Following an analysis that could be characterized as almost Talmudic (though not as mere* 'pilpul'*), my conclusions about Mill's methods are considerably more severe.*

Preamble

Below, I list John Stuart Mill's five "Methods of Experimental Inquiry"[56]; then I try to expose and evaluate them. It should be noted that though my approach is at times critical, my main intent is to clarify; I am more interested in Mill's achievements, than in his apparent mistakes. (All symbols used below are mine – introduced to facilitate and clarify discussion.)

Mill's **terminology** is a bit obscure, but can be interpreted with some effort.

In the paradigm (the first method), he seems to be looking out at the world, or a specific domain of it, and observing something (say, X) occurring in some things or events, in scattered places and times, and not occurring in others; and also observing some second thing (say, Y) occurring in some things or events, in scattered places and times, and not occurring in others; and he wonders at how two such events can be causally related.

In the first three methods, Mill verbally differentiates the two things under study by naming one "the *phenomenon*" (X, for us) and the other "the *circumstance*[57]" (Y, for us), suggesting that in his mind's eye the former is the effect and the latter its cause, although note well in his conclusions he rightly (usually) considers the two items interchangeable, so that either might be the cause or effect of the other. In the fourth method, Y is viewed as a "part" of the "phenomenon" X. In the last method, Mill refers to both items with the same word, viz. "phenomenon". Whatever the words used for X and Y, it is clear that Mill has no intent to prejudice the conclusion. These terms are intended very broadly to mean *any thing or event*, i.e. (since he is considering experimental inquiry) any object of perception. (I prefer the very neutral – purely logical – term "item" for this.)

Now, these items (X, Y, or their negations) are found scattered in the world, or some segment thereof, in various things or events, in scattered places and times – this is what Mill means by "*instances*". Wherever X, Y, or their negations occur, that is one of the "instances" or cases under consideration. Thus, the instances might be instances of a *kind* of thing (e.g. humans or

56 *System of Logic* (1843), chapter VIII. The full text is available online at: http:// books.google.com/books?id=y4MEAAAAQAAJ. Note that I have placed his "third canon" first to facilitate critique.

57 Literally: 'standing around' – suggesting something found to accompany the object in some way, a condition or situation.

time's arrow or the degree of abstraction) must be specified before we can identify a direction of causation.

Now, let us turn to ***criticism*** of Mill's formula.

Mill's first inexplicable complication is his requirement that the "circumstance in common" (viz. Y or not-Y) in the premises be *exclusive*. In the first premise, he says Y is the "only one"; and in the second, there is "nothing save" not-Y. Moreover, these circumstances must "alone" differentiate the two sets of instances, for the conclusion to follow.

Mill apparently fears that some third item, say Z, might come into play and affect the projected strong relation between X and Y. However, this fear is formally unjustified. Let us consider the extreme case where three items X, Y, Z are constantly conjoined, and their negations not-X, not-Y, not-Z likewise always occur in tandem. In such a situation, *all* the following propositions (and their respective contraposites) are true:

If X then Y, and if not X then not Y.
If X then Z, and if not X then not Z.
If Y then Z, and if not Y then not Z.

The truth of the latter two propositions does not impinge upon the truth of the first one. The causative relation between X and Y *remains the same, even if* some third factor like Z comes into play. The same can be argued if only one of these extra propositions is true. In such situations, we would simply conclude that there are parallel causations, or again causative chains.

Mill apparently failed to develop these concepts, and inserted an extraneous requirement of exclusivity in a vague attempt to insure against possible third-factor interference. In truth, the relation between any two variables X and Y can be determined without reference to any other variables.

If – as indeed does occur – the two variables under consideration are affected by others, to the extent that their relation is weaker than here concluded, we will soon notice the fact by observing that X is *not always with* Y, and/or that not-X is *not always with* not-Y. But in such case, the stated premise(s) about constant conjunction will simply not be true! In other words, in such case, Mill's conception of the premise(s) would be self-contradictory.

Perhaps, someone might interject, Mill was here trying to account for the scientific methodology of "*keeping all other things equal*"? No – because: this refers to a situation where there are two or more partial causes to an effect, and to establish each of the partial causes as such, we have to consider each one in turn without the other – and in such case, complete causation could not be a putative conclusion for any of the partial causes.

The second inexplicable complication in Mill's formula is his reference in the conclusion to a third alternative, viz. that Y might be "*an indispensable part of the cause*" of X. This clause is interesting, first of all, because it indicates that when Mill initially states that Y might be "the effect, or the cause" of X, he has in mind complete causation (as distinct from the partial causation in the third alternative).

With regard to this third alternative, let us first notice that Mill does not mention that X might equally be "an indispensable part of the cause of" Y, even though he has granted that X and Y are interchangeable in the first two alternatives. Why this asymmetry? I suspect it is not intended to convey some radical insight, but merely reflects Mill's terminology and the gradual development of his formula.

He started by referring to Y as a "circumstance", suggesting that he viewed it as the precondition or cause of X, "the phenomenon" under investigation. Then, it probably occurred to him that he could not formally distinguish between X and Y, as to which is the cause and which is the effect – so he added the possibility that Y might be the effect of X. Then, he got to thinking Y could be a partial (necessary) cause of X, so he added that in; but he simply forgot to recover symmetry and suggest the reverse to be possible.

Now, the big issue: the phrase "*an indispensable part of the cause*" clearly refers to **partial necessary causation**. Given that X and Y are indeed constantly conjoined and that their negations are constantly conjoined, no conclusion is formally permissible other than complete necessary causation. It follows that it was an error for Mill to insert this additional disjunct in his conclusion.

Note parenthetically, *Mill does not anywhere give us a clue as to how partial necessary causation might be distinguished from complete necessary causation.* Supposing such alternative conclusion had been correct, he would have been obliged to a detail practical methodology for resolving the issue.

I suspect that Mill resorted to the said third alternative conclusion due to his lingering doubt concerning some possible third factor (which we above labeled Z) weakening the relation between X and Y. Apparently, Mill considered that Z might diminish the degree of causation of X by Y from complete to partial; i.e. he viewed Z as a complementary partial cause imbedded with Y in some larger cause.

This explanation is appealing, because it suggests a correlation between the said complications in premises and conclusion. However, as already shown, Z might equally well be a parallel or concatenated complete cause – so we must still fault Mill for imprecision and confusion. In any case, logically, Mill could not have his cake and eat it too. If in the premises he has firmly excluded circumstances besides Y, there is no reason for him to make allowance in his conclusion for an eventual complement Z!

Another objection we could raise here is: if Mill considered the possibility here of partial necessary causation, why not equally that of complete contingent causation, or for that matter, the possibility of partial contingent causation? If he felt (perhaps because of their inductive basis) his premises were shaky, then why did he not foresee all possible modifications of the main conclusion (complete necessary causation)?

The answer to the latter question(s) is simply that although Mill conceived of partial causation, *he apparently never grasped the inverse concept of contingent causation.* This will become evident as we continue our analysis of his methods, and find no mention anywhere of that weak alternative to necessary causation. Mill's omission suggests that, in his mind, only "indispensable" things could be causatives (although if asked the question he might well have denied it).

Another deficiency in Mill's viewpoint is his failure to consider that in some cases, though X and Y and their negations exhibit perfect regularities of conjunction as described in the premises, we (i.e. people in general) do not conclude that Y causes X or X causes Y, but conclude that "*X and Y are both effects of some third thing*". This alternative conclusion is admittedly inexplicable formally, just as the distinction between cause and effect is difficult to pinpoint. But there may in practice be indices that encourage the former, just as there are indices for the latter. Granting this, it would have been more appropriate for Mill to use that clause as his third alternative.

To sum up: what is manifest from all our above analysis is that Mill had an unclear idea of causation, mixing its paradigm up with its possible variations. He failed to first clearly distinguish and separately consider all the determinations of causation (both generically and specifically). Consequently, when he faced the inductive issue – the issue of how in practice to identify causation – his confusion was compounded by the need to consider the fact of generalization and the possibility of particularization.

2. The Method of Agreement

Mill stated:

> *If two or more instances (A, B...) of the phenomenon (X)... have only one circumstance (Y) in common, the circumstance (Y) in which alone all the instances agree is the cause (or effect) of the given phenomenon (X).*

Let X be the phenomenon, and A, B... be instances in which it occurs; and let Y be the only circumstance they have in common. Then, according to Mill:

Instances A, B... have X and have Y (*exclusively*); and

Therefore: Y is the cause, or the effect, of X.

This may be considered as an inductive argument, with a compound premise and a disjunctive conclusion (i.e. a set of two possible conclusions). In view of the name given to this method, the conclusion may be taken to refer to the positive aspect of causation, i.e. **complete causation**. I have here put in brackets and in italics the 'exclusive' demand of the premise, which I consider mistaken for reasons to be presently given.

The essence of this argument is *generalization*, from the constant conjunction of two items, X and Y, wherever and whenever they are observed to occur (the instances A, B…), to *all* existing or possible instances. The conclusion from such universal repetition is either that "if Y, then X" (whence, Y completely causes X) or that "if X, then Y" (whence, X completely causes Y).

Such generalization is logically possible, note well, provided that the "two or more instances" (A, B…) are *all* the encountered instances of X and of Y. Mill obviously intended that, but he should have made it clear – e.g. by saying *the* two or more – to preempt his formula being construed as allowing for unspecified instances in which X occurs without Y or Y occurs without X.

Mill should have mentioned this to show his awareness of the formalities involved, notably that the form "if X, then Y" means "X is impossible without Y" (and similarly, "if Y, then X" means "Y is impossible without X"). The most significant aspect (for a causative conclusion) of the constant conjunction of the two items is the implied denial of possible conjunction between one item and the negation of the other.

We could offer a generous reading Mill's statement to cover this issue. We could suppose that Mill confused circumstances *other than* Y with circumstances *contrary to* Y, and suggest that the clause "only one circumstance (Y) in common" is intended to mean that there are no instances with X accompanied by *some negation of Y*. Likewise, the exclusive word "alone" could be taken to refer to X rather than Y, meaning that the two or more instances involving X, are *the only ones among "all the instances"* to have Y, implying that there are no instances without X that have Y. However, I do not seriously think Mill intended this interpretation.

Another tacit proviso for drawing our conclusion is that each of the items X and Y be *contingent*. Strictly speaking, a conditional proposition like "if X, then Y" or "if Y, then X" can be taken to imply causation *only if* we know that "X is possible, *but unnecessary*" and "Y is possible, *but unnecessary*"[59]. In Mill's statement, here (unlike in the joint method), the occurrence of X and Y is implied in the premise, but their non-occurrence is not mentioned. This omission is noteworthy, suggesting that Mill was not fully aware of these requirements for validity.

It should be said, too, that once the contingency of the theses is granted, a hypothetical proposition could be contraposited. That is, "if X, then Y" would imply "if not Y, then not X"; similarly, "if Y, then X" would imply "if not X, then not Y". Thus, although the intent of Mill's formula (judging by its title) was an inference of complete causation, strictly speaking his formula allows for one of necessary causation. That is, the valid conclusion from his premise is a disjunction of *four* possible conclusions.

Thus, Mill's formula leaves us uncertain, not only as to which item is the cause and which is the effect (as he admits), but also as to whether we are dealing with complete or necessary causation (which he fails to notice). One thing is sure, however, is that the conclusion is a *strong* determination. This is tacitly suggested by Mill in his use of the definite article "the" in "the cause" or in "the effect". If he had had in mind weak determination (i.e. partial or contingent causation), he would have probably written "a cause" and "an effect".

This brings us to Mill's requirement that the instances where the phenomenon (X) have "only one" circumstance (Y) in common, which he repeats when we says that the latter is "alone" that in which the instances agree. Why such exclusiveness? We have seen a similar, mystifying concern in Mill's joint method. In the present case, again, Mill seems worried that there may be circumstances other than Y that will weaken the causative relation between Y and X; i.e. he is trying to preempt any possibility of partial (or contingent) causation.

In his mind's eye, apparently, if some other circumstance (say, Z) was also (like Y) constantly conjoined with the phenomenon (X), a doubt would arise as to which of the two circumstances, Y or Z, caused X. But this is formally unjustified: the possible truth of "if Y, then X" would not be affected by the eventual truth of any other proposition like "if Z, then X"; if X, Y and Z are compatible, as our premise confirms, the two hypotheticals are quite compatible. Mill here again has apparently not considered the possibility of parallel causations or causative chains.

We might add that Mill's attempt to limit the number of accompanying circumstances to just 'one' is ontologically open to doubt. Are there anywhere in the world two or more things (instances in which X occurs) having *literally* only 'one' circumstance (Y) in common? I very much doubt it! If there is such a set of things, it must be very exceptional. Most things have many (innumerable) common factors. There are always large predicates like existence, location in space and time, size, shape, etc. to consider, for a start.

Usually, when we say something so exclusive, we do not really mean it. For example, saying "the only similarity between these two individuals is their wealth" – we do not really mean to imply that the individuals do not both have a spinal cord, a heart, a brain, etc. Such misleading

---

59 This is at least true in the logical mode of conditioning, where hypothetical propositions may be true in cases where one or both of the theses are necessary. Note that in "if X, then Y", X and Y are both implied possible anyway, and all we need to add is that Y is unnecessary, for the unnecessity of X then formally follows; similarly, "if Y, then X" only requires addition that X is unnecessary, to infer causation. In the natural or extensional modes, the issue does not arise, because unless both antecedent and consequent are contingent, we would not be formally allowed to construct a conditional proposition let alone infer causation. This may be highlighted using categoricals: "All X are Y" is equivalent to "No X is not-Y"; similarly, for "All not-X are not-Y" and "No not-X is Y".

language is not accurate in scientific statements; at least, we should think twice before ever using it or taking it literally.

3. The Method of Difference

Mill stated:

> *If an instance (A) in which the phenomenon (X)... occurs, and an instance (B) in which it (X) does not occur, have every circumstance in common save one (Y)... [, that circumstance] (Y) is the effect, or the cause, or an indispensable part of the cause... [of the given phenomenon] (X).*

Let X be the phenomenon, and A be an instance in which it occurs and B be an instance in which it does not occur; and let Y be the only circumstance they do not have in common. Then, according to Mill:

Instance A has X and has Y; and

Instance B lacks X and lacks Y and

(Instances A and B, have every other circumstance in common;)

Therefore: Y is the effect, or the cause (*or an indispensable part of the cause*), of X.

This was intended as an inductive argument, with two compound premises and a disjunctive conclusion (i.e. a set of three possible conclusions). As we shall demonstrate below, this argument is a rather gauche depiction of **necessary causation**. I have here put in brackets and in italics those parts of the premises (here treated as a third premise) and the conclusion that I consider mistaken, for reasons I shall presently discuss.

It should first be noted that Mill's formulation does not make clear whether the presence of X is accompanied by the presence or absence of Y, and inversely what the absence of X is accompanied by. I have assumed symmetry, i.e. presence with presence, and absence with absence, in order that the conclusion be expressed wholly in positive terms. It is not a very important issue, but still a puzzling imprecision on Mill's part.

Next, let us notice that Mill's formula mentions only *one* instance (A) of X's occurrence (presumably with Y) and only *one* instance (B) of X's (and Y's) non-occurrence – without this time in any way suggesting plurality, let alone universality. Mill's wording as it stands does not exclude the possibility of some third instance where X occurs with not-Y, and of some fourth instance where not-X occurs with Y. In such cases, how would Mill dare claim a causative relation?

This is very intriguing[60]: I find it hard to suppose that Mill considers that causation can be induced from single instances. One may from single occurrences *deny* that some causation is applicable, but one could in nowise affirm it. In order for the premises to allow the conclusion he proposes, we would have to replace "an instance" with "*all (known)* instances" in at least one of the premises. Causation is about patterns of conjunction, not about coincidences. Mere *occasional* agreement or difference does not establish a pattern.

One wonders what Mill possibly had in mind! (I suspect he had eaten or drunk too much the day he wrote this.)

---

60 This would be a typical case of the fallacy, known already to Aristotle, *post hoc ergo propter hoc*. An example of it would be racist "reasoning".

Perhaps Mill considered the constancy of surrounding circumstances as the requisite pattern, somehow. Why does he at all refer to the two instances (A and B) having "every circumstance in common"[61] save one? This is a redundancy: the very uniformity of surrounding circumstances makes them irrelevant. In any case, uniformity in only two instances is hardly significant.

I presume, here again, he imagined that if the surrounding circumstances had *not* been uniform, they would have somehow impinged on the causative relation between X and Y. For this reason, he insists on their distinctive uniformity whether X or not-X is the case. He is apparently not aware of the possibility of parallel causations or of causative chains.

In any case, there are always innumerable surrounding circumstances, behaving in quite random fashion, that are totally unconnected with the phenomena at hand; non-uniformity is not proof of causation. And moreover, the circumstances that are here uniform (in the instances A, B) might behave more erratically in other instances.

Is the exceptive ("save one") clause in Mill's formula, i.e. the contrasting behavior of Y, his main focus, perhaps? The given fact that one circumstance (Y) *differs* from all other circumstances in that it is *un*common, i.e. present in one instance (say, A) but absent in the other (say, B), just makes Y stand out from the rest; it does not signify a causative relation to X. This is all the more true when, as here, only a couple of instances are under consideration.

But finally, it occurs to me that there is one way we can at least in part redeem Mill's statement. That is by supposing that, when he here referred to "*an* instance" he subconsciously had in mind "*a kind of* instance"! In that case, A and B are each a set of instances, corresponding respectively to the occurrences of X (with Y) and those of not-X (with not-Y). From these (experimentally) encountered instances, we may by generalization assume the same regularities hold universally.

Granting this supposition, and ignoring the extraneous mention of uniform surrounding conditions and insistence that Y be the only non-uniform circumstance, a causative relation between X and Y can indeed be inferred. However, in such case the premises and conclusion of this method would seem identical to those of the joint method! This is obviously not Mill's intention.

Considering the title of the 'method of difference', we can safely suppose that it refers to something found in part in the 'joint method of agreement and difference' and not found in the 'method of agreement'. Mill was apparently struggling to split necessary complete causation (the 'joint method') into its two components, complete causation (agreement) and necessary causation (difference). He managed to formulate the former, positive aspect readily enough, but had considerable trouble putting his finger on the latter, negative aspect.

A further confirmation of our supposition is to be found by comparison of the conclusions of the three methods. Note first that whereas the method of agreement concludes that Y is "the cause (or effect) of" X, the other two methods conclude in reverse order that Y is "the effect, or the cause… of" X. Moreover, the joint method and the method of difference, distinctively from the method of agreement, propose as an alternative conclusion that Y might be "*an indispensable part of the cause*" of X.

---

61 Note in passing that "every" implies general knowledge – which is empirically impossible without generalization (except with regard to finite sets). We can never in practice be sure to have identified all existing circumstances; and though we may assume we have done so, as a working hypothesis, we have to remain vigilant and continue to look for still unidentified factors that might also be relevant.

This latter possibility obviously refers to partial necessary causation, as earlier pointed out. "Indispensable" means that one cannot do without it, it is a *sine qua non*, a necessity; and "part of the cause" means a fraction of the sufficient cause. All this suggests that, in Mill's mind, the causation found by the method of agreement is essentially positive and whole, whereas that found in the other two ways may be negative and fractional.

But since, as already said, the joint method and the method of difference cannot be identical, the latter must be assumed to focus on necessary causation only. We should, by combination of the methods of agreement and difference, arrive at the same result as with the joint method. So, our task is to isolate the 'difference' component (necessary causation) from the 'agreement' component (complete causation).

Mill might have achieved this by proposing some sort of negative 'mirror image' of his formula for the method of agreement, one about "two or more instances (A, B) in which the phenomenon (X) does *not* occur" having "*the absence of* one circumstance (Y) in common". Some such more analogous statement could be constructed for the method of difference, but I will not even try, because of all the difficulties in the earlier statements already discussed.

Moreover, if we attempt such a reconstruction, we soon realize the title "method of difference" to be a misnomer, in view of the use of the term "agree" within Mill's formula for the method of agreement. His method of difference is really just another application of the method of agreement, except that we focus in it on the absences, instead of presences, of the items (X, Y) concerned. "Difference" (i.e. *disagreement*) can only really be claimed in the joint method, where we switch from presence to absence or vice-versa. In this perspective, the titles 'method of agreement of positives' and 'method of agreement of negatives' might be more appropriate.

Whatever the name used for it, and the language used to formulate it, it is evident for reasons of symmetry that the method of difference aims at the negative aspect of causation, i.e. necessary causation. It follows that the premise(s) must be such that by generalization we can ideally conclude that "if not-X, then not-Y" or "if not-Y, then not-X". This would in practice be based on observed constant conjunction between not-X and not-Y. The matter is that simple!

Mill realizes this at some level, but goes quite astray in his attempt to put it in words. His statement of the method of difference is incredibly garbled. He not only repeats some of the mistakes he made in formulating the preceding two methods, but also makes many more.

Before leaving this topic, it should be added that the said constant conjunction of negations only formally implies causation after generalization if the terms concerned are known contingent, i.e. if X is possible and Y is possible. Moreover, given such contingency, the inferred conditional propositions can be contraposited to "if Y, then X" and "if X, then Y"; so that strictly speaking, the conclusion formally allows for complete causation as well necessary causation (whether of X by Y, or of Y by X).

Observed constant conjunction of negations does *not*, however, formally allow as alternative conclusion partial necessary causation – or for that matter, complete contingent causation or partial contingent causation. Mill's proposition that Y may be "an indispensable part of the cause" of X is artificial and erroneous. Needless to say, reversing its direction would also be erroneous, as would inverting the polarities of the terms. Anyway, as already pointed out, Mill apparently completely misses out on the possibility of contingent causation.

I have already discussed the issue of partial causation with regard to the joint method, and will not repeat my comments here. These are commendable attempts by Mill to insert it in his analyses, but his approach so far is unequal to the task. He makes arbitrary claims in his conclusions, which are incompatible with his premises; and even supposing consistency, he provides no means to decide between his alternative conclusions. He does, however, offer some more precise means for identifying partial causes in his next method, that of 'residues'.

4. The Method of Residues

Mill stated:

> *Subduct from any phenomenon (F) such part (D) as is known by previous inductions to be the effect of certain antecedents (A), and the residue (E) of the phenomenon (F) is the effect of the remaining antecedents (B).*

Here, Mill is attempting to deal with **partial causation**. He is saying:

Suppose: D is a part of F; and E is the rest of F (i.e. D + E = F).

And suppose: A causes D (i.e. presumably, If A, then D, etc.)

It follows that: B causes E (i.e. presumably, If B, then E, etc.)

Note that a tacit assumption, here (suggested by the reference in the conclusion to "remaining" antecedents), which we can readily grant, is that A and B together (as C, say) cause F (the compound of D and E), i.e. that:

(A + B) = C; and C causes F (i.e. presumably, If C, then F, etc.)

Note also that I presume that the kind of causation by A of D, and by B of E, intended by Mill, is complete causation[62], i.e. a relation including positive implication by the cause of the effect (i.e. if the cause, then the effect), plus strictly speaking a negation of the inverse implication (i.e. if not the cause, not-then not the effect).

The causations mentioned and tacit in Mill's statement are considered as already established, as he admits by saying "as is known by previous inductions". The means of induction used is not specified; he presumably intends one of the other four 'methods' (probably the second). His formula is only intended to infer a causation from within other, given causations. This is a purely deductive argument.

Moreover, Mill appeals to the relation between whole and parts without really defining it. We could briefly express that relation by saying that D and E together imply and are implied by F. But to fully clarify this relation, we ought to mention that D without E or E without D, as well as not-D + not-E, amount to not-F. Similarly, with regard to A + B versus C.

Mill's process of "*subduction*" is thus essentially based on the following reasoning:

If A+B (= C), then D+E (= F) – call this the major premise.

But: If A, then D – call this the minor premise.

Therefore, If B, then E – the putative conclusion.

This argument is, I hasten to add, formally *invalid*, although a common error of inference! This can be seen by splitting the major premise into the two hypotheticals:

If A + B, then D

If A + B, then E

Clearly, the minor premise "if A, then D" overrides the first proposition, "if A + B, then D", which has the same consequent, showing the component "B" of the antecedent to be

---

62 We could reexamine the whole argument, based on the opposite assumption, that necessary causation is intended throughout this argument. But I anticipate the overall result would be the same, for the underlying process of subduction is the basic issue at stake. It is just as erroneous if the elements we focus on are of negative polarity, i.e. not-A, not-B, etc.

extraneous. However, the second proposition, "if A + B, then E", whose consequent is different, is unaffected by the minor premise; i.e. its antecedent remains compound. We can, if we wish, "nest" this eduction, putting our result in the form "if A, then if B, then D". But this inference still leaves "A" conditional.

Whence, Mill's putative conclusion "if B, then E" is pretentious. The only way we could draw it would be to confirm "A" to be *categorically* true. It does not suffice to mention the element "A" conditionally, as in "if A, then D". Thus, Mill's present account of partial causation is not strictly correct.

Partial causation can readily be defined and in practice identified, but the appropriate formula for it is a bit more complicated than Mill suggests. It requires a more radical understanding and more systematic treatment of causation. There is no need to go into it here, since I treat it in detail in my main text on the subject.

The notion of a "residue" (or remainder or leftover) is a mathematical one, rooted in the relation of whole and part: if you have a basket with three fruits and you remove one, you still have two left. A similar idea can be used in causation – but only to say:

> If one of the partial causes is found to be *present*, then we can anticipate that as soon as the remaining partial causes are also found to be present, all their collective effects will follow on their heels.

Mill's 'method of residues' subconsciously appeals to this obvious truth. But he confuses the issue, when he considers that things (like D) that the present phenomenon (A) causes *by itself* (i.e. things it alone suffices to bring about) can be counted as among the *collective* effects (like E) of all the causal phenomena under consideration (A and B). This is his essential error. Here again (as with his previous attempts to infer partial causations), his premises and conclusion are not consistent with each other.

Note finally that Mill's language is positive, suggesting that he had in mind specifically partial causation. Here again, as in the preceding methods, he does not apparently consider the other form of weak causation, that involving negative theses, viz. contingent causation.

Moreover, even supposing that Mill had successfully identified partial causatives, he does not here specify that such causes might be necessary or contingent. Perhaps, having spoken (although out of place) about necessary partial causation in the joint method and the method of difference (mentioning "an indispensable part of the cause"), he might be supposed here to be focusing on contingent partial causation. But this would be reading into Mill's treatment something he has given no sign he has awareness of.

One more point worth adding, concerning the appeal to "residues" in reasoning about causes. There is indeed a method that can be so named, one commonly used by scientists and ordinary thinkers. This method was known to Francis Bacon already, long before Mill. It consists simply of disjunctive apodosis – i.e. *the gradual elimination of alternative hypotheses*. Such reasoning has the form:

> Either P or Q or R or… is the cause of S;
> these (P, Q, R,…) are *all* the conceivable causes of S.
> The cause of S is *not* …; and it is *not* R; and it is *not* Q.
> Therefore, the cause of S must be P (i.e. the *only remaining* alternative).

For example, Sherlock Holmes might say: "the culprit is either Jack or Jill; it can't be Jack, since he has an alibi; therefore, it has to be Jill."

5. The Method of Concomitant Variations

Mill stated:

> *Whatever phenomenon (X) varies in any manner whenever another phenomenon (Y) varies in some particular manner, (X) is either a cause or an effect of that phenomenon (Y), or is connected with it through some fact of causation.*

Let X be "whatever phenomenon", and Y be "another phenomenon"; let X1, X2, X3... be variants of X, and Y1, Y2, Y3... be corresponding variants of Y. Then:

- Whenever Y varies from Y1 to Y2, X varies from X1 to X2;
- Whenever Y varies from Y2 to Y3, X varies from X2 to X3;
- etc.;
- therefore, X is "either a cause or an effect of" Y, "or is connected with it through some fact of causation".

Notice Mill's use of "whenever": he is correctly referring to *unvarying* relations, not mere random coincidences. That is, we may suppose he was implying that if the variations of the two phenomena (the kinds of events we labeled X and Y) are *not* concomitant, they may be assumed independent of each other.

Mill does not explicitly tell us what degree of causation may be inferred – whether complete and necessary, or only the one or the other, or neither. He is seemingly open to all possibilities, since he vaguely mentions that "some fact of causation" may in some cases be the best conclusion we can draw. Granting this phrase refers to the weaker determinations, we may suppose that when he refers to "a cause or an effect" he means a stronger determination. However, since he here uses the indefinite article "a", instead of his usual definite article "the", this supposition is debatable. In sum, Mill concludes some sort of causation to be inferable, but is vague as to which sort and when.

Whereas in the first three methods, changes from presence to absence or vice versa are concerned – in concomitant variations, *every incremental change in measure or degree of the cause is accompanied by a corresponding incremental change in the measure or degree of the effect; and/or vice-versa.* In some cases, the correspondences between two phenomena are in this way very regular; but *in other cases, additional phenomena have to be taken into consideration to clarify the more complex relationship involved.*

In any case, the fact of concomitant variation may be considered an ontological derivative of that of causation, dealing with quantitative instead of merely qualitative relationships between two or more phenomena. That is, in Mill's terms, this fifth method is a corollary, or frequent further development, of the preceding four.

In the case of 'agreement' (interpreted as complete causation), we would expect changes in the cause to be invariably followed by concomitant variations in the effect. In the case of 'difference' (interpreted as necessary causation), we would expect changes in the effect indicative of predictable concomitant variations in the cause. In the 'joint' case (i.e. the strongest possible causative relation), both these directions of inference would be applicable.

Note that these alternatives are not made clear in Mill's formula, where X's variations follow Y's variations, yet X is concluded to be "either a cause or an effect" of Y. Given regular variation of X with Y, the more probable conclusion would be that Y is a complete cause of X; although a second possible conclusion would be that X is a necessary cause of Y. Mill presumably does not mention the reverse case, where Y varies with X, simply because he considers that in such case we would just place each term in the other's position in his formula. Fair enough, but then he should at least have mentioned in his formula that in some cases variations are concomitant in one direction only, and in others in both directions!

In the last case ('residues' – interpreted as partial and/or contingent causation), we would have to use more a elaborate technique: *to identify and monitor all the factors involved, and observe how and how much each varies with whatever changes occur, or are experimentally produced, in the other factors*. This is generally achieved using the cunning method of *"keeping all other things equal" while investigating just two factors at a time, until all the factors have successively been played off against each other and we obtain a full picture of their multilateral quantitative relationship.*

In my view, Mill should have mentioned all that explicitly in his formula. It is reasonable to assume he knew it, since the method was oft used in scientific experiments in his day. Why didn't he, then? Let us go on, anyway, and analyze these matters a bit more.

We can theoretically express concomitant variations by means of series of causal propositions, either through statements mentioning *changes* (as above initially done) or more radically through statements mentioning *states*, like:

> If A=A1, then B=B1;
> if A=A2, then B=B2;
> **etc.**

However, often in practice, these innumerable, point-by-point correlations between various *quantities* are plotted on a graph and then *summarized* in a mathematical equation. For example, if B is directly proportional to A, we would write (where k is some constant):

> **B = kA.**

Actually, we have to be careful in this matter, because such a mathematical *equation* implies/presupposes *fully convertible* relations. Thus, the following would also have to be true:

> If A=A1, then B=B1; if B=B1, then A=A1; where B1=kA1.
> If A=A2, then B=B2; if B=B2, then A=A2; where B2=kA2.
> etc.

This does not have to imply the causation involved to be *reversible*, only that A be a complete and necessary cause of B. Thus, while in common language we can readily express concomitant variation between a merely complete cause and its effect (or conceivably between a merely necessary cause and its effect) - in the language of mathematical equations, necessary as well as complete causation is implied (although, I believe, modern mathematics can readily overcome this difficulty).

Note well that if we just say "If A=A1, then B=B1", it does not exclude that for another value of A (say, A8), B may have the same value (B1) again. Such reiterations of value will translate mathematically into more complex formulas than mere proportionality.

More complex relationships may, but do not in all cases, signify partial and/or contingent causation, involving more than two items (at least two causes and one effect). Thus, note well, Mill's statement of this method need not be limited to two variables; he presumably had this in mind when he wrote the alternative conclusion "or is connected with it through some fact of causation".

Note, finally: the idea of comparing variations between two or more variables was proposed long before Mill, by Francis Bacon.

Concluding Remarks

John Stuart Mill (1806-73) was an English philosopher, a highly educated man whose interests ranged very widely, including all aspects of logic. He published the work in which he presents the above 'methods of experimental inquiry', *A System of Logic*, when he was 37. He sought for a pragmatic, empiricist, inductive approach to knowledge; an updated logic, but one that would "supplement and not supersede" Aristotle's.

Mill's five methods have generally been well received, and I acknowledge them as having been an inspiration to me. However, as the above analysis shows, though his intentions were laudable, his performance was often woefully inadequate. I take no pleasure in saying this; but I am somewhat consoled by the knowledge that others have before me also sharply criticized him.

If these methods had been developed before the dawn of modern science – say before the publication of Isaac Newton's *Principia* (1687) – I would have congratulated their author for having provided researchers with potentially valuable cognitive tools. But Mill's work is dated 1843 – almost the mid-19th Century!

At that late date in modern science and philosophy, one could no longer discover these research tools, but one could at least give an *ex post facto* exposé and validation of them. Mill's effort in that direction was, in the last analysis, surprisingly confused, considering his broad knowledge of science and philosophy till his day.

As we have seen, Mill's methods could just as well be characterized as techniques 'for identifying causation', because that is the form of their conclusions; and also, because experimental data is not essential to them, i.e. they can be applied as well to passive observations. His method of residues, unlike the others, is deductive rather than inductive. Whether this list of methods, without regard to its internal imperfections, constitutes an exhaustive summary of actual scientific techniques is open to debate.

What is clear, anyhow, is that Mill did not fully understand the relations of causation. Flaws are evident in his treatment of each of his five methods. Briefly put:

- In the 'joint method', he seemingly tries and succeeds defining or identifying the paradigm of causation, complete necessary causation. However, his understanding is put in doubt by his mention of irrelevant conditions (exclusiveness of the circumstances) in the premises, and his drawing of an alternative conclusion ("an indispensable part of the cause") logically contrary to the given premises.
- In the 'agreement method', he seemingly tries and succeeds defining or identifying complete causation. However, his understanding is put in doubt by his mention of irrelevant conditions (exclusiveness) in the premises, and his failure to specify the unnecessity of the theses (as needed to infer causation from constant conjunction).
- In the 'difference method', he seemingly tries but quite fails defining or identifying necessary causation. His understanding is put in doubt by his appeal to single instances (instead of kinds), his mention of extraneous conditions (the uniformity of surrounding circumstances), and his drawing of an alternative conclusion ("an indispensable part of the cause") logically contrary to the seemingly intended premises.
- In the 'residues method', he seemingly tries to deduce a partial causation from two complete causations. His understanding is here again put in doubt, by his proposing conflicting premises (the same thing cannot be both a complete and a partial cause of a

given phenomenon), and his suggesting an excessive conclusion (i.e. more than the givens allow).
- In the 'concomitant variations method', he seemingly tries and vaguely succeeds defining or identifying the quantitative aspect of causation. This is logically his soundest method, but he fails to mention and distinguish the various degrees of causation that may be involved.

Mill obviously had difficulty with the concept of plurality of causes; i.e. distinguishing between parallelism and composition of causes. The inclusion of redundancies concerning surrounding circumstances in some of his statements indicates that he did not have an entirely accurate picture of causation. His resort to seemingly last minute inserts at the tail end of certain conclusions leads to the same suspicion. Moreover, in none of the five methods does he so much as hint he has heard of contingent causation.

Mill's first four methods may be taken to essentially refer to the causative forms **mn**, **m**, **n**, and **p**, respectively. The first and third methods *mention* the specific determination **np**, but give us no clue as to *how* such causation might be established, i.e. concluded *rather than* **mn** or **n**, respectively. Since he apparently ignores the generic determination **q**, he misses the specific determinations **mq** and **pq**. His treatment is thus neither symmetrical nor exhaustive.

I should also point out that Mill does not clearly distinguish between generic and specific determinations. I assume he does not intend the generic determinations that he separately as well as jointly affirms (namely: **m**, **n**) as absolute *lone* determinations; but the issue is not to my knowledge explicitly raised by him so we cannot be sure what he imagined.

As we have seen, Mill's formulations are open to further criticism. His language is often ambiguous and its intent difficult to fathom. He did not always manage to capture in words what he was trying to say. His logic is in places dubious, if not downright self-contradictory. He may propose mutually incoherent premises and/ conclusions that contradict explicit premises.

The main reason for the weaknesses in Mill's treatment is perhaps *his attempt to deal with definition and induction simultaneously*. He would have been more successful if he had, more systematically, first defined the various forms of causation (*ratio essendi*) and then investigated how their contents may be induced (*ratio cognoscendi*). Perhaps due to his association with the Utilitarian school of philosophy, he was ideologically inclined towards a rather heuristic approach, eschewing a more theoretical treatment of causation.

The logician's main task is to describe *and validate* forms of reasoning. While Mill took some pains to describe causal arguments, he made little effort to validate them. At times, his treatment seems like a sham – not out of malice, but due to negligence. He does not seem to intentionally lie (as some do); but one gets the impression he has not really done his best to do a good job, and he does not expect anyone to notice or care.

The logician's role is also to provide methodological aids for scientists, students, and indeed thinking people in general. Whether Mill's contributions to causal logic ever actually affected anyone's investigation of nature in a positive or negative way is hard to say. Nevertheless, some of his thoughts on the subject were misleading, and the fact should be made public.

This is all very disappointing, considering J. S. Mill's status in British intellectual history. How could a man of his social standing and educational caliber have made such mistakes, and moreover gotten away with them, one wonders.

After all, causation and its varieties were pretty well known to the ancients; this is even evident in commonly used Latin legal terms, like *causa sufficiens* or *sine qua non*. And Mill was very well read in ancient thought; he was brought up with it by his father, James.

Major British philosophers had already discussed causality at considerable length. John Locke (1632-1704), in *An Essay Concerning Human Understanding* (1690), put forward a theory of induction based on regularities of sequence between phenomena. David Hume (1711-1776), for all his avowed skepticism in *An Enquiry Concerning Human Understanding* (1748), had clearly expounded constant conjunction. Mill's views about causation were frankly influenced by Hume's.

Most shocking, is the realization that Mill's logical treatise (1843) was published *238 years* after the founding father of British Empiricism, Francis Bacon (1561-1626), published his *Novum Organon* (1605). Mill was aware of Bacon's work, too, since he (rightly) criticized Bacon's view of causation as simplistic in various respects. But he manifestly failed to notice and learn the important lessons taught by this unsung (or insufficiently sung) hero of the modern scientific method; namely, Bacon's programme of **adduction** and **matricial analysis** (to use my terminology).

Suffices to quote the *Encyclopaedia Britannica* (2004) description of Bacon's "new method" for this failure of Mill's to be clear:

> The crucial point, Bacon realized, is that induction must work by elimination not, as it does in common life and the defective scientific tradition, by simple enumeration. Thus he stressed "the greater force of the negative instance"—the fact that while "all A are B" is only very weakly confirmed by "this A is B," it is shown conclusively to be false by "this A is not B." He devised tables, or formal devices for the presentation of singular pieces of evidence, in order to facilitate the rapid discovery of false generalizations. What survives this eliminative screening, Bacon assumes, may be taken to be true.
>
> Bacon presents tables of presence, of absence, and of degree. Tables of presence contain a collection of cases in which one specified property is found. They are then compared to each other to see what other properties are always present. Any property not present in just one case in such a collection cannot be a necessary condition of the property being investigated. Second, there are tables of absence, which list cases that are as alike as possible to the cases in the tables of presence except for the property under investigation. Any property that is found in the second case cannot be a sufficient condition of the original property. Finally, in tables of degree proportionate variations of two properties are compared to see if the proportion is maintained.

AVI SION

THE LOGIC OF CAUSATION:

PHASE TWO: MICROANALYSIS

Phase II: Microanalysis. Seeing various difficulties encountered in the first phase, and the fact that some issues were left unresolved in it, a more precise method is developed in the second phase, capable of systematically answering most outstanding questions. This improved matricial analysis (microanalysis) is based on tabular prediction of all logically conceivable combinations and permutations of conjunctions between two or more items and their negations (grand matrices). Each such possible combination is called a 'modus' and is assigned a permanent number within the framework concerned (for 2, 3, or more items). This allows us to identify each distinct (causative or other, positive or negative) propositional form with a number of alternative moduses.

This technique greatly facilitates all work with causative and related forms, allowing us to systematically consider their eductions, oppositions, and syllogistic combinations. In fact, it constitutes a most radical approach not only to causative propositions and their derivatives, but perhaps more importantly to their constituent conditional propositions. Moreover, it is not limited to logical conditioning and causation, but is equally applicable to other modes of modality, including extensional, natural, temporal and spatial conditioning and causation. From the results obtained, we are able to settle with formal certainty most of the historically controversial issues relating to causation.

Chapter 11. PIECEMEAL MICROANALYSIS.

1. Binary Coding and Unraveling.

We have developed a theory of causative propositions and arguments (eductions and syllogisms) by means of an analysis of the possibilities and impossibilities implied for the various combinations of the items concerned. This was characterized as 'matricial analysis', because of our recourse to tables for assessing and recording results.

But thus far we have only really engaged in elementary matricial analysis, which may be called **macroanalysis**. We shall now introduce a more advanced approach, which may be called **microanalysis**. They are not different methods. Microanalysis is based on macroanalysis; it is merely a more detailed examination, digging deeper into the issues concerned, in an attempt to solve outstanding problems.

As we have seen, the determinations of causation are best expressed through a *matrix*, a table composed of 'items' and 'moduses'. The *items* are the terms or theses related by the causative proposition concerned. Each conceivable conjunction of these items, in positive or negative form, defines a row of the matrix. The *modus* for each such conjunction is a statement regarding its logical possibility or impossibility, or 'openness' (the latter in cases where the conjunction is in some unspecified contexts possible and in others impossible, so that an uncertainty remains). The moduses for the various conjunctions of items together constitute an additional column of the matrix.[63]

If we array the items of a matrix in a conventional arrangement (presenting the same row always in the same place), then the modus columns of all matrices will be comparable. By such standardization, we can express a determination of causation by merely writing down a string of moduses (i.e. its modus column), which we may call the modus of the determination concerned as a whole, or (for reasons we shall see presently) its *summary modus*.

To simplify things, we may revert to *binary codes*. We may express the presence or absence of each item in the matrix by a 1 or 0 notation. Similarly, we may code the modus for each row by a 1, 0 or · (dot - meaning blank). *The zeros or ones have different meanings in the items and modus cells of the matrix, note well*:

Binary codes:
In the *items* columns: **1** = present **0** = absent
In the *modus* column(s): **1** = possible **0** = impossible · = open

Such notation is merely convenient abbreviation, allowing us to express the summary modus of any determination as a relatively short string of digits and see the whole matrix in one sweep of the eyes. It is also, obviously, useful for computer programming purposes. Of course, if we are dealing with two items (say, P, R), the modus string will have $2^2 = 4$ digits; if with

63 The word 'modus' was chosen to highlight the modal character of such statements. The plural form should perhaps be *modera* (just as genera is plural of genus); but we shall use moduses, anyway.

three items (say, P, Q, R), it will have $2^3 = 8$ digits; and so forth. Whether the string of digits is *distinctive* for each determination, we shall look into further on[64].

As we said, the rows of a matrix are defined and (conventionally) located by combinations of items. Thus, for two items, P and R, the four possible *PR sequences* are 11, 10, 01, 00, which may be labeled a, b, c, d if need be. We may choose this order of combinations as our standard arrangement (any other permutation is equally conceivable, but we conventionally settle on this one[65]). Similarly, for three items, P, Q and R, there are eight possible *PQR sequences*, which may be labeled a-h if need be. And so forth, for more items.

We may thus, to begin with, present the matrices of the generic determinations of causation as in the following tables. These include (in the first two or three columns) the items in positive (1) or negative (0) forms, arrayed in standard combinations; followed by the summary modus for each propositional form (shaded column, symbol Σ), *which you will recall we developed at the beginning of our research* (in Phase 1, chapter 2) by analyzing the meaning of each of its constituent clauses and assessing the result of their interactions.

New columns are then introduced, which present all the conceivable realizations of the summary modus. These realizations, called *alternative moduses*, are obtained simply by substituting, successively, a 0 (for 'impossible') or 1 (for 'possible') for each dot ('open' position) encountered in the summary modus, so that no dots are leftover. This process can be called **unraveling**. The alternative moduses thus make explicit all cases inherent in the summary modus; and conversely, the latter is a summary of all the information contained in the former.

Note that the alternative moduses are themselves, ultimately, summaries, too. For while a zero (for impossibility) signifies that the combination of items concerned is in every context or *always absent*, a one (for possibility) signifies that it is in some contexts or *sometimes present*[66]. Thus, to remove all implicit modality, and consider only actualities, we would have to dissect each such modus into an unspecifiable number of actualizations, where '0' means absent and '1' means present, simply. However, such further analysis is not needed for our purposes; the moduses as above defined are sufficiently informative.[67]

Consideration of a summary modus constitutes *macro*analysis; that of alternative moduses, *micro*analysis. That is all the difference between these two methods of matricial analysis: one of degree of detail. In the former, we have a rough idea of the relations involved; in the latter, it is as if we scrutinize them under a microscope.

The similar strings of zeros and ones used by computer programmers to code letters of the alphabet and symbols (I am thinking of ASCII codes) were arbitrary, pure conventions. But

---

64 The answer to that question is no.

65 The labeling of columns (1-16) would change meaning in other permutations, but the meaning of the moduses would be unchanged. Our present study is in language 'abcd'; 23 other languages could express the same information (since the rows might be ordered in 24 different ways). It might be interesting to compare these competing languages, in search for the most attractive; but we have adopted this one. (Note that the columns, also, could be ordered in umpteen different ways.)

66 'Sometimes present', remember, means 'either always or only sometimes so' - i.e. it allows for necessity as well as contingency.

67 There are thus four senses of the word *modus*: the modus of a single conjunction of items (a cell in the grand matrix); the modus of all conceivable conjunctions of those items (a column, referring to the *summary* modus); the modus(es) which are the conceivable realizations of the summary modus (one or more columns, called the *alternative* modus(es)); and lastly, the actualizations underlying the possibilities inherent in the modal definition of the 0 and 1 codes (subsidiary columns, further subdividing each alternative modus), which we might call *radical* modus(es).

here, note well, once the meanings of zeros and ones, and the order of their presentation, are decided, there is nothing conventional about the string for each determination; it is a *logical* property of it, *objectively given* information.

2. The Generic Determinations.

In the four tables below, the precise significance of the numbers heading the columns of alternative moduses will be made clear in the next chapter; for now, just consider them as arbitrary labels. It should be stressed at the outset that these *modus numbers* are not to be confused with the determination numbers or mood numbers used in earlier chapters. Note also that Tables 11.1 and 11.2 concern two items (P, R), whereas Tables 11.3 and 11.4 concern three items (P, Q, R)[68]; the summary moduses of these two sets are therefore not directly comparable, the former being within a 'two-item framework', the latter within a 'three-item framework'.

The two-item modus of **complete** causation of form PR (symbolized by **m,** or more precisely $\mathbf{m}_{PR}$) was previously established to be "**10.1**". This is, through the following table, worked out to have two conceivable realizations, namely "1001" or "1011" (labeled respectively Nos. 10, 12).

Table 11.1. Matrix of "P is a complete cause of R".

Items		Σ	2 alternative moduses	
P	**R**	**m**	**10**	**12**
1	1	1	1	1
1	0	0	0	0
0	1	·	0	1
0	0	1	1	1

In contrast, the two-item modus, summarily put, of **necessary** causation of form PR (symbolized by **n,** or most precisely $\mathbf{n}_{PR}$) was previously established to be "**1.01**". This is, through the following table, worked out to have two conceivable realizations, namely "1001" or "1101" (labeled respectively Nos. 10, 14).

Table 11.2. Matrix of "P is a necessary cause of R".

Items		Σ	2 alternative moduses	
P	**R**	**n**	**10**	**14**
1	1	1	1	1
1	0	·	0	1
0	1	0	0	0
0	0	1	1	1

---

68 We can, of course, symbolize the two or three items concerned by any letters we like. I have here chosen PR for two items to facilitate comparisons with P(Q)R for three items. The items could just as well have been labeled PQ and P(R)Q. These are mere matters of convention.

The three-item summary modus of *relative* **partial** causation of form P(Q)R (symbolized, according to context, by **p** or $\mathbf{p}_{rel}$, or $\mathbf{p}_Q$ or most precisely $\mathbf{p}_{PQR}$) was previously established to be "**10.1.1..**" . This is, through the following table, worked out to have sixteen conceivable realizations, as shown below (labeled respectively Nos. 149-152, 157-160, 181-184, 189-192). Note well that this is true *relative to complement Q*; we shall consider absolute partial causation further on.

Table 11.3. Matrix of "P (complemented by Q) is a partial cause of R".

Items			Σ	16 alternative moduses															
P	(Q)	R	p	149	150	151	152	157	158	159	160	181	182	183	184	189	190	191	192
1	1	1	1	1	1	1	1	1	1	1	1	1	1	1	1	1	1	1	1
1	1	0	0	0	0	0	0	0	0	0	0	0	0	0	0	0	0	0	0
1	0	1	·	0	0	0	0	0	0	0	0	1	1	1	1	1	1	1	1
1	0	0	1	1	1	1	1	1	1	1	1	1	1	1	1	1	1	1	1
0	1	1	·	0	0	0	0	1	1	1	1	0	0	0	0	1	1	1	1
0	1	0	1	1	1	1	1	1	1	1	1	1	1	1	1	1	1	1	1
0	0	1	·	0	0	1	1	0	0	1	1	0	0	1	1	0	0	1	1
0	0	0	·	0	1	0	1	0	1	0	1	0	1	0	1	0	1	0	1

The three-item summary modus (column Σ) of **contingent** causation of form P(Q)R (symbolized, according to context, by **q** or $\mathbf{q}_{rel}$, or $\mathbf{q}_Q$ or most precisely $\mathbf{q}_{PQR}$) was previously established to be "**..1.1.01**". This is, through the following table, worked out to have sixteen conceivable realizations, as shown below (labeled respectively Nos. 42, 46, 58, 62, 106, 110, 122, 126, 170, 174, 186, 190, 234, 238, 250, 254). Note well that this is true *relative* to complement Q; we shall consider absolute contingent causation further on.

Table 11.4. Matrix of "P (complemented by Q) is a contingent cause of R".

Items			Σ	16 alternative moduses															
P	(Q)	R	q	42	46	58	62	106	110	122	126	170	174	186	190	234	238	250	254
1	1	1	·	0	0	0	0	0	0	0	0	1	1	1	1	1	1	1	1
1	1	0	·	0	0	0	0	1	1	1	1	0	0	0	0	1	1	1	1
1	0	1	1	1	1	1	1	1	1	1	1	1	1	1	1	1	1	1	1
1	0	0	·	0	0	1	1	0	0	1	1	0	0	1	1	0	0	1	1
0	1	1	1	1	1	1	1	1	1	1	1	1	1	1	1	1	1	1	1
0	1	0	·	0	1	0	1	0	1	0	1	0	1	0	1	0	1	0	1
0	0	1	0	0	0	0	0	0	0	0	0	0	0	0	0	0	0	0	0
0	0	0	1	1	1	1	1	1	1	1	1	1	1	1	1	1	1	1	1

As above stated, we shall presently look into the summary moduses of *absolute* partial and contingent causation. As we shall see, they are much wider than those relative to a given complement, dealt with above. This is natural, since absolute weak causations are vaguer forms than relative weak causations.

Comparing the summary moduses of complete and necessary causation with two identical items, 10.1 and 1.01, we see more clearly in what sense they are 'mirror images' of each other: the strings are identical, viewing one from left to right and the other from right to left. Similarly for the summary moduses of partial and contingent causation with three identical items, 10.1.1.. and ..1.1.01.

It should also be noted that *a weak cause and its complement have the same summary modus.* That is, partial causation of forms P(Q)R and Q(P)R have the same 10.1.1.. summary; and contingent causation of forms P(Q)R and Q(P)R have the same ..1.1.01 summary. This was obvious from the original definitions of these determinations, in which P and Q had the same relations to each other and to R; the distinction of one or the other of P, Q as complement was purely one of convenience or focus.[69]

Another observation we can make at this stage is that *a generic determination and its (appropriate) converse would have one and the same summary modus.*

That is true for the strong and absolute weak[70] determinations, which all concern two items. For instance, "P is a complete cause of R" and "R is a necessary cause of P" (note the change of determination, as well as that of item positions) are both here described by the string "10.1". This can be ascertained by reading Table 11.2 in a different order, starting with the first row, then the third, then the second, then the fourth.

That is also true for the weak determinations, which involve three items. For instance, "P (complemented by Q) is a partial cause of R" is convertible to "R (complemented by notQ) is a contingent cause of P" (note well the change of complement polarity, as well as of determination and item positions) are both here described by the string "10.1.1..". Again, we can prove this by rereading Table 11.4 in a different order, starting with the third row, then the seventh, then the first, then the fifth, then the fourth, then the eighth, then the second, then the sixth.

Indeed, we can say that convertibility is to be explained by such identity of moduses. Clearly, it follows that we cannot express direction of causation by reference to summary moduses. The orientations "from P to R" and "from R to P" must have some meaning - they are not empty verbal distinctions - but that meaning is not apparent in the way of a difference between moduses. It has to be sought in other properties, as already argued.

3. Contraction and Expansion.

Now, the above account does not allow us to compare the moduses of the strong determinations with those of the weak ones, nor tell us how to distinguish absolute from relative weak determinations. To enable such comparisons, we need to develop two processes: (a) contraction of a three-item modus into a two-item modus, and (b) expansion of a two-item modus into a three-item modus....

---

69 The reader can ascertain this by taking the matrix of partial or contingent causation (Table 11.3 or 11.4) and reordering the rows: the columns are found to be in different order but have the same overall content.

70 As we shall see further down, the summary modus of absolute partial causation is "11.1" and that of absolute contingent causation is "1.11".

Let us first consider **contraction** of the three-item moduses of **p** or **q** (in their relative forms). Take first the case of partial causation by P of R, with reference to Table 11.3, above.

- The conjunction (P + Q + R) is possible, since the first row is always coded 1, whereas (P + notQ + R) is open, since the third row is sometimes coded 0 and sometimes 1. Nevertheless, it follows that the conjunction (P + R) is possible (i.e. to be coded 1), since "(P + Q + R) is possible" implies that "(P + R) is possible". Note that if we regarded (P + R) as merely open, we would fail to record that **there is *no* column with 0s in *both* the first and third cells**.

  Note this well: *it is a finding we altogether missed in macroanalysis, and which may therefore affect some of our results.*
- The same reasoning applies for the conjunctions (P + notR), comprising the second and fourth rows, and (notP + notR), comprising the sixth and eighth rows. They are both possible conjunctions, and not merely open.
- On the other hand, the conjunction (notP + R), comprising the fifth and seventh rows, must be declared open (i.e. be coded ·), since it is conceivable for both (notP + Q + R) and (notP + notQ + R) to be found impossible (as in the columns numbered 149, 150, 181, 182).

In this way the three-item modus for relative partial causation "10.1.1.." becomes the two-item modus "11.1". Similarly with contingent causation: its three-item modus "..1.1.01" becomes the two-item modus "1.11".

Thus, in case of need, we can contract a three-item modus into a two-item one, by changing a combination of 1 and 0, or 1 and ·, in corresponding locations, into a 1. Also, a combination of two dots yields one dot. Note well the rule of contraction:

1. **Where there is a 1 in the 3-item modus, there must be a 1 in the 2-item modus.**

Additionally note, though we have not yet encountered cases:

2. **The only way we could obtain a 0 in a two-item modus, from a three-item modus, would be to find *only* 0s along both rows of the latter.**
3. **If we find cases of '11','10' and/or '01' mixed with cases of '00' in the three-item modus, we must conclude a dot (·) in the two-item modus.**

Now, what have we found here? We started with weak causations by P of R, relative to some complement Q specifically, and ended with weak causations by P or R, without specification of Q, i.e. *absolutely*. The three-item modus for **p** or **q** relative to Q could not be equated to the same relative to some other complement, say Q_1; their matrices are superficially similar, but the items involved (namely PQR and PQ_1R) are quite different. But if we contract both kinds to two-item moduses, they would be indistinguishable, since the items involved (namely PR) are exactly identical.

Thus, the two-item modus of *absolute* (which includes relative) **partial** causation of form PR (symbolized, according to context, by **p** or $\mathbf{p}_{abs}$, or most precisely $\mathbf{p}_{PR}$) is by contraction found to be "**11.1**". This is, through the following table, worked out to have two conceivable realizations, namely "1101" or "1111" (labeled respectively Nos. 14, 16).

Table 11.5. Matrix of "P is a partial cause of R".

Items		Σ	2 alternative moduses	
P	**R**	**p**	**14**	**16**
1	1	1	1	1
1	0	1	1	1
0	1	•	0	1
0	0	1	1	1

In contrast, the two-item modus, summarily put, of *absolute* (which includes relative) **contingent** causation of form PR (symbolized, according to context, by **q** or $\mathbf{q}_{abs}$, or most precisely $\mathbf{q}_{PR}$) is by contraction found to be "**1.11**". This is, through the following table, worked out to have two conceivable realizations, namely "1011" or "1111" (labeled respectively Nos. 12, 16).

Table 11.6. Matrix of "P is a contingent cause of R".

Items		Σ	2 alternative moduses	
P	**R**	**q**	**12**	**16**
1	1	1	1	1
1	0	•	0	1
0	1	1	1	1
0	0	1	1	1

Notice the similarity between the summary moduses of **m** (10.1) and $\mathbf{p}_{abs}$ (11.1), or those of **n** (1.01) and $\mathbf{q}_{abs}$ (1.11). Where for the strong determination we have a '0' code, in the corresponding absolute weak determination we have a '1'; the remaining codes being identical. This, as we shall see in a later chapter[71], allows us to define the absolute weak determinations in formal terms.

Also note that the absolute weak determinations are convertible, just like the strong ones (as we pointed out in the previous section). For instance, "P is a partial cause of R" converts to "R is a contingent cause of P" (note the change in determination, as well as that of item positions), since these two forms have the same summary modus "11.1".

The next question to ask is: what are the three-item moduses of **m** or **n**, or of **p** or **q** in their absolute forms? We can answer this question by means of **expansion**, as follows.

Consider, to begin with, the strong determinations. In the case of complete causation by P of R, the following can be said:

- Knowing the conjunction (P + notR) is impossible, it follows that *both* (P + Q + notR) and (P + notQ + notR) are impossible conjunctions (whence the initial modus 0 becomes two moduses 0).

---

71 See "Some More Microanalyses", last section.

- Whereas, since the conjunction (P + R) is possible, it follows only that *at least one* of (P + Q + R) and (P + notQ + R) is a possible conjunction (i.e. they cannot *both* be impossible) - but we cannot predict *which* one is possible, so both conjunctions must be declared open (whence, the initial modus 1 becomes two moduses ·). Similarly, *mutatis mutandis*, for (notP + notR).
- Lastly, since the conjunction (notP + R) is open, so will a fortiori its two derivatives be (i.e. the initial modus · becomes two moduses ·).

In this way the two-item modus "10.1" for **complete** causation becomes the three-item modus "**.0.0....**". Similarly with **necessary** causation: its two-item modus "1.01" becomes the three-item modus "**....0.0.**". Notice the loss of information occasioned by the change in each case, due to the fact that ones become dots; the results of such expansions are vaguer than their sources.

Thus, in case of need, we can expand a two-item modus into a three-item one, by changing a zero into two zeros in the appropriate locations, and a one or dot into two dots as appropriate. Here, note well the 'appropriate locations' are *not* adjacent rows: they are the first and third, the second and fourth, the fifth and sixth, etc., reflecting a correspondence in combination of items - such as (P + notR) becoming (P + Q + notR) or (P + notQ + notR), which means moving from a PR sequence 10 to the PQR sequences 110 and 100, which signify the second and fourth rows of the matrix.

However, note well the restrictions implied in the following rules of expansion:

1. **Where there is a 0 in the 2-item modus, there *must* be two 0s in the 3-item modus.**
2. **Where there is a 1 in the 2-item modus, there *cannot* be two 0s in the 3-item modus.**
3. **Where there is a dot in the 2-item modus, there might be any combinations of 0s and/or 1s in the 3-item modus.**

With regard to 'zero' moduses (impossibility), they are universalized as it were from the initial row to the corresponding expanded rows. With regard to 'ones' (possibility), what is universalized from the single initial row to the two subsumed rows is the interdiction of zeros: just as in the two-item modus 0 is excluded by 1, so the three-item expansion cannot include columns (moduses) having 0s in both the corresponding rows. A fortiori, in the case of 'dots' (which might include zeros or ones), we cannot predict combinations in the two cells concerned, since all pairs are allowed, i.e. 0 and 0, 0 and 1, 1 and 0, 1 and 1.

Consider now the weak determinations, in accord with the rules of expansion just ascertained. If we similarly expand the two-item modus of absolute (or relative) **partial** causation, namely 11.1, into a three-item modus, we obtain "**........**", since all ones or zeros become dots. Likewise, by expansion of the two-item modus of absolute (or relative) **contingent** causation, namely 1.11, into a three-item modus, we obtain "**........**".

Note well that the result in both these cases is a string of dots, signifying complete uncertainty, the least possible amount of information. Each initial one or dot was expanded into two dots, so that all remaining specificity in the initial string was dissolved in its derivatives.

Note also the marked difference between the three-item strings of the absolute weak determinations "**........**", and those of the corresponding relative forms, namely "10.1.1.." and "..1.1.01" respectively.

Clearly, for all four generic determinations, expansion of a two-item modus "1" (possible) into two three-item moduses "•" (open) results in a *loss of data*; i.e. the information that 'at least one of the two conjunctions concerned must be possible: i.e. they cannot both be impossible' is no longer coded in our table. A calculus of causation should be so designed as to avoid all loss of information due to mere linguistic inadequacies[72]. Thus, we have to find a way to express, through a special code in the modus, say μ (Gk. letter mu), that *at least one of the two (or more) conjunctions so coded is implicitly a "1"*.[73]

Complete causation	= μ0μ0.μ.μ
Necessary causation	= μ.μ.0μ0μ
Partial causation (absolute)	= μμμμ.μ.μ
Contingent causation (absolute)	= μ.μ.μμμμ

This measure by itself is not enough; to save all available information, we would have to specify the rows concerned, say by labeling them a-h. For instance, if at least one of rows 'a' and 'c' has to have modus 1, each would have to be coded more specifically as μ_{ac}. Such coding means that the PQR sequences signified by the labels a and c (namely, 111 and 101) may have moduses with 0 and 1, 1 and 0, 1 and 1 - but they may not have the pair 0 and 0. It follows that:

- if a = μ_{ac} and c = 0, then a = 1 (since 00 is inconceivable), and
- if a = μ_{ac} and c = 1, then a = • (since 01 and 11 are both conceivable);

and likewise, of course, if c = μ_{ac} and a = 0 then c = 1, and if c = μ_{ac} and a = 1 then c = •.
All this information can be considered implicit in a table like the following (the relative weak determinations of form PQR are included for comparison):

Table 11.7. Summary moduses for the six generic determinations of form PR or PQR.

Row label	Items			m	n	p_{abs}	q_{abs}	p_{rel}	q_{rel}
	P	Q	R	PR	PR	PR	PR	PQR	PQR
a	1	1	1	μ_{ac}	μ_{ac}	μ_{ac}	μ_{ac}	1	•
b	1	1	0	0	•	μ_{bd}	•	0	•
c	1	0	1	μ_{ac}	μ_{ac}	μ_{ac}	μ_{ac}	•	1
d	1	0	0	0	•	μ_{bd}	•	1	•
e	0	1	1	•	0	•	μ_{eg}	•	1
f	0	1	0	μ_{fh}	μ_{fh}	μ_{fh}	μ_{fh}	1	•
g	0	0	1	•	0	•	μ_{eg}	•	0
h	0	0	0	μ_{fh}	μ_{fh}	μ_{fh}	μ_{fh}	•	1

All this may seem pretty complicated, but as we shall see it simplifies a lot of things. Through the summary moduses in the above table, we can identify precisely the alternative moduses in a three-item framework implied by each of the four determinations (for the items PR or PQR).

---

[72] I do not doubt that a better symbolic or mathematical logician than myself could develop neater approach. This is not my forte.

[73] When μ occurs, it occurs in pairs, note well. In contrast, within that notation, when a pair of dots occur, it means that both these positions may well be 0s.

As we shall see in the next chapter, the strong determinations **m** and **n** turn out to have 36 such alternative moduses each, while the weak determinations **p** and **q** in absolute form (as here), have 108 alternative moduses each (to compare to the 16 moduses of relative weaks). We shall list these moduses in the next chapter, so no need to do so here; they are easy to unravel by substituting zeros and ones for dots as previously explained.

4. Intersection, Nullification and Merger.

We shall now consider certain inferences from the above data.

The joining of generic determinations can be considered as the **intersection** of their respective summary moduses. By such conjunction of two propositions (or more), two classes (generic determinations) are used to express a more restrictive class (a joint determination), with whatever they have *in common.*

By this process, in a *two-item* framework (where the weak determinations are absolute), the joint determinations are found to have the following summary moduses:

- complete-necessary causation, **mn** = 10.1 + 1.01 = 1001 (modus No. 10);
- complete-contingent causation, **mq** = 10.1 + 1.11 = 1011 (modus No. 12);
- necessary-partial causation, **np** = 1.01 + 11.1 = 1101 (modus No. 14);
- partial-contingent causation, **pq** = 11.1 + 1.11 = 1111 (modus No. 16).

As can be seen, the result in each case is a single alternative modus (mentioned in brackets), which represents what the joined generics have in common. Thus, for instance, **m** has moduses 10 and 12, and **n** has moduses 10 and 14; therefore **mn** (meaning **m** + **n**) will have modus 10. The resulting summary modus is more defined than its sources, i.e. there are less dots, there are less uncertainties in the relation between the items.

This operation is merely an application of the well-known rule of class logic, that the logical product of two classes (such as **m** and **n**, each of which subsumes two subclasses, namely 10 and 12 for **m** and 10 and 14 for **n**) is the elements they have in common (namely, modus 10, in the case of **mn**). This can be seen for example in an Euler diagram, comprising two circles which overlap: their *common* area is the outcome of their product, and usually smaller than the circles (in our example, modus 10).

Note that by 'logical product' logicians mean that the two (or more) classes are conjoined together (i.e. **mn** means **m** + **n**)[74]. It must be stressed that modus lists are disjunctive not conjunctive, so that underlying this formula is another one (namely **mn** = 'modus 10 or modus 12' and 'modus 10 or modus 14', which means 'in any event, modus 10', i.e. it refers to the leftover after removing from consideration the elements 'modus 12' and 'modus 14', which are exclusive in either disjunct.

Similarly, in a *three-item* framework (where the weak determinations may be absolute or relative), intersection of the generic determinations yields the joint determinations, with the following summary moduses:

- complete-necessary causation:

74 This process is therefore, despite its name, in some ways more akin to addition. See *Future Logic*, chapter 28, on logical compositions.

$\mathbf{mn}$ = μ0μ0.μ.μ + μ.μ.0μ0μ = μ0μ00μ0μ (9 alternative moduses)

- complete-contingent causation,
  - $\mathbf{mq}_{abs}$ = μ0μ0.μ.μ + μ.μ.μμμμ = μ0μ0μμμμ (27 alternative moduses)
  - $\mathbf{mq}_{rel}$ = μ0μ0.μ.μ + ..1.1.01 = .0101.01 (4 alternative moduses)
- necessary-partial causation:
  - $\mathbf{np}_{abs}$ = μ.μ.0μ0μ + μμμμ.μ.μ = μμμμ0μ0μ (27 alternative moduses)
  - $\mathbf{np}_{rel}$ = μ.μ.0μ0μ + 10.1.1.. = 10.1010. (4 alternative moduses)
- partial-contingent causation:
  - $\mathbf{p}_{abs}\mathbf{q}_{abs}$ = μμμμ.μ.μ + μ.μ.μμμμ = μμμμμμμμ (81 alternative moduses)
  - $\mathbf{p}_{rel}\mathbf{q}_{rel}$ = 10.1.1.. + ..1.1.01 = 10111101 (1 alternative modus)

Here again, the result signifies the alternative moduses that the joined generics have in common; we shall not list them at this stage: the list will be given in the next chapter. In the case of $\mathbf{p}_{rel}\mathbf{q}_{rel}$, exceptionally, the result is a fully specifying summary modus, i.e. a single alternative modus (that labeled #190, as we shall see later). The resulting summary modus fuses together the most definite elements of the initial summary moduses; some dots become μ's, and some dots or μ's become more specifically a 0 or a 1. The μ's concerned are in pairs like μ_{ac} remember; the subscripts are not mentioned here for brevity.

Some of these results correspond to those obtained by macroanalysis, note. To grasp the rules of intersection, let us review the examples shown above:

1. The summary moduses are never conflicting in a given position (1 in one case and 0 in the other); this simply means that the determinations joined are *compatible*.
2. For each position identical in both generic summary moduses, or more definite (μ or 1 or 0) in one and indefinite (· or μ) in the other, the resulting corresponding position in the joint summary modus has that equal or more definite value.

We *cannot* join two determinations whose summary moduses have conflicting elements in the same position (a 0 in one and a 1 in the other): they are incompatible propositions, it is an impossible conjunction. In alternative modus terms, it means that these determinations do not have even one modus in common; in class logic terms, it means that the given classes (generic determinations) do not overlap: they have no intersection. Such logically empty concepts are known as null classes; we might therefore refer to the act of judging a class to be null as **nullification**.

For instances, the conjunctions **mp**, **nq** are null classes. Since **m** has moduses 10, 12 and **p** has moduses 14, 16, they have no common ground, no modus in which to coexist. Similarly for **n** and **q**, *mutatis mutandis*. This we know already from macroanalysis. More interesting, is *the capacity nullification gives us to judge the feasibility of lone determinations*, as we shall see in the next chapter.

Let us now consider another logical composition, that of **merger**, which disjoins two (or more) propositions, to obtain a single, vaguer proposition. In alternative modus terms, this process puts together all the alternative moduses listed for the given propositions in a larger list for the merged proposition. In class logic terms, this means that the two (or more) classes together become a single class covering all the areas they have exclusively as well as those they have in common.

This corresponds to the 'logical sum' of classes, where the two initial classes merge into a larger class by inclusive disjunction (expressed by operator *or*, which here means 'and/or', i.e. 'not both not'; this is often symbolized by a 'v', or in some computer languages by a '|')[75]. Inclusive disjunction means that all the elements subsumed by the given classes are to be included in the larger class; if a subclass subsumes x elements and another involves y elements, then the larger class covers $(x + y)$ elements. In contrast, in conjunction, only the elements subsumed by all the given classes are selected, forming a narrower class.[76]

We can, for instance, merge joint determinations into generics; thus, "**mn v mq**" is equivalent to just "**m**", "**mn v np**" becomes "**n**", "**np v pq**" becomes "**p**", and "**mq v pq**" results in "**q**". We can likewise merge generic determinations into broader concepts, such as strong or weak causation or causation, as shown below. Merger is easy if we work directly with alternative moduses; but it becomes very complicated if we refer to summary moduses, due to the inadequacies of the *ad hoc* notation system we have used so far.

In a *two-item* framework, it is feasible if we introduce an additional symbol, say λ (Gk. letter lambda), signifying that the two positions in the formula where it occurs cannot both be coded '1' (in contrast to μ, which signifies that they cannot both be '0', remember). In such case, we can predict the summary moduses of the following vague propositions (**s**, **w**, **c**) on the basis of the generics merged in them:

- **strong causation**, symbol **s** = **m or n** = 10.1 v 1.01 = 1λλ1 (moduses Nos. 10, 12, 14)[77];
- absolute **weak causation**, symbol $\mathbf{w}_{abs}$ = $\mathbf{p}_{abs}$ **or** $\mathbf{q}_{abs}$ = 11.1 v 1.11 = 1μμ1 (moduses Nos. 12, 14, 16);
- relative **weak causation**, symbol $\mathbf{w}_{rel}$ = $\mathbf{p}_{rel}$ **or** $\mathbf{q}_{rel}$ = same *two-item* summary modus as for absolute weak causation;
- **causation**, symbol **c** = **m or n or** $\mathbf{p}_{abs}$ **or** $\mathbf{q}_{abs}$ = 10.1 v 1.01 v 11.1 v 1.11 = 1..1 (moduses Nos. 10, 12, 14, 16).

Note that 'causation' here means *some* causation, causation of *any* determination whatever, whether **m**, **n**, $\mathbf{p}_{abs}$ or $\mathbf{q}_{abs}$. As we will show in the next chapter, 'contributory causation' (**m or p**) and 'possible causation' (**n or q**) are different from it only with reference to relatives; in absolute terms, they are identical to each other and to causation (because **m** implies **not-**$\mathbf{p}_{abs}$, and **n** implies **not-**$\mathbf{q}_{abs}$).

The same four operations in a *three-item* system all apparently yield one and the same conclusion, namely "μ.μ..μ.μ" (try and see) - which is the summary modus of causation, covering 144 alternative moduses, as we shall see. This is of course an absurd result, because, as we shall see in the next chapter, strong causation in fact covers 63 moduses; absolute weak causation, 135 moduses; and relative weak causation, 31 moduses! It follows that our notation system is inadequate for merger operations other than:

---

75 This process turns out, despite its name, in some ways more akin to multiplication. See *Future Logic*, chapter 28, on logical compositions.

76 Exclusive disjunction, note in passing, refers to the results of inclusive disjunction less those of conjunction, i.e. to the subsumptions of the given classes *not* common to them all.

77 The two middle positions of the merged summary modus have to be λ, because in the given summary moduses they may only be 00 or 01 (in the first) or 00 or 10 (in the second); i.e. the remaining possibility 11 is excluded.

c = μ0μ0.μ.μ v μ.μ.0μ0μ v μμμμ.μ.μ v μ.μ.μμμμ = μ.μ..μ.μ (144 moduses).

What this means is that the symbolic language developed so far is too simple to express more complex relations than those intended by a 0, 1, · or μ (or even λ, just introduced to enable merger of the two-item summary moduses of strong determinations[78]). It does not generate a distinctive summary modus for each and every form. However, I will not bother to attempt improving on it, not wishing to get bogged down in inessential matters. For our primary goal here is not to develop a calculus of summary moduses, but to ascertain how generic propositions can be merged into vaguer forms. And this we can readily do with reference to the underlying alternative moduses, which is good enough.

5. Negation.

Before moving on, let us review the ground covered thus far. We started with *binary coding* of the summary moduses of the generic determinations known to us thanks to macroanalyses performed at the very start of our research into causative propositions. We saw that these summary moduses involved uncertainties (coded ·). To eliminate these information gaps, we had to *unravel* the summary moduses, that is, identify the underlying alternative moduses (involving 0 or 1 codes exclusively). We thus introduced microanalysis.

However, the strong determinations **m**, **n** were expressed in a two-item (PR) framework, while the relative weaks $\mathbf{p}_{rel}$, $\mathbf{q}_{rel}$ were expressed in a three-item (PQR) framework - so these two sets of forms were not comparable. We therefore had to work out the means for *contraction* and *expansion* of their summary moduses (the latter process required that we introduce a fourth code, μ). This also allowed us to ascertain the two- and three- item summary moduses of absolute weak determinations $\mathbf{p}_{abs}$, $\mathbf{q}_{abs}$ - first by contracting those of the relative weaks $\mathbf{p}_{rel}$, $\mathbf{q}_{rel}$, then by expanding these results.

Having thus obtained both the two- and three- item summary moduses of all six generic determinations, we had all the information we need to work out the matrices of all derivative propositions. Indeed, by means of *intersection* we can readily identify the alternative moduses of any conjunction of determinations: they are the alternative moduses the latter have in common. A special case of this is *nullification*: if the propositions we wish to conjoin have no alternative moduses in common, they are incompatible. And by means of *merger* we can readily identify the alternative moduses of any disjunction of determinations: they are the alternative moduses the latter have all taken together.

We thus dispose of the basic data and logical processes we need for microanalysis of all positive forms, be they generic, joint (i.e. narrower than the generics) or vague (i.e. broader than the generics). But we still lack the alternative moduses of negative forms of whatever breadth. We cannot obtain their summary moduses by macroanalysis, as we did for the generic positive forms, because of the underlying complexity of negative causative propositions. So we must look for more profound means.

Thus far, we have engaged in microanalysis that may be characterized as *piecemeal*. In the next chapter, we shall approach this topic with a more holistic perspective, which we may

---

[78] For three items, we would have to introduce, as well as the concepts 'not 00' (μ) and not '11' (λ), 'not 01' and 'not 10', among others (supposedly). All of which becomes more complicated than useful.

refer to as *systematic* microanalysis. That consists in considering all conceivable alternative moduses in a given framework (fixed by the number of items under consideration), and then locating the determination(s) under consideration within this full range of possibilities.

The alternative moduses of negative forms become easy to identify thereby. Having the list of all conceivable alternative moduses in a given framework, and the alternative moduses of a positive form, we can readily infer those of the corresponding negative form: they are all the remaining alternative moduses! This process, which we shall simply call **negation**[79], is akin to subtraction. If a class subsumes *x* elements and a subclass of it involves *y* elements, then the remaining area covers (*x* - *y*) elements.

Microanalysis thus ultimately enables us to distinctively define any and every causative proposition (and other, related forms, as we shall see), with little effort. Furthermore, such detailed matricial analysis turns out to be a panacea, providing us with resolutions to all deductive issues in causation.

In particular, note that once we identify the moduses of negative generics, we can ascertain those of **lone** determinations, which conjoin one positive generic with the negations of all others. As we shall see in the next chapter, *absolute* lones are nullified. However, as we shall see in a subsequent chapter, *relative* lones are not nullified. Let us here mention for the record their *summary* moduses, which may be constructed knowing their alternative moduses, there identified (check and see for yourself that these summaries give rise to the correct alternatives):

- $\mathbf{m\text{-}alone}_{rel}$ = μ0μ0μμμμ
- $\mathbf{n\text{-}alone}_{rel}$ = μμμμ0μ0μ
- $\mathbf{p\text{-}alone}_{rel}$ = 10.1μ1μ.
- $\mathbf{q\text{-}alone}_{rel}$ = .μ1μ1.01

If we compare these to the summary moduses of **m**, **n**, $\mathbf{p}_{rel}$ and $\mathbf{q}_{rel}$, respectively (which are given in Table 11.7 above), as well as to those of joint determinations **mn**, $\mathbf{mq}_{rel}$, $\mathbf{np}_{rel}$, $\mathbf{p}_{rel}\mathbf{q}_{rel}$ (given in the previous section), we may observe the following mutations.

A code 0 or 1 for a generic is retained in a joint or lone including it. A μ found in **m** (or **n**, as the case may be) is retained in **mn**, and in $\mathbf{m\text{-}alone}_{rel}$ (or $\mathbf{n\text{-}alone}_{rel}$), but not in $\mathbf{mq}_{rel}$ (or $\mathbf{np}_{rel}$), because in the latter one μ is superseded by the 1 found in the corresponding position in $\mathbf{p}_{rel}$ (or $\mathbf{q}_{rel}$), so that the remaining μ becomes a dot. There are no dots left in $\mathbf{p}_{rel}\mathbf{q}_{rel}$ because all the dots in $\mathbf{p}_{rel}$ or $\mathbf{q}_{rel}$ have all been superseded by a 1 or 0. A dot in **m** or **n** becomes a μ in $\mathbf{m\text{-}alone}_{rel}$ or $\mathbf{n\text{-}alone}_{rel}$, respectively. As for $\mathbf{p\text{-}alone}_{rel}$ or $\mathbf{q\text{-}alone}_{rel}$, the dots in $\mathbf{p}_{rel}$ or $\mathbf{q}_{rel}$ *not* paired-off with a 1 become μ, whereas those paired-off with a 1 remain dots.

As already explained, a μ signifies that the pair of cells containing it (the first and third, the second and fourth, the fifth and seventh, or the sixth and eighth) may separately be 0 or 1, but cannot together be 0. No such restriction occurs where there are mere dots. Thus, what the above teaches us, especially, is that a relative lone determination has a *slightly more restrictive* modus than the corresponding generic determination, but is in all other respects identical.

---

79 Not to be confused with nullification, dealt with in the previous section.

Chapter 12. SYSTEMATIC MICROANALYSIS.

1. Grand Matrices.

Our study of causative propositions, in a first phase, consisted in conception of positive forms, their dissection into defining clauses, and their matricial analysis, or more precisely their macroanalysis. That provided us with the means to solve various problems, including many syllogistic issues; but it left us without practical means to answer questions concerning negative forms. We consequently, in a second phase, opted for a more detailed and deep method of study, microanalysis. We thus somewhat improved our predictive abilities; but serious difficulties remained, due to our approach being piecemeal.

To resolve outstanding issues, we must approach microanalysis in a more systematic manner. Instead of constructing matrices for each propositional form, we shall proceed in the opposite direction and conceive a *grand matrix* for the items concerned in which each and every propositional form can be located. A grand matrix tabulates *all conceivable moduses for a given number of items*, and assigns a numerical label (an address, as it were) to each such logical possibility. Once this is developed, we can identify the places of the various determinations within such a broad framework, and easily predict all their interactions.

Through grand matrices, we have an overview of all possible relations between the items concerned. We can then focus on particular segments of the matrix as signifying this or that specific relation.

Two items (P, R) give rise to a table with $2^2 = 4$ rows (with PR sequences 11, 10, 01, 00, conventionally so ordered), and $2^4 = 16$ modus columns (conventionally ordered with the maximum number of 'zeros' on the left and the maximum number of 'ones' on the right, then numbered 1-16). Such a table defines the *general* relation of any pair of items, and is the same whatever they happen to be.

A *specific* relation proposed for two particular items is then expressed by highlighting the modus column(s) corresponding to that specific relation (or by stating their numerical labels). The degree of determination involved is visually represented by the pattern of zeros and ones which stand out against the background of the grand matrix in which they are imbedded.

The grand matrix prefigures all 'potential' configurations for the number of items involved; while the highlighted alternative(s) depict the apparent or supposed 'actual' configuration for the particular items under scrutiny, which constitutes the distinctive determination relating them with each other.

In the case of three items (P, Q, R), the table has $2^3 = 8$ rows and $2^8 = 256$ modus columns, conventionally ordered in a similar manner. For four items (P, Q, R, S) we can expect a table with $2^4 = 16$ rows and $2^{16} = 65,536$ modus columns. And so forth. Note well that the concrete content of the items is irrelevant to the structure of the grand matrix; it looks the same for any given number of items.

From an epistemological and ontological point of view, a grand matrix depicts **the universe of imaginable relations between any two (or more) items in the world or in knowledge taken at random**. In reality, i.e. in the experienced world or at a given stage of knowledge development, only some of these relations (alternative moduses, i.e. conjunctions of presences and absences) will be found applicable to the items under scrutiny.

Thus, we can visualize the 'distance' (their separation in space-time, or their conceptual difference) between any two or more items in the world or in knowledge as inhabited by a belt[80] with strips of zeros and ones (a grand matrix with alternative moduses), of which some are highlighted or potent in the case concerned, and the rest are neutralized or inactive. We thus propose a very binary structure for the world and for knowledge, appealing by its universality and simplicity.

Indeed, in this perspective, we can even conceive of a 'universal matrix', comprising the umpteen items in the world or in knowledge, and an enormous tapestry of logically possible relations with zillions of zeros and ones in their every combination and permutation. For x items, this matrix would have $y = 2^x$ rows and $z = 2^y$ columns.

With this image in mind, the pursuit of knowledge can be considered as an attempt to pinpoint - on the basis of sensory and other experience, as well as of mental speculation and logical insight - the applicable moduses within such broad ranges, for the items concerned. A specific relation like 'causation' or 'complete causation' is thus a selection of moduses proposed as applicable to the concrete items concerned. The applicable alternative moduses constitute the 'bond' (of some degree) between the items in a given case.

Identification of applicable moduses proceeds gradually, *inductively* (with deduction as but a tool of induction). They are not known immediately, without residual doubts. Intellectual work is required.

We start with a mass of phenomena in flux. Appearances are presented to consciousness, perceptually (concretes) or conceptually (abstracts). We stratify some as 'given' (pure) and others as 'speculative' (mental projections about the pure), and try through logical insight to judge the hypotheses most fitting for the overall context of currently available data.

Much of our 'thinking' in relation to causation consists simply in trying to encapsulate the data available in the different forms of causation. This is a trial and error process, which may be characterized as successive formulation and (if need arise) elimination of hypotheses. Our approach may be passive, unconscious; or proactive, purposeful.

Normally, we first try out the strongest form of causation (**mn**), then lesser forms (**mq** or **np**), and finally the weakest (**pq**); if none of these work, we conclude with non-causation. Alternatively, we may proceed on a deeper level, with reference to if-then statements or, more cautiously, to moduses, before we build up comprehensive causative propositions.

As the empirical context changes, growing and becoming more focused, our opinion may vary. We may also discover, through deductive reasoning, inconsistencies between different conclusions. What seemed previously a successful summary of information then has to be reviewed. But eventually things seem to settle down and solidify, and we may presume that our opinion at last corresponds to (or more closely than ever approaches) the 'real' state of affairs, and may be regarded as knowledge.

Logic, after working out matricial configurations, immediately imposes **one universal restriction: the alternative modus in any grand matrix consisting only of zeros, with no**

---

80 To stress this image, we could place the items at opposite ends of the matrix. For two items, the 'belt' would be flat; for three items voluminous in three dimensions; and so forth. Another idea is to imagine the matrix as somehow enveloping the items, with varying force of cohesion. Each alternative modus indeed signifies a centripetal or centrifugal force relating the items concerned.

ones, cannot be true. Whatever the grand matrix, i.e. for any number and content of items, only alternative moduses involving at least one '1' code are at all credible; in every such matrix, the first modus, composed entirely of '0' codes, has no credibility.

This is just a restatement with regard to matricial analysis of the Laws of Non-Contradiction and of the Excluded Middle. Since the rows of our matrix *already* predict every conceivable combination of the items in their positive and negative forms, at least one of these rows has to possibly exist; if a column means that none of these combinations may occur, it contradicts that setup and lays claim to *yet another* combination of items. Such a claim would be absurd, and may be rejected at the outset.

All other moduses are logically sound *per se*, though they might well be excluded *within a given context*. Indeed, the knowledge enterprise may be viewed as a search for good reasons for the elimination of as many moduses as we can, so as to be left with a limited number of moduses which signify an interesting specific relation like causation. We thus move from the vaguely conceivable, to a more focused and pondered evaluation.

We cannot say at the outset which relation (expressed by one or more moduses) applies in a given case. There is bound to be *some* relation, but as we shall soon see logic does not insist on a specifically *causative* relation, it allows for a non-causative relation. *Ab initio*, all logic stipulates is that the modus consisting only of zeros can never apply.

This is the nearest thing to a 'law of causation' we can foresee at this stage; which by itself implies that there is no law of causation in the traditional senses, or that if there is one it must be sought for in other ways. We shall, of course, return to this topic in more detail, in a later chapter.

2. Moduses in a Two-Item Framework.

We shall first consider a two-item framework, and catalogue all its conceivable moduses, then enumerate those applicable to each category of proposition. In the following table, P is looked upon as a putative cause, while R is looked upon as a putative effect. Their conceivable combinations define rows, and columns refer to all initially conceivable alternative moduses for them.

In a two-item grand matrix, there are 4 rows and 16 columns, as we have seen, and therefore 64 cells. Each cell may equally be coded 0 (impossible) or 1 (possible), so that each code will occur a total of 32 times. The matrix is constructed by coding: in the first row, 0 in the first 8 cells then 1 in the last 8 cells; for the second row, 0 in the first and third set of 4 cells then 1 in the second and fourth set of 4 cells; in the third row, we have a succession of pairs, 00, 11, 00, 11, and so forth; finally, in the fourth row, we coded 0, 1, 0, 1, in succession. We are thus sure to have foreseen every possible interplay of 0 and 1 codes.

Take the time to notice that we have ordered the alternative moduses in a progressive manner, starting with a maximum number of 0s in a column (no cell coded 1) and ending with a maximum number of 1s in a column (no cell coded 0). We then conventionally number (or label) the columns so ordered, 1-16. The rows, note well, are also in a conventional arrangement, with four PR sequences 11, 10, 01, 00, respectively (labeled a-d, if need be).

Now, the column labeled No. 1 is an *impossible* modus, since at least one row has to have a '1', by the Laws of Non-Contradiction and of the Excluded Middle. Significantly, this is the only combination excluded universally by those logical laws, as already explained.

Concerning the remaining 15 *possible* moduses, they are *exhaustive* (one of them must be true) and *mutually exclusive* (no more than one may be true at once).
Here, then, is the grand matrix for two items, a catalogue of all conceivable alternative moduses for any two items, like P, R:

Table 12.1. Catalogue of moduses for the four conjunctions of two items (P, R).

Row label	Items		**	Possible moduses, labeled 2-15														
	P	**R**	**1**	**2**	**3**	**4**	**5**	**6**	**7**	**8**	**9**	**10**	**11**	**12**	**13**	**14**	**15**	**16**
a	**1**	**1**	0	0	0	0	0	0	0	0	1	1	1	1	1	1	1	1
b	**1**	**0**	0	0	0	0	1	1	1	1	0	0	0	0	1	1	1	1
c	**0**	**1**	0	0	1	1	0	0	1	1	0	0	1	1	0	0	1	1
d	**0**	**0**	0	1	0	1	0	1	0	1	0	1	0	1	0	1	0	1

** *Column labeled No. 1 is an impossible modus.*

The following table interprets the preceding, by enumeration of the alternative moduses of the main causative forms. It is based on the known characteristics of positive strong and absolute weak generics, i.e. the moduses given in Tables 1, 2, 5 and 6 of the previous chapter. From this initial information, we can, using the processes of negation, intersection and merger, infer the alternative moduses of derivative forms, i.e. negatives, as well as joints and vaguer forms (**s**, **w**, **c**), and their negations.
Note that relative weak determinations are not dealt with here, because, in a two-item framework, they have the same moduses as absolutes. They can only be distinguished as of a three-item framework, so we cannot analyze them and their derivatives till we get there.

Table 12.2. Enumeration of two-item moduses for the strong or absolute weak determinations and their derivatives (form PR).

Determination	Column number(s)	Comment
Strongs and their negations:		
m	10, 12	2 alternatives, by macroanalysis.
n	10, 14	2 alternatives, by macroanalysis.
not-m	2-9, 11, 13-16	All alternatives but those of **m**; i.e. 13 cases.
not-n	2-9, 11-13, 15-16	All alternatives but those of **n**; i.e. 13 cases.
Absolute weaks and their negations:		
$\mathbf{p}_{abs}$	14, 16	2 alternatives, by macroanalysis of $\mathbf{p}_{rel}$ and contraction.
$\mathbf{q}_{abs}$	12, 16	2 alternatives, by macroanalysis of $\mathbf{q}_{rel}$ and contraction.
not-$\mathbf{p}_{abs}$	2-13, 15	All alternatives but those of $\mathbf{p}_{abs}$; i.e. 13 cases.
not-$\mathbf{q}_{abs}$	2-11, 13-15	All alternatives but those of $\mathbf{q}_{abs}$; i.e. 13 cases.

Table 12.2 continued.

Joints (absolute) and their negations:		
mn	10	Their one common alternative, by intersection.
mq$_{abs}$	12	Their one common alternative, by intersection.
np$_{abs}$	14	Their one common alternative, by intersection.
p$_{abs}$**q**$_{abs}$	16	Their one common alternative, by intersection.
not(mn)	2-9, 11-16	All alternatives but that of **mn**; i.e. 14 cases.
not(mq$_{abs}$**)**	2-11, 13-16	All alternatives but that of **mq**$_{abs}$; i.e. 14 cases.
not(np$_{abs}$**)**	2-14, 15-16	All alternatives but that of **np**$_{abs}$; i.e. 14 cases.
not(p$_{abs}$**q**$_{abs}$**)**	2-15	All alternatives but that of **p**$_{abs}$**q**$_{abs}$; i.e. 14 cases.
Strong causation and its negation:		
s = m or n	10, 12, 14	All their 3 alternatives, by merger.
not-s = not-m + not-n	2-9, 11, 13, 15-16	All alternatives but the preceding; i.e. 12 cases.
Absolute weak causation and its negation:		
w$_{abs}$ = **p**$_{abs}$ **or q**$_{abs}$	12, 14, 16	All their 3 alternatives, by merger.
not- w$_{abs}$ = **not-p**$_{abs}$ + **not-q**$_{abs}$	2-11, 13, 15	All alternatives but the preceding; i.e. 12 cases.
Causation (absolute) and its negation:		
c$_{abs}$ = **m or n or p**$_{abs}$ **or q**$_{abs}$	10, 12, 14, 16	All their four alternatives, by merger.
not-c$_{abs}$ = **not-m + not-n** + **not-p**$_{abs}$ + **not-q**$_{abs}$	2-9, 11, 13, 15	All alternatives but the preceding; i.e. 11 cases.

Let us highlight some of the information in the above table. First, take note of the ease with which we are now able to define any negative form, given the moduses of the corresponding positive form, by simply listing the leftover moduses. We can also readily define vaguer positive forms, like **s**, **w**, **c**, by merging the modus lists of their components. These forms were until here very difficult to define, remember.

Second, we can see at a glance that compatible forms are those which have a common modus (or more); for instance, **m** and **n**, **m** and **q**$_{abs}$, **n** and **p**$_{abs}$, **p**$_{abs}$ and **q**$_{abs}$ can be joined, because they share a modus (respectively, 10, 12, 14 and 16). Incompatibilities are also made evident by such a table; thus, **m** and **p**$_{abs}$ have no common modus, nor do **n** and **q**$_{abs}$; so these are incompatible pairs and give rise to no form.

Third, certain compounds of positives and negatives have not been listed in the above table, because they are equivalent to already listed forms, i.e. all their moduses are the same. ***Implication*** signifies that *every* modus of the implying form is a modus of the implied form; this is not mere overlap, note, but full inclusion of one form in the other.

Two *one-way* implications (and their contraposites) must be noted:

- that **p**$_{abs}$ implies **not-m** (or **m** implies **not-p**$_{abs}$), and
- that **q**$_{abs}$ implies **not-n** (or **n** implies **not-q**$_{abs}$).

This is because the moduses Nos. 14, 16 of **p**$_{abs}$ are both also moduses of **not-m**, and the moduses Nos. 12, 16 of **q**$_{abs}$ are both also moduses of **not-n**. Given that **m** implies **not-p**$_{abs}$, it follows that **(m + not-p**$_{abs}$**)** is identical to **m**. Similarly, **(n + not-q**$_{abs}$**) = n**; **(not-m + p**$_{abs}$**) = p**$_{abs}$; and **(not-n + q**$_{abs}$**) = q**$_{abs}$. There is therefore no need to list these four conjunctions separately.

Mutual implication or equivalence occurs when the forms compared have the very same alternative modus list. Thus,

- **(m + not-q**$_{abs}$**) = (n + not-p**$_{abs}$**) = mn** (modus 10);
- **(m + not-n) = (not-p**$_{abs}$ **+ q**$_{abs}$**) = mq**$_{abs}$ (modus 12);
- **(not-m + n) = (p**$_{abs}$ **+ not-q**$_{abs}$**) = np**$_{abs}$ (modus 14); and
- **(not-m + q**$_{abs}$**) = (not-n + p**$_{abs}$**) = p**$_{abs}$**q**$_{abs}$ (modus 16).

There is therefore no need to list these various conjunctions separately. In contrast, for instance, **m** and **n** do not imply each other, though they have one modus in common (No. 10), because each has a modus the other lacks. Likewise for **p**$_{abs}$ and **q**$_{abs}$, they overlap only in one of their moduses (No. 16) and both have a distinct additional modus.

Fourth, some compositions have not been listed in the above table, because they do not constitute an interesting concept. Falling in this category are **m or q**$_{abs}$ (moduses 10, 12, 16) and its negation **(not-m + not-q**$_{abs}$**)**, or again **n or p**$_{abs}$ (moduses 10, 14, 16) and its negation **(not-n + not-p**$_{abs}$**)**.

Fifth, certain conjunctions of positives and negatives have not been listed in the above table, because they give rise to no forms. Note especially that **(absolute) lone determinations are excluded** from consideration (or nullified) by this technique. That is, we cannot form the following conjunctions of positive and negatives, because they do not share a single common alternative modus:

- **m-alone**$_{abs}$ = **m + not-n + not-p**$_{abs}$ **+ not-q**$_{abs}$ = null-class;
- **n-alone**$_{abs}$ = **n + not-m + not-p**$_{abs}$ **+ not-q**$_{abs}$ = null-class;
- **p-alone**$_{abs}$ = **p**$_{abs}$ **+ not-m + not-n + not-q**$_{abs}$ = null-class;
- **q-alone**$_{abs}$ = **q**$_{abs}$ **+ not-m + not-n + not-p**$_{abs}$ = null-class.

Thus, for instance, **m** shares modus 12 with **not-n** and (needless to say, since it implies it) with **not-p**$_{abs}$, but this modus is absent in **not-q**$_{abs}$. And so forth, for the other absolute lones. These symbolically contrived conjunctions are therefore impossible in fact: *by reference to the moduses we can definitively establish this fact and understand it.*

This is an important formal principle, which may be looked upon as a 'law of causation' (among others)[81]. Had (absolute) lone determinations been possible, our view of the causative

81 The expression 'law of causation' is traditionally used with reference to general statements such as "everything has a cause", for which we have so far not found formal justification, though they might eventually be adopted as inductive principles. Here, the phrase is used in a more open sense,

relation would have been much less deterministic. Before microanalysis, we could not ascertain whether or not the generic determinations **m**, **n**, $\mathbf{p}_{abs}$ or $\mathbf{q}_{abs}$ may logically exist without intersection; now we know for sure that they can only exist within joint determinations.

The following equations follow from the nullification of lones:

- **m** = (**mn or mq**$_{abs}$), and **n** = (**mn or np**$_{abs}$);
- $\mathbf{p}_{abs}$ = (**np**$_{abs}$ **or** $\mathbf{p}_{abs}\mathbf{q}_{abs}$), and $\mathbf{q}_{abs}$ = (**mq**$_{abs}$ **or** $\mathbf{p}_{abs}\mathbf{q}_{abs}$).

Again, **s** = (**mn or mq**$_{abs}$ **or np**$_{abs}$), and $\mathbf{w}_{abs}$ = (**mq**$_{abs}$ **or np**$_{abs}$ **or** $\mathbf{p}_{abs}\mathbf{q}_{abs}$). Consequently, $\mathbf{c}_{abs}$ = (**mn or mq**$_{abs}$ **or np**$_{abs}$ **or** $\mathbf{p}_{abs}\mathbf{q}_{abs}$); and it is equivalent to (**m or p**$_{abs}$) and to (**n or q**$_{abs}$). Also, by negation, **not-c**$_{abs}$ is equivalent to (**not-m** + **not-p**$_{abs}$) and to (**not-n** + **not-q**$_{abs}$).

These various compounds are therefore implicit in the above table, and need not be listed.

Lastly, we should notice the *genus-species* relations between forms. Thus, **mn** is a species of **m** and a species of **n**, because it shares a modus (No. 10) with each of them, and has none they lack; the latter forms are more generic or less definite, since they involve additional alternatives. Similarly, **s** is vaguer or broader in possibilities than **m** or **n**, and therefore a genus of theirs; likewise, $\mathbf{p}_{abs}$ and $\mathbf{q}_{abs}$ are species of $\mathbf{w}_{abs}$. Causation (**c**) is clearly the summum genus for all the positive forms. Negatives can be examined in the same perspective.

It is also worth noticing what underlies the *relative strengths of determinations*. Note that the alternative moduses of the strong determinations (10, 12, 14) involve more zeros than those of the weaks (12, 14, 16). In particular, ignoring the common moduses (12, 14), compare modus 10 (two 0s) with modus 16 (no 0s). Clearly, **m** and **n** are stronger than **p** and **q**, because they involve more impossibility (two extra zeros); zeros more firmly delimit a relation. Similarly, comparing joints with each other; the more zeros in the modus, the stronger the determination.

3. Catalogue of Moduses, for Three Items.

Let us now consider a three-item framework. We shall here catalogue all its conceivable moduses; and in the next section, we shall enumerate those applicable to each category of proposition. In the following table, P and Q are looked upon as putative causes, while R is looked upon as a putative effect. Their conceivable combinations define rows, and columns refer to all initially conceivable alternative moduses for them.

In a three-item grand matrix, there are 8 rows and 256 columns, as we have seen, and therefore 2048 cells. Each cell may equally be coded 0 (impossible) or 1 (possible), so that each code will occur a total of 1024 times. This matrix is constructed in the same manner as the preceding one, by coding 0s and 1s progressively throughout it, so symmetrically that we can be sure it is exhaustive.

The columns (representing the alternative moduses), so ordered, are then numbered (or labeled) 1-256. Since the order of the rows is also fixed conventionally, with eight PQR

---

reflecting the usual usage of the term 'law'. In this sense, as we saw earlier, the fact that alternative modus No. 1 (consisting only of zeros) is impossible is a law; and likewise the fact that absolute lone determinations do not exist. Indeed, in this sense, all formal processes about causation - including all oppositions, eductions, syllogisms - are laws.

sequences 111, 110, 101, 100, 011, 010, 001, 000 (which can, if need be, be labeled a-h, respectively), the modus number suffices to symbolize the modus concerned.[82]

Now, the column labeled No. 1 is an *impossible* modus, since at least one row has to have a '1', by the Laws of Non-Contradiction and of the Excluded Middle. Significantly, this is the only combination excluded universally by those logical laws, as already explained. Concerning the remaining 255 *possible* moduses, they are *exhaustive* (one of them must be true) and *mutually exclusive* (no more than one may be true at once).

Here, then, is the grand matrix for three items, a catalogue of all conceivable alternative moduses for any three items, such as P, Q, R:

Table 12.3. Catalogue of moduses for the eight conjunctions of three items (P, Q, R).

Items			**	Possible moduses, labeled 2-16														
P	**(Q)**	**R**	**1**	**2**	**3**	**4**	**5**	**6**	**7**	**8**	**9**	**10**	**11**	**12**	**13**	**14**	**15**	**16**
1	**1**	**1**	0	0	0	0	0	0	0	0	0	0	0	0	0	0	0	0
1	**1**	**0**	0	0	0	0	0	0	0	0	0	0	0	0	0	0	0	0
1	**0**	**1**	0	0	0	0	0	0	0	0	0	0	0	0	0	0	0	0
1	**0**	**0**	0	0	0	0	0	0	0	0	0	0	0	0	0	0	0	0
0	**1**	**1**	0	0	0	0	0	0	0	0	1	1	1	1	1	1	1	1
0	**1**	**0**	0	0	0	0	1	1	1	1	0	0	0	0	1	1	1	1
0	**0**	**1**	0	0	1	1	0	0	1	1	0	0	1	1	0	0	1	1
0	**0**	**0**	0	1	0	1	0	1	0	1	0	1	0	1	0	1	0	1

** *Column labeled No. 1 is an impossible modus.*

Same table continued.

Items			Moduses, labeled 17-32															
P	**(Q)**	**R**	**17**	**18**	**19**	**20**	**21**	**22**	**23**	**24**	**25**	**26**	**27**	**28**	**29**	**30**	**31**	**32**
1	**1**	**1**	0	0	0	0	0	0	0	0	0	0	0	0	0	0	0	0
1	**1**	**0**	0	0	0	0	0	0	0	0	0	0	0	0	0	0	0	0
1	**0**	**1**	0	0	0	0	0	0	0	0	0	0	0	0	0	0	0	0
1	**0**	**0**	1	1	1	1	1	1	1	1	1	1	1	1	1	1	1	1
0	**1**	**1**	0	0	0	0	0	0	0	0	1	1	1	1	1	1	1	1
0	**1**	**0**	0	0	0	0	1	1	1	1	0	0	0	0	1	1	1	1
0	**0**	**1**	0	0	1	1	0	0	1	1	0	0	1	1	0	0	1	1
0	**0**	**0**	0	1	0	1	0	1	0	1	0	1	0	1	0	1	0	1

[82] Needless to say, one should not confuse the modus numbers 1-16 in a two-item framework, with the first 16 of 256 modus numbers used for a three-item framework. These are mere homonyms. The framework concerned should always be specified, if not implicitly clear. (See Table 12.6 below for precise correspondences.)

Table 12.3 continued.

Items			Moduses, labeled 33-48															
P	(Q)	R	33	34	35	36	37	38	39	40	41	42	43	44	45	46	47	48
1	**1**	**1**	0	0	0	0	0	0	0	0	0	0	0	0	0	0	0	0
1	**1**	**0**	0	0	0	0	0	0	0	0	0	0	0	0	0	0	0	0
1	**0**	**1**	1	1	1	1	1	1	1	1	1	1	1	1	1	1	1	1
1	**0**	**0**	0	0	0	0	0	0	0	0	0	0	0	0	0	0	0	0
0	**1**	**1**	0	0	0	0	0	0	0	0	1	1	1	1	1	1	1	1
0	**1**	**0**	0	0	0	0	1	1	1	1	0	0	0	0	1	1	1	1
0	**0**	**1**	0	0	1	1	0	0	1	1	0	0	1	1	0	0	1	1
0	**0**	**0**	0	1	0	1	0	1	0	1	0	1	0	1	0	1	0	1

Same table continued.

Items			Moduses, labeled 49-64															
P	(Q)	R	49	50	51	52	53	54	55	56	57	58	59	60	61	62	63	64
1	**1**	**1**	0	0	0	0	0	0	0	0	0	0	0	0	0	0	0	0
1	**1**	**0**	0	0	0	0	0	0	0	0	0	0	0	0	0	0	0	0
1	**0**	**1**	1	1	1	1	1	1	1	1	1	1	1	1	1	1	1	1
1	**0**	**0**	1	1	1	1	1	1	1	1	1	1	1	1	1	1	1	1
0	**1**	**1**	0	0	0	0	0	0	0	0	1	1	1	1	1	1	1	1
0	**1**	**0**	0	0	0	0	1	1	1	1	0	0	0	0	1	1	1	1
0	**0**	**1**	0	0	1	1	0	0	1	1	0	0	1	1	0	0	1	1
0	**0**	**0**	0	1	0	1	0	1	0	1	0	1	0	1	0	1	0	1

Same table continued.

Items			Moduses, labeled 65-80															
P	(Q)	R	65	66	67	68	69	70	71	72	73	74	75	76	77	78	79	80
1	**1**	**1**	0	0	0	0	0	0	0	0	0	0	0	0	0	0	0	0
1	**1**	**0**	1	1	1	1	1	1	1	1	1	1	1	1	1	1	1	1
1	**0**	**1**	0	0	0	0	0	0	0	0	0	0	0	0	0	0	0	0
1	**0**	**0**	0	0	0	0	0	0	0	0	0	0	0	0	0	0	0	0
0	**1**	**1**	0	0	0	0	0	0	0	0	1	1	1	1	1	1	1	1
0	**1**	**0**	0	0	0	0	1	1	1	1	0	0	0	0	1	1	1	1
0	**0**	**1**	0	0	1	1	0	0	1	1	0	0	1	1	0	0	1	1
0	**0**	**0**	0	1	0	1	0	1	0	1	0	1	0	1	0	1	0	1

Same table continued.

Items			Moduses, labeled 81-96															
P	(Q)	R	81	82	83	84	85	86	87	88	89	90	91	92	93	94	95	96
1	**1**	**1**	0	0	0	0	0	0	0	0	0	0	0	0	0	0	0	0
1	**1**	**0**	1	1	1	1	1	1	1	1	1	1	1	1	1	1	1	1
1	**0**	**1**	0	0	0	0	0	0	0	0	0	0	0	0	0	0	0	0
1	**0**	**0**	1	1	1	1	1	1	1	1	1	1	1	1	1	1	1	1
0	**1**	**1**	0	0	0	0	0	0	0	0	1	1	1	1	1	1	1	1
0	**1**	**0**	0	0	0	0	1	1	1	1	0	0	0	0	1	1	1	1
0	**0**	**1**	0	0	1	1	0	0	1	1	0	0	1	1	0	0	1	1
0	**0**	**0**	0	1	0	1	0	1	0	1	0	1	0	1	0	1	0	1

Table 12.3 continued.

Items			Moduses, labeled 97-112															
P	(Q)	R	97	98	99	100	101	102	103	104	105	106	107	108	109	110	111	112
1	**1**	**1**	0	0	0	0	0	0	0	0	0	0	0	0	0	0	0	0
1	**1**	**0**	1	1	1	1	1	1	1	1	1	1	1	1	1	1	1	1
1	**0**	**1**	1	1	1	1	1	1	1	1	1	1	1	1	1	1	1	1
1	**0**	**0**	0	0	0	0	0	0	0	0	0	0	0	0	0	0	0	0
0	**1**	**1**	0	0	0	0	0	0	0	0	1	1	1	1	1	1	1	1
0	**1**	**0**	0	0	0	0	1	1	1	1	0	0	0	0	1	1	1	1
0	**0**	**1**	0	0	1	1	0	0	1	1	0	0	1	1	0	0	1	1
0	**0**	**0**	0	1	0	1	0	1	0	1	0	1	0	1	0	1	0	1

Same table continued.

Items			Moduses, labeled 113-128															
P	(Q)	R	113	114	115	116	117	118	119	120	121	122	123	124	125	126	127	128
1	**1**	**1**	0	0	0	0	0	0	0	0	0	0	0	0	0	0	0	0
1	**1**	**0**	1	1	1	1	1	1	1	1	1	1	1	1	1	1	1	1
1	**0**	**1**	1	1	1	1	1	1	1	1	1	1	1	1	1	1	1	1
1	**0**	**0**	1	1	1	1	1	1	1	1	1	1	1	1	1	1	1	1
0	**1**	**1**	0	0	0	0	0	0	0	0	1	1	1	1	1	1	1	1
0	**1**	**0**	0	0	0	0	1	1	1	1	0	0	0	0	1	1	1	1
0	**0**	**1**	0	0	1	1	0	0	1	1	0	0	1	1	0	0	1	1
0	**0**	**0**	0	1	0	1	0	1	0	1	0	1	0	1	0	1	0	1

Same table continued.

Items			Moduses, labeled 129-144															
P	(Q)	R	129	130	131	132	133	134	135	136	137	138	139	140	141	142	143	144
1	**1**	**1**	1	1	1	1	1	1	1	1	1	1	1	1	1	1	1	1
1	**1**	**0**	0	0	0	0	0	0	0	0	0	0	0	0	0	0	0	0
1	**0**	**1**	0	0	0	0	0	0	0	0	0	0	0	0	0	0	0	0
1	**0**	**0**	0	0	0	0	0	0	0	0	0	0	0	0	0	0	0	0
0	**1**	**1**	0	0	0	0	0	0	0	0	1	1	1	1	1	1	1	1
0	**1**	**0**	0	0	0	0	1	1	1	1	0	0	0	0	1	1	1	1
0	**0**	**1**	0	0	1	1	0	0	1	1	0	0	1	1	0	0	1	1
0	**0**	**0**	0	1	0	1	0	1	0	1	0	1	0	1	0	1	0	1

Same table continued.

Items			Moduses, labeled 145-160															
P	(Q)	R	145	146	147	148	149	150	151	152	153	154	155	156	157	158	159	160
1	**1**	**1**	1	1	1	1	1	1	1	1	1	1	1	1	1	1	1	1
1	**1**	**0**	0	0	0	0	0	0	0	0	0	0	0	0	0	0	0	0
1	**0**	**1**	0	0	0	0	0	0	0	0	0	0	0	0	0	0	0	0
1	**0**	**0**	1	1	1	1	1	1	1	1	1	1	1	1	1	1	1	1
0	**1**	**1**	0	0	0	0	0	0	0	0	1	1	1	1	1	1	1	1
0	**1**	**0**	0	0	0	0	1	1	1	1	0	0	0	0	1	1	1	1
0	**0**	**1**	0	0	1	1	0	0	1	1	0	0	1	1	0	0	1	1
0	**0**	**0**	0	1	0	1	0	1	0	1	0	1	0	1	0	1	0	1

Table 12.3 continued.

Items			Moduses, labeled 161-176															
P	(Q)	R	161	162	163	164	165	166	167	168	169	170	171	172	173	174	175	176
1	**1**	**1**	1	1	1	1	1	1	1	1	1	1	1	1	1	1	1	1
1	**1**	**0**	0	0	0	0	0	0	0	0	0	0	0	0	0	0	0	0
1	**0**	**1**	1	1	1	1	1	1	1	1	1	1	1	1	1	1	1	1
1	**0**	**0**	0	0	0	0	0	0	0	0	0	0	0	0	0	0	0	0
0	**1**	**1**	0	0	0	0	0	0	0	0	1	1	1	1	1	1	1	1
0	**1**	**0**	0	0	0	0	1	1	1	1	0	0	0	0	1	1	1	1
0	**0**	**1**	0	0	1	1	0	0	1	1	0	0	1	1	0	0	1	1
0	**0**	**0**	0	1	0	1	0	1	0	1	0	1	0	1	0	1	0	1

Same table continued.

Items			Moduses, labeled 177-192															
P	(Q)	R	177	178	179	180	181	182	183	184	185	186	187	188	189	190	191	192
1	**1**	**1**	1	1	1	1	1	1	1	1	1	1	1	1	1	1	1	1
1	**1**	**0**	0	0	0	0	0	0	0	0	0	0	0	0	0	0	0	0
1	**0**	**1**	1	1	1	1	1	1	1	1	1	1	1	1	1	1	1	1
1	**0**	**0**	1	1	1	1	1	1	1	1	1	1	1	1	1	1	1	1
0	**1**	**1**	0	0	0	0	0	0	0	0	1	1	1	1	1	1	1	1
0	**1**	**0**	0	0	0	0	1	1	1	1	0	0	0	0	1	1	1	1
0	**0**	**1**	0	0	1	1	0	0	1	1	0	0	1	1	0	0	1	1
0	**0**	**0**	0	1	0	1	0	1	0	1	0	1	0	1	0	1	0	1

Same table continued.

Items			Moduses, labeled 193-208															
P	(Q)	R	193	194	195	196	197	198	199	200	201	202	203	204	205	206	207	208
1	**1**	**1**	1	1	1	1	1	1	1	1	1	1	1	1	1	1	1	1
1	**1**	**0**	1	1	1	1	1	1	1	1	1	1	1	1	1	1	1	1
1	**0**	**1**	0	0	0	0	0	0	0	0	0	0	0	0	0	0	0	0
1	**0**	**0**	0	0	0	0	0	0	0	0	0	0	0	0	0	0	0	0
0	**1**	**1**	0	0	0	0	0	0	0	0	1	1	1	1	1	1	1	1
0	**1**	**0**	0	0	0	0	1	1	1	1	0	0	0	0	1	1	1	1
0	**0**	**1**	0	0	1	1	0	0	1	1	0	0	1	1	0	0	1	1
0	**0**	**0**	0	1	0	1	0	1	0	1	0	1	0	1	0	1	0	1

Same table continued.

Items			Moduses, labeled 209-224															
P	(Q)	R	209	210	211	212	213	214	215	216	217	218	219	220	221	222	223	224
1	**1**	**1**	1	1	1	1	1	1	1	1	1	1	1	1	1	1	1	1
1	**1**	**0**	1	1	1	1	1	1	1	1	1	1	1	1	1	1	1	1
1	**0**	**1**	0	0	0	0	0	0	0	0	0	0	0	0	0	0	0	0
1	**0**	**0**	1	1	1	1	1	1	1	1	1	1	1	1	1	1	1	1
0	**1**	**1**	0	0	0	0	0	0	0	0	1	1	1	1	1	1	1	1
0	**1**	**0**	0	0	0	0	1	1	1	1	0	0	0	0	1	1	1	1
0	**0**	**1**	0	0	1	1	0	0	1	1	0	0	1	1	0	0	1	1
0	**0**	**0**	0	1	0	1	0	1	0	1	0	1	0	1	0	1	0	1

Table 12.3 continued.

Items			Moduses, labeled 225-240															
P	**(Q)**	**R**	**225**	**226**	**227**	**228**	**229**	**230**	**231**	**232**	**233**	**234**	**235**	**236**	**237**	**238**	**239**	**240**
1	**1**	**1**	1	1	1	1	1	1	1	1	1	1	1	1	1	1	1	1
1	**1**	**0**	1	1	1	1	1	1	1	1	1	1	1	1	1	1	1	1
1	**0**	**1**	1	1	1	1	1	1	1	1	1	1	1	1	1	1	1	1
1	**0**	**0**	0	0	0	0	0	0	0	0	0	0	0	0	0	0	0	0
0	**1**	**1**	0	0	0	0	0	0	0	0	1	1	1	1	1	1	1	1
0	**1**	**0**	0	0	0	0	1	1	1	1	0	0	0	0	1	1	1	1
0	**0**	**1**	0	0	1	1	0	0	1	1	0	0	1	1	0	0	1	1
0	**0**	**0**	0	1	0	1	0	1	0	1	0	1	0	1	0	1	0	1

Same table continued.

Items			Moduses, labeled 241-256															
P	**(Q)**	**R**	**241**	**242**	**243**	**244**	**245**	**246**	**247**	**248**	**249**	**250**	**251**	**252**	**253**	**254**	**255**	**256**
1	**1**	**1**	1	1	1	1	1	1	1	1	1	1	1	1	1	1	1	1
1	**1**	**0**	1	1	1	1	1	1	1	1	1	1	1	1	1	1	1	1
1	**0**	**1**	1	1	1	1	1	1	1	1	1	1	1	1	1	1	1	1
1	**0**	**0**	1	1	1	1	1	1	1	1	1	1	1	1	1	1	1	1
0	**1**	**1**	0	0	0	0	0	0	0	0	1	1	1	1	1	1	1	1
0	**1**	**0**	0	0	0	0	1	1	1	1	0	0	0	0	1	1	1	1
0	**0**	**1**	0	0	1	1	0	0	1	1	0	0	1	1	0	0	1	1
0	**0**	**0**	0	1	0	1	0	1	0	1	0	1	0	1	0	1	0	1

4. Enumeration of Moduses, for Three Items.

The following table interprets the preceding, by enumeration of the alternative moduses of the main causative forms. It is based on the known characteristics of positive strong and weak generics, i.e. the moduses given in Tables 1-6 of the previous chapter. From this initial information, we can, using the processes of negation, intersection and merger, infer the alternative moduses of derivative forms, i.e. negatives, as well as joints and vaguer forms (**s**, **w**, **c**), and their negations.

We shall deal here only with the absolute weak determinations and their derivatives; relative weaks and their derivatives will be considered in the next chapter.

Table 12.4. Enumeration of three-item moduses for the generic determinations and their derivatives (form PR).

Determination	Modus numbers	Comment
Strongs and their negations:		
m	34, 36-40, 42, 44-48, 130, 132-136, 138, 140-144, 162, 164-168, 170, 172-176	36 alternatives, by macroanalysis.
n	34, 37-38, 50, 53-54, 98, 101-102, 114, 117-118, 130, 133-134, 146, 149-150, 162, 165-166, 178, 181-182, 194, 197-198, 210, 213-214, 226, 229-230, 242, 245-246	36 alternatives, by macroanalysis.
not-m	2-33, 35, 41, 43, 49-129, 131, 137, 139, 145-161, 163, 169, 171, 177-256	All alternatives but those of **m**, i.e. 219 cases.
not-n	2-33, 35-36, 39-49, 51-52, 55-97, 99-100, 103-113, 115-116, 119-129, 131-132, 135-145, 147-148, 151-161, 163-164, 167-177, 179-180, 183-193, 195-196, 199-209, 211-212, 215-225, 227-228, 231-241, 243-244, 247-256	All alternatives but those of **n**, i.e. 219 cases.
Absolute weaks and their negations:		
$\mathbf{p}_{abs}$	50, 52-56, 58, 60-64, 98, 100-104, 106, 108-112, 114, 116-120, 122, 124-128, 146, 148-152, 154, 156-160, 178, 180-184, 186, 188-192, 194, 196-200, 202, 204-208, 210, 212-216, 218, 220-224, 226, 228-232, 234, 236-240, 242, 244-248, 250, 252-256	108 alternatives, by macroanalysis of $\mathbf{p}_{rel}$ then contraction and expansion.
$\mathbf{q}_{abs}$	36, 39-40, 42, 44-48, 52, 55-56, 58, 60-64, 100, 103-104, 106, 108-112, 116, 119-120, 122, 124-128, 132, 135-136, 138, 140-144, 148, 151-152, 154, 156-160, 164, 167-168, 170, 172-176, 180, 183-184, 186, 188-192, 196, 199-200, 202, 204-208, 212, 215-216, 218, 220-224, 228, 231-232, 234, 236-240, 244, 247-248, 250, 252-256	108 alternatives, by macroanalysis of $\mathbf{q}_{rel}$ then contraction and expansion.
not-$\mathbf{p}_{abs}$	2-49, 51, 57, 59, 65-97, 99, 105, 107, 113, 115, 121, 123, 129-145, 147, 153, 155, 161-177, 179, 185, 187, 193, 195, 201, 203, 209, 211, 217, 219, 225, 227, 233, 235, 241, 243, 249, 251	All alternatives but those of $\mathbf{p}_{abs}$, i.e. 147 cases.
not-$\mathbf{q}_{abs}$	2-35, 37-38, 41, 43, 49-51, 53-54, 57, 59, 65-99, 101-102, 105, 107, 113-115, 117-118, 121, 123, 129-131, 133-134, 137, 139, 145-147, 149-150, 153, 155, 161-163, 165-166, 169, 171, 177-179, 181-182, 185, 187, 193-195, 197-198, 201, 203, 209-211, 213-214, 217, 219, 225-227, 229-230, 233, 235, 241-243, 245-246, 249, 251	All alternatives but those of $\mathbf{q}_{abs}$, i.e. 147 cases.

Table 12.4 continued.

Joints (absolute) and their negations:		
mn	34, 37-38, 130, 133-134, 162, 165-166	Their 9 common alternatives, by intersection.
$\mathbf{mq}_{abs}$	36, 39-40, 42, 44-48, 132, 135-136, 138, 140-144, 164, 167-168, 170, 172-176	Their 27 common alternatives, by intersection.
$\mathbf{np}_{abs}$	50, 53-54, 98, 101-102, 114, 117-118, 146, 149-150, 178, 181-182, 194, 197-198, 210, 213-214, 226, 229-230, 242, 245-246	Their 27 common alternatives, by intersection.
$\mathbf{p}_{abs}\mathbf{q}_{abs}$	52, 55-56, 58, 60-64, 100, 103-104, 106, 108-112, 116, 119-120, 122, 124-128, 148, 151-152, 154, 156-160, 180, 183-184, 186, 188-192, 196, 199-200, 202, 204-208, 212, 215-216, 218, 220-224, 228, 231-232, 234, 236-240, 244, 247-248, 250, 252-256	Their 81 common alternatives, by intersection.
not(mn)	2-33, 35-36, 39-129, 131-132, 135-161, 163-164, 167-256	All alternatives but those of **mn**; i.e. 246 cases.
not($\mathbf{mq}_{abs}$**)**	2-35, 37-38, 41, 43, 49-131, 133-134, 137, 139, 145-163, 165-166, 169, 171, 177-256	all alternatives but those of $\mathbf{mq}_{abs}$; i.e. 228 cases.
not($\mathbf{np}_{abs}$**)**	2-49, 51-52, 55-97, 99-100, 103-113, 115-116, 119-145, 147-148, 151-177, 179-180, 183-193, 195-196, 199-209, 211-212, 215-225, 227-228, 231-241, 243-244, 247-256	All alternatives but those of $\mathbf{np}_{abs}$; i.e. 228 cases.
not($\mathbf{p}_{abs}\mathbf{q}_{abs}$**)**	2-51, 53-54, 57, 59, 65-99, 101-102, 105, 107, 113-115, 117-118, 121, 123, 129-147, 149-150, 153, 155, 161-179, 181-182, 185, 187, 193-195, 197-198, 201, 203, 209-211, 213-214, 217, 219, 225-227, 229-230, 233, 235, 241-243, 245-246, 249, 251	All alternatives but those of $\mathbf{p}_{abs}\mathbf{q}_{abs}$; i.e. 174 cases.
Strong causation and its negation:		
s = m or n	34, 36-40, 42, 44-48, 50, 53-54, 98, 101-102, 114, 117-118, 130, 132-136, 138, 140-144, 146, 149-150, 162, 164-168, 170, 172-176, 178, 181-182, 194, 197-198, 210, 213-214, 226, 229-230, 242, 245-246	Their 63 separate and common alternatives (including overlap, i.e. **mn**), by merger.
not-s = not-m + not-n	2-33, 35, 41, 43, 49, 51-52, 55-97, 99-100, 103-113, 115-116, 119-129, 131, 137, 139, 145, 147-148, 151-161, 163, 169, 171, 177, 179-180, 183-193, 195-196, 199-209, 211-212, 215-225, 227-228, 231-241, 243-244, 247-256	All alternatives but the preceding; i.e. 192 cases.

Table 12.4 continued.

Absolute weak causation and its negation:		
$\mathbf{w}_{abs}$ = $\mathbf{p}_{abs}$ **or** $\mathbf{q}_{abs}$	36, 39-40, 42, 44-48, 50, 52-56, 58, 60-64, 98, 100-104, 106, 108-112, 114, 116-120, 122, 124-128, 132, 135-136, 138, 140-144, 146, 148-152, 154, 156-160, 164, 167-168, 170, 172-176, 178, 180-184, 186, 188-192, 194, 196-200, 202, 204-208, 210, 212-216, 218, 220-224, 226, 228-232, 234, 236-240, 242, 244-248, 250, 252-256	Their 135 separate and common alternatives (including overlap, i.e. $\mathbf{p}_{abs}\mathbf{q}_{abs}$), by merger.
not- $\mathbf{w}_{abs}$ = **not-**$\mathbf{p}_{abs}$ + **not-**$\mathbf{q}_{abs}$	2-35, 37-38, 41, 43, 49, 51, 57, 59, 65-97, 99, 105, 107, 113, 115, 121, 123, 129-131, 133-134, 137, 139, 145, 147, 153, 155, 161-163, 165-166, 169, 171, 177, 179, 185, 187, 193, 195, 201, 203, 209, 211, 217, 219, 225, 227, 233, 235, 241, 243, 249, 251	All alternatives but the preceding; i.e. 120 cases.
Causation (absolute) and its negation:		
$\mathbf{c}_{abs}$ = **m or n or** $\mathbf{p}_{abs}$ **or** $\mathbf{q}_{abs}$	34, 36-40, 42, 44-48, 50, 52-56, 58, 60-64, 98, 100-104, 106, 108-112, 114, 116-120, 122, 124-128, 130, 132-136, 138, 140-144, 146, 148-152, 154, 156-160, 162, 164-168, 170, 172-176, 178, 180-184, 186, 188-192, 194, 196-200, 202, 204-208, 210, 212-216, 218, 220-224, 226, 228-232, 234, 236-240, 242, 244-248, 250, 252-256	Their 144 separate and common alternatives (including overlap).
not-$\mathbf{c}_{abs}$ = **not-m** + **not-n** + **not-**$\mathbf{p}_{abs}$ + **not-**$\mathbf{q}_{abs}$	2-33, 35, 41, 43, 49, 51, 57, 59, 65-97, 99, 105, 107, 113, 115, 121, 123, 129, 131, 137, 139, 145, 147, 153, 155, 161, 163, 169, 171, 177, 179, 185, 187, 193, 195, 201, 203, 209, 211, 217, 219, 225, 227, 233, 235, 241, 243, 249, 251	All alternatives but the preceding; i.e. 111 cases.

The results obtained in Table 12.4 can be made to conveniently stand out by color coding each form's moduses in Table 12.3. This is left to the reader to do.

We need not repeat here what was said before, with reference to the similar table for a two-item framework (Table 12.2); the same comments apply, because the relationships there established are true irrespective of framework. We will, however, highlight something which was less visible before, namely the consistency between various results.

There are never overlaps between contradictory propositions, and their alternatives sum up to 255; also, each generic sums up to two joints (since absolute lones do not exist). For instance, **m** comprises 36 alternative moduses, the 9 of **mn** plus the 27 of $\mathbf{mq}_{abs}$; while **not-m** has the 219 remaining alternatives. Similarly, with regard to **n**. Likewise, $\mathbf{p}_{abs}$ comprises 108 alternatives, the 27 of $\mathbf{np}_{abs}$ plus the 81 of $\mathbf{p}_{abs}\mathbf{q}_{abs}$; while **not-**$\mathbf{p}_{abs}$ has the 147 remaining alternatives. Similarly, with regard to $\mathbf{q}_{abs}$.

Moreover, the number of moduses corresponding to the vaguer forms are predictable. Thus, **s** (= **m or n**) comprises the 36 moduses of **m** plus the 36 of **n**, less the 9 of **mn**[83], a total of 63 alternatives; and its negation has 255 - 63 = 192 alternatives. We can similarly predict the moduses of $\mathbf{w}_{abs}$ (= $\mathbf{p}_{abs}$ **or** $\mathbf{q}_{abs}$) to be 108 + 108 - 81 = 135; and a residue of 120 alternatives for its negation. For **c** (= **s or** $\mathbf{w}_{abs}$) we have 63 + 135 - 2*27 = 144 (the 54 subtracted being those of $\mathbf{mq}_{abs}$ and $\mathbf{np}_{abs}$ - i.e. of $\mathbf{sw}_{abs}$); for its negation, 111.

Thus, incidentally, causation in all its forms covers more than half the matrix, but still leaves a large space to non-causation.

83 So as to avoid double accounting of **mn**, which is implicit in both **m** and **n**.

5. Comparing Frameworks.

Now let us compare the results in Tables 2 and 4. They are essentially the same tables, except that each modus of the first is, as it were, further subdivided into a number of moduses in the second. However, the subdivision is evidently *not* proportional, say in the ratio 16:256; you cannot just say that to each two-item modus there corresponds 16 three-item ones. The following table makes this disproportionality clear:

Table 12.5. Numbers of Moduses for Positive Forms, in Different Frameworks.

Framework	**m**, **n**	$\mathbf{p}_{abs}$, $\mathbf{q}_{abs}$	**mn**	$\mathbf{mq}_{abs}$, $\mathbf{np}_{abs}$	$\mathbf{p}_{abs}\mathbf{q}_{abs}$	**s**	$\mathbf{w}_{abs}$	**c**
Two-Item	2	2	1	1	1	3	3	4
Three-Item	36	108	9	27	81	63	135	144

The explanation is easy. **Expansion of a two-item alternative modus into a number of three-item moduses depends on how many zero or one codes it involves.** For, as we saw in the previous chapter (with the proviso of appropriate locations), each '0' in a two-item framework has a single expression ('0 0') in the three-item framework; whereas each '1' in the former has three expressions in the latter ('0 1', '1 0' or '1 1' - i.e. *any but* '0 0').

Thus, if a two-item modus involves four 'zeros' and no 'one', its three-item equivalent will consist of 1*1*1*1 = 1 (equally impossible) modus; if the former involves three zeros and a one, the latter will consist of 1*1*1*3 = 3 moduses; if the former involves two zeros and two ones, the latter will consist of 1*1*3*3 = 9 moduses; if the former involves one zero and three ones, the latter will consist of 1*3*3*3 = 27 moduses; and if the former involves no zero and four ones, the latter will consist of 3*3*3*3 = 81 moduses.

Whence, the strongs **m**, **n**, which each involves two two-item moduses, one with two zeros (No. 10) and one with a single zero (no. 12 or 14), will have 9 + 27 = 36 three-item moduses; whereas, the weaks $\mathbf{p}_{abs}$, $\mathbf{q}_{abs}$, which each involves two two-item moduses, one with a single zero (no. 12 or 14) and one with no zero (No. 16) and will have 27 + 81 = 108 three-item moduses.

The numbers of three-item moduses for the conjunctions and disjunctions of these forms follow. The joint **mn** (two-item modus No. 10) will have 9 of them; $\mathbf{mq}_{abs}$ (modus No. 12) and $\mathbf{np}_{abs}$ (modus No. 14) will each have 27; and $\mathbf{p}_{abs}\mathbf{q}_{abs}$ (modus 16) will have 81. The vague form **s** (moduses 10, 12, 14) will have 9 + 2*27 = 63; $\mathbf{w}_{abs}$ (moduses 12, 14, 16) will have 2*27 + 81 = 135; and **c** (moduses 10, 12, 14, 16) will have 9 + 2*27 + 81 = 144.

We can proceed in a like manner to predict expansions of negative forms. Furthermore, given the two-item modus(es) of a form, we can predict not only *how many* moduses it will have in a three-item framework, but precisely *which* moduses it will have. Thus, a table of equivalencies between the two frameworks can be constructed without difficulty. In short, we have here a functioning calculus.

The precise three-item modus(es) corresponding to each two-item modus are given in the following table:

Table 12.6. Correspondences between two- and three item frameworks.

Two-item modus	No. of zeros in it	Corresponding three-item modus numbers	No. of moduses
1	4	1	1
2	3	2, 5, 6	3
3	3	3, 9, 11	3
4	2	4, 7, 8, 10, 12-16	9
5	3	17, 65, 81	3
6	2	18, 21-22, 66, 69-70, 82, 85-86	9
7	2	19, 25, 27, 67, 73, 75, 83, 89, 91	9
8	1	20, 23-24, 26, 28-32, 68, 71-72, 74, 76-80, 84, 87-88, 90, 92-96	27
9	3	33, 129, 161	3
10	2	34, 37-38, 130, 133, 134, 162, 165-166	9
11	2	35, 41, 43, 131, 137, 139, 163, 169, 171	9
12	1	36, 39-40, 42, 44-48, 132, 135-136, 138, 140-144, 164, 167-168, 170, 172-176	27
13	2	49, 97, 113, 145, 177, 193, 209, 225, 241	9
14	1	50, 53-54, 98, 101-102, 114, 117-118, 146, 149-150, 178, 181-182, 194, 197-198, 210, 213-214, 226, 229-230, 242, 245-246	27
15	1	51, 57, 59, 99, 105, 107, 115, 121, 123, 147, 153, 155, 179, 185, 187, 195, 201, 203, 211, 217, 219, 227, 233, 235, 243, 249, 251	27
16	0	52, 55-56, 58, 60-64, 100, 103-104, 106, 108-112, 116, 119-120, 122, 124-128, 148, 151-152, 154, 156-160, 180, 183-184, 186, 188-192, 196, 199-200, 202, 204-208, 212, 215-216, 218, 220-224, 228, 231-232, 234, 236-240, 244, 247-248, 250, 252-256	81
16	Total number of moduses		256

Needless to say, each modus will occur only once in the above table, making a total of 16 or 256 moduses, according to the framework. Clearly, if we had developed this table earlier, we could have derived Table 12.4 from Table 12.2.[84]

Obviously, we can follow the same procedures to expand three-item alternative moduses (of which there are 256) into four-item alternative moduses (of which there are 65,536 - as seen earlier).

The *number and configuration* of the latter will emerge from the each of the former, in accordance with the number of zero and one codes it contains and the way they are arrayed

[84] I would like to slip in an unrelated comment here, regarding summary moduses, for the record. We could predict all conceivable summary moduses, within a two-item or three-item framework. Such lists would include all alternative moduses, since summary moduses may also be free of dots, and thus constitute enlarged grand matrices with additional columns and numbers. I do not do this in view of the inadequacies, which we encountered in the previous chapter, of the notation system adopted in this work for summary moduses.

within it (i.e. the incidence, prevalence and locations of zero and one codes in it). A **table of correspondences** can thus be constructed, which details the results obtained in each case.

We have above identified the main lines of what might be called the *two-three* (2/3) table of correspondences, emerging from the operation of *expansion* of '0' into '0 0' and '1' into '0 1', '1 0', '1 1' (all pairs but '0 0'). We could thereafter, step by step, build similar tables of correspondence of size 3/4 or 4/5... and so forth on to infinity, if need arise to resolve eventual issues.

For instance, from a three-item matrix (which has 8 rows) to a four-item matrix, each combination of zeros and ones will result in a product of eight factors of 1 (for '0' codes) or 3 (for '1' codes) - e.g., a modus with 1 zero and 7 ones will become 1*3*3*3*3*3*3*3 = 2187 moduses, in various possible permutations. These are long-winded techniques, which may or may not be needed.

Chapter 13. SOME MORE MICROANALYSES.

1. Relative Weaks.

We have in the previous chapter identified the alternative moduses of the absolute weak determinations and their derivatives. We will here ascertain those of relative weaks and their derivatives. In a two-item framework, relatives are of course indistinguishable from absolutes; they arise only as of a three-item framework.

The following table may be viewed as a continuation of Table 12.4 of the previous chapter; and the modus numbers listed in it refer to the grand matrix in Table 12.3 of the previous chapter. Note well that $\mathbf{p}_{rel}$ and $\mathbf{q}_{rel}$ (and their derivatives with the same suffix), below, refer to partial or contingent causation between P and R ***relative to Q***; that is, P with complement Q are putative causes of R.

Table 13.1. Enumeration of three-item moduses for the relative weak determinations and their derivatives (form PQR).

Determination	Modus numbers	Comment
Relative weaks and their negations:		
$\mathbf{p}_{rel}$	149-152, 157-160, 181-184, 189-192	16 alternatives, by macroanalysis.
$\mathbf{q}_{rel}$	42, 46, 58, 62, 106, 110, 122, 126, 170, 174, 186, 190, 234, 238, 250, 254	16 alternatives, by macroanalysis.
not-$\mathbf{p}_{rel}$	2-148, 153-156, 161-180, 185-188, 193-256	All alternatives but those of $\mathbf{p}_{rel}$, i.e. 239 cases.
not-$\mathbf{q}_{rel}$	2-41, 43-45, 47-57, 59-61, 63-105, 107-109, 111-121, 123-125, 127-169, 171-173, 175-185, 187-189, 191-233, 235-237, 239-249, 251-253, 255-256	All alternatives but those of $\mathbf{q}_{rel}$, i.e. 239 cases.
Joints (relative) and their negations:		
mn	34, 37-38, 130, 133-134, 162, 165-166	Their 9 common alternatives.
m$\mathbf{q}_{rel}$	42, 46, 170, 174	Their 4 common alternatives.
n$\mathbf{p}_{rel}$	149-150, 181-182	Their 4 common alternatives.
$\mathbf{p}_{rel}\mathbf{q}_{rel}$	190	Their 1 common alternatives.
not(mn)	2-33, 35-36, 39-129, 131-132, 135-161, 163-164, 167-256	All alternatives but those of **mn**; i.e. 246 cases.
not(m$\mathbf{q}_{rel}$)	2-41, 43-45, 47-169, 171-173, 175-256	All alternatives but those of **m**$\mathbf{q}_{rel}$; i.e. 251 cases.
not(n$\mathbf{p}_{rel}$)	2-148, 151-180, 183-256	All alternatives but those of **n**$\mathbf{p}_{rel}$; i.e. 251 cases.
not($\mathbf{p}_{rel}\mathbf{q}_{rel}$)	2-189, 191-256	All alternatives but those of $\mathbf{p}_{rel}\mathbf{q}_{rel}$; i.e. 254 cases.

Table 13.1 continued.

Relative lones and their negations:		
m-alone$_{rel}$	36, 39-40, 44-45, 47-48, 132, 135-136, 138, 140-144, 164, 167-168, 172-173, 175-176	The 23 common alternatives of **m**, **not-n**, and **not-q**$_{rel}$.
n-alone$_{rel}$	50, 53-54, 98, 101-102, 114, 117-118, 146, 178, 194, 197-198, 210, 213-214, 226, 229-230, 242, 245-246	The 23 common alternatives of **n**, **not-m**, and **not-p**$_{rel}$.
p-alone$_{rel}$	151-152, 157-160, 183-184, 189, 191-192	The 11 common alternatives of **p**$_{rel}$, **not-n**, and **not-q**$_{rel}$.
q-alone$_{rel}$	58, 62, 106, 110, 122, 126, 186, 234, 238, 250, 254	The 11 common alternatives of **q**$_{rel}$, **not-m**, and **not-p**$_{rel}$.
not(m-alone$_{rel}$**)**	2-35, 37-38, 41-43, 46, 49-131, 133-134, 137, 139, 145-163, 165-166, 169-171, 174, 177-256	All alternatives but those of **m-alone**$_{rel}$; i.e. 232 cases.
not(n-alone$_{rel}$**)**	2-49, 51-52, 55-97, 99-100, 103-113, 115-116, 119-145, 147-177, 179-193, 195-196, 199-209, 211-212, 215-225, 227-228, 231-241, 243-244, 247-256	All alternatives but those of **n-alone**$_{rel}$; i.e. 232 cases.
not(p-alone$_{rel}$**)**	2-150, 153-156, 161-182, 185-188, 190, 193-256	All alternatives but those of **p-alone**$_{rel}$; i.e. 244 cases.
not(q-alone$_{rel}$**)**	2-57, 59-61, 63-105, 107-109, 111-121, 123-125, 127-185, 187-233, 235-237, 239-249, 251-253, 255-256	All alternatives but those of **q-alone**$_{rel}$; i.e. 244 cases.
Relative weak causation and its negation:		
w$_{rel}$ = **p**$_{rel}$ **or q**$_{rel}$	42, 46, 58, 62, 106, 110, 122, 126, 149-152, 157-160, 170, 174, 181-184, 186, 189-192, 234, 238, 250, 254	Their 31 separate and common alternatives (including overlap, i.e. **p**$_{rel}$**q**$_{rel}$ = 1).
p$_{rel}$ + **not-q**$_{rel}$	149-152, 157-160, 181-184, 189, 191-192	Their 15 common alternatives.
not-p$_{rel}$ + **q**$_{rel}$	42, 46, 58, 62, 106, 110, 122, 126, 170, 174, 186, 234, 238, 250, 254	Their 15 common alternatives.
not-w$_{rel}$ = **not-p**$_{rel}$ + **not-q**$_{rel}$	2-41, 43-45, 47-57, 59-61, 63-105, 107-109, 111-121, 123-125, 127-148, 153-156, 161-169, 171-173, 175-180, 185, 187-188, 193-233, 235-237, 239-249, 251-253, 255-256	All alternatives but those of **w**$_{rel}$; i.e. 224 cases.
Contributory causation (relative) and its negation:		
m or p$_{rel}$	34, 36-40, 42, 44-48, 130, 132-136, 138, 140-144, 149-152, 157-160, 162, 164-168, 170, 172-176, 181-184, 189-192	Their 52 separate alternatives (no overlap).
not-m + not-p$_{rel}$	2-33, 35, 41, 43, 49-129, 131, 137, 139, 145-148, 153-156, 161, 163, 169, 171, 177-180, , 185-188, 193-256	All alternatives but the preceding; i.e. 203 cases.
Possible causation (relative) and its negation:		
n or q$_{rel}$	34, 37-38, 42, 46, 50, 53-54, 58, 62, 98, 101-102, 106, 110, 114, 117-118, 122, 126, 130, 133-134, 146, 149-150, 162, 165-166, 170, 174, 178, 181-182, 186, 190, 194, 197-198, 210, 213-214, 226, 229-230, 234, 238, 242, 245-246, 250, 254	Their 52 separate alternatives (no overlap).

Table 13.1 continued.

not-n + not-q_{rel}	2-33, 35-36, 39-41, 43-45, 47-49, 51-52, 55-57, 59-61, 63-97, 99-100, 103-105, 107-109, 111-113, 115-116, 119-121, 123-125, 127-129, 131-132, 135-145, 147-148, 151-161, 163-164, 167-169, 171-173, 175-177, 179-180, 183-185, 187-189, 191-193, 195-196, 199-209, 211-212, 215-225, 227-228, 231-233, 235-237, 239-241, 243-244, 247--249, 251-253, 255-256	All alternatives but the preceding; i.e. 203 cases.
Causation (relative) and its negation:		
c_{rel} = **m or n or p_{rel} or q_{rel}**	34, 36-40, 42, 44-48, 50, 53-54, 58, 62, 98, 101-102, 106, 110, 114, 117-118, 122, 126, 130, 132-136, 138, 140-144, 146, 149-152, 157-160, 162, 164-168, 170, 172-176, 178, 181-184, 186, 189-192, 194, 197-198, 210, 213-214, 226, 229-230, 234, 238, 242, 245-246, 250, 254	Their 86 separate and common alternatives (including overlap).
not-c_{rel} = **not-m + not-n** **+ not-p_{rel} + not-q_{rel}**	2-33, 35, 41, 43, 49, 51-52, 55-57, 59-61, 63-97, 99-100, 103-105, 107-109, 111-113, 115-116, 119-121, 123-125, 127-129, 131, 137, 139, 145, 147-148, 153-156, 161, 163, 169, 171, 177, 179-180, 185, 187-188, 193, 195-196, 199-209, 211-212, 215-225, 227-228, 231-233, 235-237, 239-241, 243-244, 247-249, 251-253, 255-256	All alternatives but the preceding; i.e. 169 cases.

Now, let us compare the above results for relative weaks to those for absolute weaks in Table 12.4 of the previous chapter. The logical properties of these forms are quite distinct. When we unravel the summary modus μμμμ.μ.μ of $\mathbf{p}_{abs}$, we obtain 108 alternative moduses; similarly, the summary modus μ.μ.μμμμ of $\mathbf{q}_{abs}$ yields 108 alternative moduses. In contrast, the summaries of $\mathbf{p}_{rel}$ and $\mathbf{q}_{rel}$ - namely, 10.1.1.. and ..1.1.01 - give rise to 16 alternatives each.

The first thing to note is that the 16 moduses of $\mathbf{p}_{rel}$ are all *included in* the 108 of $\mathbf{p}_{abs}$; and likewise, the 16 of $\mathbf{q}_{rel}$ are *among* the 108 of $\mathbf{q}_{abs}$. Look at the tables, and see this for yourself. What this means is that the positive relative weaks *imply* and are *species of* the positive absolute weaks.

Moreover, note that the latter are more than twice as broad in possibilities than the former. This reveals to us that $\mathbf{p}_{PR}$ is not merely the sum of $\mathbf{p}_{Q}$ and $\mathbf{p}_{notQ}$, i.e. that "P (with whatever complement) is a partial cause of R" means more than "P (whether with complement Q or notQ) is a partial cause of R"; similarly, regarding **q**. We shall list the precise moduses of $\mathbf{p}_{notQ}$ and $\mathbf{q}_{notQ}$ further on; but we can predict at the outset that they will be 16 in number in each case, by the demands of symmetry. Therefore, absolute weak causation between P and R can occur with complements other than Q or notQ; and ***we cannot engage in dilemmatic arguments, saying that if Q is not the complement, notQ must be it***. It is wise to keep that in mind.

Consequently, the negations of the relative weaks are broader than those of the corresponding absolute weaks; the former involve 239 (255 - 16) alternative moduses each, the latter only 147 (255 - 108) among these.[85]

---

[85] We can see here why relative weaks should not be listed in a two-item framework. In their positive generic forms, they would have the same alternative moduses as the absolute weaks (though in fact, as we know with reference to the three-item framework, covering *only part of* these moduses). However, when such two-item moduses are negated, the similarity between relatives and absolutes

Consider now the **relative joint** determinations: $\mathbf{mq}_{rel}$ and $\mathbf{np}_{rel}$ have only 4 moduses each, while the corresponding absolute joints $\mathbf{mq}_{abs}$ and $\mathbf{np}_{abs}$ have 27 each; and $\mathbf{p}_{rel}\mathbf{q}_{rel}$ has only 1 modus, in contrast to the 81 of $\mathbf{p}_{abs}\mathbf{q}_{abs}$. Thus, as we move from absolute to relative determination, we narrow down the possibilities, we get more specific. On the negative side, the possibilities are broadened, from 228 to 251 or 174 to 254.

We saw in the previous chapter that absolute lone determinations do not exist, for the simple reason that their constituents have no common modus. On the other hand, as can be seen above, **relative lone** determinations do indeed exist, since their constituents have common moduses, 23 for the strongs and 11 for the weaks.

But the latter concepts are of course not as significant as the former. For as we can see with reference to the moduses involved, the relative lones - together with the relative joints - are merely species of (i.e. are all included in) the absolute joints; that is:

- $\mathbf{m\text{-}alone}_{rel}$ + $\mathbf{mq}_{rel}$ (23 + 4) = $\mathbf{mq}_{abs}$ (27, i.e. the 36 of **m** less the 9 of **mn**);
- $\mathbf{n\text{-}alone}_{rel}$ + $\mathbf{np}_{rel}$ (23 + 4) = $\mathbf{np}_{abs}$ (27, i.e. the 36 of **n** less the 9 of **mn**);
- $\mathbf{p\text{-}alone}_{rel}$ + $\mathbf{q\text{-}alone}_{rel}$ + $\mathbf{p}_{rel}\mathbf{q}_{rel}$ (11 + 11 + 1 = 23) *imply* $\mathbf{p}_{abs}\mathbf{q}_{abs}$ (81).

Thus, whereas $\mathbf{w}_{abs}$ = $\mathbf{mq}_{abs}$ **or** $\mathbf{np}_{abs}$ **or** $\mathbf{p}_{abs}\mathbf{q}_{abs}$, we must equate $\mathbf{w}_{rel}$ to $\mathbf{mq}_{rel}$ **or** $\mathbf{np}_{rel}$ **or** $\mathbf{p}_{rel}\mathbf{q}_{rel}$ **or** $\mathbf{p\text{-}alone}_{rel}$ **or** $\mathbf{q\text{-}alone}_{rel}$; check it out with reference to the moduses involved. Note that $\mathbf{w}_{rel}$ involves only 31 moduses, the 15 of $\mathbf{p}_{rel}$ + $\mathbf{not\text{-}q}_{rel}$, the 15 of $\mathbf{not\text{-}p}_{rel}$ + $\mathbf{q}_{rel}$, and the 1 of $\mathbf{p}_{rel}\mathbf{q}_{rel}$. This is in contrast to $\mathbf{w}_{abs}$ which has 135 (the same 31, and 103 more besides). Consequently, $\mathbf{not\text{-}w}_{rel}$ has 224 moduses, including all 120 of $\mathbf{not\text{-}w}_{abs}$.

We saw in the previous chapter that contributory causation, possible causation and causation *tout court* are one and the same concept with regard to absolute weaks, all with the same 144 moduses. But with regard to relative weaks, they are different concepts, as the above table clearly shows.

The relative form of contributory causation "**m or** $\mathbf{p}_{rel}$" has 52 moduses, and that of possible causation "**n or** $\mathbf{q}_{rel}$" has 52, while relative causation "**m or n or** $\mathbf{p}_{rel}$ **or** $\mathbf{q}_{rel}$" involves 86. The latter 86 moduses comprise the preceding 52 + 52, *minus* the 18 moduses of the four relative joint determinations (their overlaps); and all these moduses are of course included in the list of 144 for absolute causation.

The moduses of the negations of these three relative forms follow, as shown in our table. Note especially that negation of relative causation, $\mathbf{not\text{-}c}_{rel}$ (169 moduses), does *not* imply negation of absolute causation, $\mathbf{not\text{-}c}_{abs}$ (111 moduses); but instead, the latter implies and is a species of the former, including all its moduses and more.

We need not mention in the above table the combinations (**m** + $\mathbf{not\text{-}p}_{rel}$), (**n** + $\mathbf{not\text{-}q}_{rel}$), (**not-m** + $\mathbf{p}_{rel}$), (**not-n** + $\mathbf{q}_{rel}$), because, as can be seen with reference to the common moduses of the positive and negative forms constituting them, they are respectively equivalent to **m**, **n**, $\mathbf{p}_{rel}$, $\mathbf{q}_{rel}$.

The remaining combinations are not mentioned because they are not particularly interesting. This refers to (**m or** $\mathbf{q}_{rel}$), comprising the 4 moduses of $\mathbf{mq}_{rel}$ plus the 32 of "**m** + $\mathbf{not\text{-}q}_{rel}$" plus the 12 of "**not-m** + $\mathbf{q}_{rel}$", a total 48 alternatives; and to "**n or** $\mathbf{p}_{rel}$", comprising the 4 moduses of $\mathbf{np}_{rel}$ plus the 32 of "**n** + $\mathbf{not\text{-}p}_{rel}$" plus the 12 of "**not-n** + $\mathbf{p}_{rel}$", a total 48 alternatives; as well as to their respective negations, "**not-m** + $\mathbf{not\text{-}q}_{rel}$" and "**not-n** + $\mathbf{not\text{-}p}_{rel}$", which involve 207 moduses each.

---

would cease, and we would be led astray, unaware that negative relatives are *broader* than negative absolutes.

2. Items of Negative Polarity in Two-Item Framework.

The grand matrices, in which the various forms of causative propositions are embedded, are equally the habitats of similar propositions involving like items but of negative polarity. Such propositions need also to be microanalyzed, for reasons which will be become apparent after we do so. The job is rather easy, involving a mere reshuffling of the summary moduses of propositions with items of positive polarity.

Let us consider, to begin with, the positive generic forms in a two-item framework (strongs or absolute weaks only - relative weaks being indistinguishable here), with reference to Table 12.1 of the previous chapter (turn to it, and note well that it has P and R as column headings for items).

We have previously ascertained the summary moduses of generics with items 'P.R'; our task here is to find out those for the same forms with items 'notP.notR', 'P.notR' and 'notP.R'. Symbolically, such forms can be distinguished by changes in suffix. Thus, for complete causation, symbol **m**, we would write $\mathbf{m}_{PR}$, $\mathbf{m}_{notPnotR}$, $\mathbf{m}_{PnotR}$, and $\mathbf{m}_{notPR}$, according to the sequence of items intended; similarly for **n**, **p**, **q** - each form gives rise to four.

Now, if we changed the column headings of the said table from P.R to some other combination (notP.notR, P.notR or notP.R), the modus numbers (labels) applicable to each form would remain the same but change meanings (i.e. refer to different arrays of an equal number of 0 and 1 codes), and we would not be able to compare same forms with different suffixes.

What we need to do, rather, is retain **the same grand matrix** (*the one for positive items P.R*), and locate within it the moduses of the forms we want to compare. This grand matrix has four rows, which we may label a-d, in which the PR sequences are 11 (both present), 10 (P present, R absent), 01 (P absent, R present), and 00 (both absent).

If we wish to refer to this same matrix as our **standard framework**, for forms with an item of different polarity, we must refer to a different rows. Clearly, notP = 1 is the same as P = 0, and notP = 0 is the same as P = 1; similarly with respect to notR. Thus, the reshuffling of rows is therefore predictable, as follows:

Table 13.2. Row references in a standard (PR) matrix for different polarities of items.

Row in PR matrix	Row label	Sequences for different polarities of items							
		PR		notPnotR		PnotR		notPR	
PR	a	11	a	00	d	10	b	01	c
PnotR	b	10	b	01	c	11	a	00	d
notPR	c	01	c	10	b	00	d	11	a
notPnotR	d	00	d	11	a	01	c	10	b

Consider **m**, for instance. Whereas the summary modus for $\mathbf{m}_{PR}$ is abcd = 10.1 (as previously ascertained by macroanalysis, yielding alternative moduses Nos. 10, 12 after unraveling) - for $\mathbf{m}_{notPnotR}$ it will be the mirror image dcba = 1.01 (moduses 10, 14); for $\mathbf{m}_{PnotR}$ it will be badc = 011. (moduses 7, 8); and for $\mathbf{m}_{notPR}$ it will be the mirror image cdab = .110 (moduses 7, 15). That is, knowing the summary modus for $\mathbf{m}_{PR}$ to be 10.1 (1 in row a, 0 in row b, • in row c,

and 1 in row d), we can predict it for all the other forms of **m** by merely reshuffling the rows as indicated in the above table. Similarly, with regard to **n**, **p**, **q**.
We can in this manner, without much effort, identify the summary and alternative moduses in a standard two-item grand matrix of the positive generic forms (and thence, if need be, of all other forms, using the processes of negation, intersection and merger). The following table presents the desired information without further ado:

Table 13.3. Enumeration of moduses of positive generic forms with different polarities of items, with reference to standard two-item (PR) grand matrix.

		Causation		Prevention	
Determination	Moduses	PR	notPnotR	PnotR	notPR
m	summary	10.1	1.01	011.	.110
	alternative	10, 12	10, 14	7, 8	7, 15
n	summary	1.01	10.1	.110	011.
	alternative	10, 14	10, 12	7, 15	7, 8
$\mathbf{p}_{abs}$	summary	11.1	1.11	111.	.111
	alternative	14, 16	12, 16	15, 16	8, 16
$\mathbf{q}_{abs}$	summary	1.11	11.1	.111	111.
	alternative	12, 16	14, 16	8, 16	15, 16

All the above table is inferable from the preceding table, given the summary moduses of **m** and $\mathbf{p}_{abs}$. Notice the identities between the moduses of pairs of forms with different suffixes. Thus, $\mathbf{m}_{PR}$ and $\mathbf{n}_{notPnotR}$ are identical; as are $\mathbf{m}_{notPnotR}$ and $\mathbf{n}_{PR}$; likewise, $\mathbf{m}_{PnotR} = \mathbf{n}_{notPR}$, and $\mathbf{m}_{notPR} = \mathbf{n}_{PnotR}$. Similarly with regard to the weaks, $\mathbf{p}_{PR}$ and $\mathbf{q}_{notPnotR}$, etc. These identities simply signify that, as we already know, these pairs of forms are inverses of each other. Notice also the mirror images (same string in opposite directions), like for example $\mathbf{m}_{PR}$ and $\mathbf{n}_{PR}$, which have the same significance.
These equations allow us to see that forms in PR and notPnotR are closely associated, by mirroring; and similarly for forms in PnotR and notPR. Furthermore, that the former and latter pairs are in turn associated, in another sense, insofar as the first and last digits of the summary modus for the one are identical to the middle digits of it for the other, and vice-versa. Clearly, whatever the respective polarities of the items, their relations remain essentially causative.
All these forms therefore embody similar concepts in different guises, signifying various types and degrees of bondage or cohesion between the items concerned; they have common aspects and are all logically or structurally interrelated. They form a *family* of propositions. We have so far in our study concentrated on items PR or notPnotR, but given little attention to items PnotR or notPR in view of their similarities and the derivability of their logical properties. But now let us look upon them as distinct paradigms.
All these forms may be classified as 'causative relations', in the broad sense we ultimately understand for this term. Yet we have in the present study gotten used to a more restrictive sense of the term 'causation', as meaning specifically PR or notPnotR relations. Granting this, we need another term to refer specifically to PnotR and notPR relations; and yet another term to refer to the broad, all-inclusive sense.
Therefore, I propose the following convention, in the appropriate contexts. PR or notPnotR causative relations will be called causation (restrictive sense), while PnotR and notPR

causative relations will be called **prevention**[86]. Thus, "P *prevents* R" is to mean "P causes notR" (still in the restrictive sense of causation). Both causation and prevention are species of causative relations in a broad sense; but when we want to avoid confusion let us call the latter genus of both, say, **connection**[87].

We would thus say that two items P and R are *connected*, if either item or its negation causes (in the restrictive sense) or prevents the other item or its negation. And just as causation may vary in determination, i.e. be complete, necessary, partial or contingent - so may prevention be subdivided.

My purpose here is to make the reader aware that when we speak of causation in a wide sense, we must mentally include both causation in a narrow sense and its family relative prevention. Similarly, note well, if we speak of *non*causation, we must know whether we mean negation of causation in a restrictive sense (which does not imply negation of prevention) or negation of all causative relation, i.e. of connection (which implies negation of both causation and prevention).

However, before we adopt such loaded terminology, let us examine the relationships involved more closely. As will be seen, we will have to qualify our statement somewhat.

As we stressed from the word go, causation (and similarly, of course, prevention) formally implies the *contingency* of the items it involves: i.e. each of the items considered separately must be possible but not necessary[88] for a causative relation between them to be conceivable. If one or more of the items involved is/are not contingent, the other item(s) cannot be causing or caused by it. But it does not follow that any two contingent items are causatively related.

Now, according to our analysis so far, the two-item moduses of causation are four, viz. Nos. 10, 12, 14, 16 (and of noncausation are eleven: Nos. 2-9, 11, 13, 15), those of prevention are four, viz. Nos. 7-8, 15-16 (and of nonprevention are eleven: Nos. 2-6, 9-14. Note that these positives have **one common modus, No. 16 (1111)**, which means that causation and prevention are, in this instance (namely, $\mathbf{p}_{abs}\mathbf{q}_{abs}$, i.e. *absolute* **pq**, note well), overlapping and compatible. It follows that the two-item moduses of connection are seven, viz. Nos. 7-8, 10, 12, 14-16 (and of nonconnection are eight: Nos. 2-6, 9, 11, 13).

Next, look again at Table 12.1 of the previous chapter. The question may well be asked: *what is so special about* the above-mentioned moduses of connection (as tentatively defined)? That is, *what distinguishes* them from the moduses of nonconnection? Let us look for an answer in ***the number of cells coded 1 or 0*** in their alternative moduses.

Connection refers to moduses with four 1s (No. 16), three 1s and one 0 (Nos. 8, 12, 14-15), or two 1s and two 0s (Nos. 7, 10). Nonconnection has moduses with two 1s and two 0s (Nos. 4, 6, 11, 13), or one 1 and three 0s (Nos. 2, 3, 5, 9). Thus, though connection is distinguishable by its comprising moduses with three or four 1s, and nonconnection through moduses with only one 1, they both have moduses with two 1s!

However, we need not be surprised or alarmed. For moduses #s 2, 3, 4 mean that P is impossible (they have code 0 for it, with or without R), and moduses #s 5, 9, 13 mean that P is necessary (i.e. that notP is impossible). Similarly, moduses #s 2, 5, 6 mean that R is

---

86 Any synonym, like hindrance, obstruction, forestalling, inhibition, counteraction, etc., would do as well; though some of these have slightly different connotations - more active or passive, or psychological or ethical, rather than natural, and so forth. Prevention is to be understood in a very general sense, here.

87 I use this term in another (though not unrelated) sense in *Future Logic* (see p. 124), with reference to conditional propositions.

88 In the mode concerned.

impossible (coded 0, whether P is present or absent), and moduses #s 3, 9, 11 mean that R is necessary (i.e. that notR is impossible).

Thus, all the moduses of nonconnection refer to situations where one or two items is/are *incontingent*, which means present or absent (as the case may be) *independently of* any other item. In its moduses with three zeros (Nos. 2, 3 5, 9), two items are incontingent; in those with two zeros (Nos. 4, 6, 11, 13), one item is incontingent. In contrast, connection never involves an incontingent item.

Therefore, by this reasoning, connection could be conceptually distinguished from nonconnection with reference to the contingency of both items or to the incontingency of one or the other of them, respectively. But this is nonsensical: *it would mean that any two contingent items are necessarily causatively related*! Clearly, we must have misinterpreted some relevant fact.

It is this: *the last modus of any grand matrix, i.e. the modus involving only 1s*, i.e. modus #16 in a two-item framework (similarly, modus #256 for three items, or #65,536 for four items), *does not necessarily signify causation* (or prevention or connection). For no matter whether the items concerned or their negations are together or apart, the combination is always 'possible' (i.e. coded 1) in this modus. ***So we cannot in fact tell with reference to this uniform modus alone whether the items concerned have any impact on each other.***

It follows that in this special case, we must interpret the modus as indicative of *possible* causation (or prevention or connection); but there *may* also in some cases turn out to be neither causation nor prevention (i.e. nonconnection). That is to say, the last modus (with all 1s) is ***indefinite*** with regard to connection (or causation or prevention) or nonconnection (or noncausation or non prevention). The last modus is in all frameworks included in the form $\mathbf{p}_{abs}\mathbf{q}_{abs}$, and indeed in $\mathbf{c}_{abs}$, but when we consider more than two items, it is not part of $\mathbf{p}_{rel}\mathbf{q}_{rel}$, or of $\mathbf{c}_{rel}$.

This new finding is in agreement with common sense. Taking any two items at random, we cannot reasonably say that they are either (a) both contingent and causatively connected or (b) one or both incontingent and therefore not causatively connected. There is still another possibility: that (c) they are both contingent and *yet not* causatively connected. This possibility is inherent, as already stated, in the 'last modus' of any matrix, which being composed only of 1s, cannot be definitely interpreted one way or the other.

This realization leaves us a window of opportunity for eventual development of a concept of **spontaneity** (i.e. chance, and perhaps also freewill). For if we are unable to find for some contingent item any other contingent item with which we may causatively relate it in some way, we may be in the long run allowed to inductively generalize from this "failure to find despite due diligence in searching" to a presumed "spontaneity". Obviously, if we opt for the postulate of a "law of universal causation", such a movement of thought becomes illicit. But granting that such a law is itself a product of generalization, we have some freedom of choice in the matter. These important insights will naturally affect our later investigations.

(See discussion in Chapter 16.2, including Table 16.1)

3. Items of Negative Polarity in Three-Item Framework.

All the above can be repeated in a three-item framework. In following table, which concerns strongs and absolute weaks (relative weaks will be dealt with further on), the summary moduses are obtained from those given in Table 13.3 above, by expansion[89]; and the alternative moduses are derived from those given in that table, by applying the correspondences between two- and three- item frameworks developed in Table 12.6 of the previous chapter.

Table 13.4. Enumeration of moduses of strong and absolute weak determinations with different polarities of items, with reference to standard three-item (PQR) grand matrix.

	Causation		Prevention	
Determination	PR	notPnotR	PnotR	notPR
m	.0.0....	....0.0.	0.0.....	.....0.0
	34, 36-40, 42, 44-48, 130, 132-136, 138, 140-144, 162, 164-168, 170, 172-176	34, 37-38, 50, 53-54, 98, 101-102, 114, 117-118, 130, 133-134, 146, 149-150, 162, 165-166, 178, 181-182, 194, 197-198, 210, 213-214, 226, 229-230, 242, 245-246	19-20, 23-32, 67-68, 71-80, 83-84, 87-96	19, 25, 27, 51, 57, 59, 67, 73, 75, 83, 89, 91, 99, 105, 107, 115, 121, 123, 147, 153, 155, 179, 185, 187, 195, 201, 203, 211, 217, 219, 227, 233, 235, 243, 249, 251
n	....0.0.	.0.0....	.....0.0	0.0.....
	34, 37-38, 50, 53-54, 98, 101-102, 114, 117-118, 130, 133-134, 146, 149-150, 162, 165-166, 178, 181-182, 194, 197-198, 210, 213-214, 226, 229-230, 242, 245-246	34, 36-40, 42, 44-48, 130, 132-136, 138, 140-144, 162, 164-168, 170, 172-176	19, 25, 27, 51, 57, 59, 67, 73, 75, 83, 89, 91, 99, 105, 107, 115, 121, 123, 147, 153, 155, 179, 185, 187, 195, 201, 203, 211, 217, 219, 227, 233, 235, 243, 249, 251	19-20, 23-32, 67-68, 71-80, 83-84, 87-96
p_{abs}	....	....	....	....
	50, 52-56, 58, 60-64, 98, 100-104, 106, 108-112, 114, 116-120, 122, 124-128, 146, 148-152, 154, 156-160, 178, 180-184, 186, 188-192, 194, 196-200, 202, 204-208, 210, 212-216, 218, 220-224, 226, 228-232, 234, 236-240, 242, 244-248, 250, 252-256	36, 39-40, 42, 44-48, 52, 55-56, 58, 60-64, 100, 103-104, 106, 108-112, 116, 119-120, 122, 124-128, 132, 135-136, 138, 140-144, 148, 151-152, 154, 156-160, 164, 167-168, 170, 172-176, 180, 183-184, 186, 188-192, 196, 199-200, 202, 204-208, 212, 215-216, 218, 220-224, 228, 231-232, 234, 236-240, 244, 247-248, 250, 252-256	51-52, 55-64, 99-100, 103-112, 115-116, 119-128, 147-148, 151-160, 179-180, 183-192, 195-196, 199-208, 211-212, 215-224, 227-228, 231-240, 243-244, 247-256	20, 23-24, 26, 28-32, 52, 55-56, 58, 60-64, 68, 71-72, 74, 76-80, 84, 87-88, 90, 92-96, 100, 103-104, 106, 108-112, 116, 119-120, 122, 124-128, 148, 151-152, 154, 156-160, 180, 183-184, 186, 188-192, 196, 199-200, 202, 204-208, 212, 215-216, 218, 220-224, 228, 231-232, 234, 236-240, 244, 247-248, 250, 252-256

[89] The dots in all the summary moduses of this table are of course meant as μ - as explained in the chapter on piecemeal analysis, in the section on expansion and contraction.

Table 13.4 continued.

q_{abs}	……..	……..	……..	……..
	36, 39-40, 42, 44-48, 52, 55-56, 58, 60-64, 100, 103-104, 106, 108-112, 116, 119-120, 122, 124-128, 132, 135-136, 138, 140-144, 148, 151-152, 154, 156-160, 164, 167-168, 170, 172-176, 180, 183-184, 186, 188-192, 196, 199-200, 202, 204-208, 212, 215-216, 218, 220-224, 228, 231-232, 234, 236-240, 244, 247-248, 250, 252-256	50, 52-56, 58, 60-64, 98, 100-104, 106, 108-112, 114, 116-120, 122, 124-128, 146, 148-152, 154, 156-160, 178, 180-184, 186, 188-192, 194, 196-200, 202, 204-208, 210, 212-216, 218, 220-224, 226, 228-232, 234, 236-240, 242, 244-248, 250, 252-256	20, 23-24, 26, 28-32, 52, 55-56, 58, 60-64, 68, 71-72, 74, 76-80, 84, 87-88, 90, 92-96, 100, 103-104, 106, 108-112, 116, 119-120, 122, 124-128, 148, 151-152, 154, 156-160, 180, 183-184, 186, 188-192, 196, 199-200, 202, 204-208, 212, 215-216, 218, 220-224, 228, 231-232, 234, 236-240, 244, 247-248, 250, 252-256	51-52, 55-64, 99-100, 103-112, 115-116, 119-128, 147-148, 151-160, 179-180, 183-192, 195-196, 199-208, 211-212, 215-224, 227-228, 231-240, 243-244, 247-256

The negations, intersections and mergers of these forms can easily be worked out, if need arise.

Notice repetitions (there are only eight sets of moduses for sixteen forms); they signify inversions (with change in polarity of both items and change in determination). But more broadly, note well all the compatibilities and incompatibilities between these various forms, which tell us which of them can occur in tandem and which cannot. The following tables, derived from the above, highlight these oppositions for **m** and $\mathbf{p}_{abs}$; needless to say, similar tables can be constructed for **n** and $\mathbf{q}_{abs}$, *mutatis mutandis*.

Table 13.5. Oppositions between $\mathbf{m}_{PR}$ and the other generic forms.

Forms compared		Compatibility	Common moduses
PR	PR		
m	**m**	yes	all 36
m	**n**	yes	the 9 of **mn**
m	$\mathbf{p}_{abs}$	no	None
m	$\mathbf{q}_{abs}$	yes	the 27 of $\mathbf{mq}_{abs}$
PR	notPnotR		
m	**m**	yes	the 9 of **mn**
m	**n**	yes	all 36
m	$\mathbf{p}_{abs}$	yes	the 27 of $\mathbf{mq}_{abs}$
m	$\mathbf{q}_{abs}$	no	None
PR	PnotR		
m	**m**	no	None
m	**n**	no	None
m	$\mathbf{p}_{abs}$	no	None
m	$\mathbf{q}_{abs}$	no	None

Table 13.5 continued.

PR	notPR		
m	**m**	no	None
m	**n**	no	None
m	$\mathbf{p}_{abs}$	no	None
m	$\mathbf{q}_{abs}$	no	None

Similarly for **n**, *mutatis mutandis*. Notice that the forms of strong causation and of prevention have no moduses in common, and are therefore incompatible. But *within* either causation or prevention, there are certain compatibilities.

Table 13.6. Oppositions between $\mathbf{p}_{PR}$ and the other generic forms.

Forms compared		Compatibility	Common moduses
PR	PR		
$\mathbf{p}_{abs}$	**m**	no	None
$\mathbf{p}_{abs}$	**n**	yes	the 27 of $\mathbf{np}_{abs}$
$\mathbf{p}_{abs}$	$\mathbf{p}_{abs}$	yes	all 108
$\mathbf{p}_{abs}$	$\mathbf{q}_{abs}$	yes	the 81 of $\mathbf{p}_{abs}\mathbf{q}_{abs}$
PR	notPnotR		
$\mathbf{p}_{abs}$	**m**	yes	the 27 of $\mathbf{np}_{abs}$
$\mathbf{p}_{abs}$	**n**	no	None
$\mathbf{p}_{abs}$	$\mathbf{p}_{abs}$	yes	the 81 of $\mathbf{p}_{abs}\mathbf{q}_{abs}$
$\mathbf{p}_{abs}$	$\mathbf{q}_{abs}$	yes	all 108
PR	PnotR		
$\mathbf{p}_{abs}$	**m**	no	None
$\mathbf{p}_{abs}$	**n**	no	None
$\mathbf{p}_{abs}$	$\mathbf{p}_{abs}$	yes	the 81 of $\mathbf{p}_{abs}\mathbf{q}_{abs}$
$\mathbf{p}_{abs}$	$\mathbf{q}_{abs}$	yes	the 81 of $\mathbf{p}_{abs}\mathbf{q}_{abs}$
PR	notPR		
$\mathbf{p}_{abs}$	**m**	no	None
$\mathbf{p}_{abs}$	**n**	no	None
$\mathbf{p}_{abs}$	$\mathbf{p}_{abs}$	yes	the 81 of $\mathbf{p}_{abs}\mathbf{q}_{abs}$
$\mathbf{p}_{abs}$	$\mathbf{q}_{abs}$	yes	the 81 of $\mathbf{p}_{abs}\mathbf{q}_{abs}$

Similarly for $\mathbf{q}_{abs}$, *mutatis mutandis*. Notice that the weak forms of causation and prevention have moduses in common, always the same 81, which are none other than the three-item moduses corresponding to the two-item modus No. 16 (see Table 12.6 of the previous chapter). This is consistent with our earlier finding, that $\mathbf{p}_{abs}\mathbf{q}_{abs}$ has the same modus whatever the polarities of its two items (except where the two forms involved are equivalent).

Now let us consider *relative* weak determinations, which only arise as of a three-item framework. For each PR sequence, and each determination, there are two complements to consider: both Q and notQ. To identify the alternative moduses of each form, we may proceed by consideration of their summary moduses.

We know, from Tables 11.3 and 11.4 of the chapter on piecemeal microanalysis, the summary modus of $\mathbf{p}_{PQR}$ to be "10.1.1.." and that of $\mathbf{q}_{PQR}$ to be "..1.1.01". These are mirror images of each other, note.

Now, the summary moduses of $\mathbf{p}_{PnotQR}$ and $\mathbf{q}_{PnotQR}$ are bound to have the same numbers of zeros, ones and dots; only they will be in a different order, such that Q = 1 (i.e. Q) and Q = 0 (i.e. notQ) are in each other's place. If the eight rows of our matrix are labeled a-h, then keeping the values (1 or 0) of P and R constant, row a will be replaced by c, row b will swap places with d, and likewise e with g and f with h. Thus, we can infer the summary moduses of $\mathbf{p}_{PnotQR}$ and $\mathbf{q}_{PnotQR}$ to be respectively ".110...1" and "1...011."; once again these are mirrors, notice.

Next consider forms with items PQnotR. Using similar reasoning with regard to the change from R to notR, we can predict the pairs of rows which replace each other to be: a b, c d, e f, and g h. Thus, the summary modus of $\mathbf{p}_{PQnotR}$ has to be "011.1..." and that of $\mathbf{q}_{PQnotR}$ "...1.110". Concerning forms with items PnotQnotR, it follows that the summary modus of $\mathbf{p}_{PnotQnotR}$ has to be "1.01..1." and that of $\mathbf{q}_{PnotQnotR}$ ".1..10.1".

Similarly arguing with regard to a change from PQR to notPQR, the pairs are seen to be a e, b f, c g, and d h, so that the summary modus for $\mathbf{p}_{notPQR}$ is ".1..10.1" and that of $\mathbf{q}_{notPQR}$ is "1.01..1.". Concerning forms with items notPnotQR, it follows that the summary modus for $\mathbf{p}_{notPnotQR}$ is "...1.110" and that of $\mathbf{q}_{notPnotQR}$ is "011.1...".

Finally, the forms $\mathbf{p}_{notPQnotR}$ and $\mathbf{q}_{notPQnotR}$ may be derived from, say, those with suffix notPQR (by transposition of adjacent rows); which yields summary moduses "1...011." and ".110...1". We may thence infer the summary moduses of the forms with items notPnotQnotR, to be "..1.1.01" in the case of $\mathbf{p}_{notPnotQnotR}$ and "10.1.1.." for $\mathbf{q}_{notPnotQnotR}$.

We have thus obtained the summary moduses of all forms of **p** and **q** for the items concerned, and can now readily unravel and list their respective alternative moduses. The following table, which may be viewed as a continuation of the preceding, is thereby obtained with reference to the three-item grand matrix (see Table 12.3 of the previous chapter).

Table 13.7. Enumeration of moduses of relative weak determinations with different polarities of items, with reference to standard three-item (PQR) grand matrix.

	Causation		Prevention	
Determination	PR	notPnotR	PnotR	notPR
$\mathbf{p}_{Q}$	10.1.1.. 149-152, 157-160, 181-184, 189-192	1...011. 135-136, 151-152, 167-168, 183-184, 199-200, 215-216, 231-232, 247-248	011.1... 105-112, 121-128	.1..10.1 74, 76, 90, 92, 106, 108, 122, 124, 202, 204, 218, 220, 234, 236, 250, 252
$\mathbf{q}_{notQ}$	1...011. 135-136, 151-152, 167-168, 183-184, 199-200, 215-216, 231-232, 247-248	10.1.1.. 149-152, 157-160, 181-184, 189-192	.1..10.1 74, 76, 90, 92, 106, 108, 122, 124, 202, 204, 218, 220, 234, 236, 250, 252	011.1... 105-112, 121-128
$\mathbf{q}_{Q}$	..1.1.01 42, 46, 58, 62, 106, 110, 122, 126, 170, 174, 186, 190, 234, 238, 250, 254	.110...1 98, 100, 102, 104, 106, 108, 110, 112, 226, 228, 230, 232, 234, 236, 238, 240	...1.110 23, 31, 55, 63, 87, 95, 119, 127, 151, 159, 183, 191, 215, 223, 247, 255	1.01..1. 147-148, 151-152, 155-156, 159-160, 211-212, 215-216, 219-220, 223-224

Table 13.7 continued.

$\mathbf{p}_{notQ}$	.110...1	..1.1.01	1.01..1.	...1.110
	98, 100, 102, 104, 106, 108, 110, 112, 226, 228, 230, 232, 234, 236, 238, 240	42, 46, 58, 62, 106, 110, 122, 126, 170, 174, 186, 190, 234, 238, 250, 254	147-148, 151-152, 155-156, 159-160, 211-212, 215-216, 219-220, 223-224	23, 31, 55, 63, 87, 95, 119, 127, 151, 159, 183, 191, 215, 223, 247, 255

The negations, intersections and mergers of these forms, with each other and with strongs, can easily if need arise be worked out.

Notice repetitions (there are eight sets for sixteen forms); they signify inversions (with change in polarity of all three items and change in determination). But more broadly, note well all the compatibilities or incompatibilities between the various forms of relative weak connection, *which tell us which of them can occur in tandem and which cannot.* The following table shows, for example, which forms can be conjoined or not with $\mathbf{p}_{PQR}$.

Table 13.8. Oppositions between $\mathbf{p}_{PQR}$ and the other relative weaks.

Forms compared		Compatibility	Common moduses
PR	PR		
$\mathbf{p}_Q$	$\mathbf{p}_Q$	yes	all
$\mathbf{p}_Q$	$\mathbf{q}_Q$	yes	190
$\mathbf{p}_Q$	$\mathbf{p}_{notQ}$	no	none
$\mathbf{p}_Q$	$\mathbf{q}_{notQ}$	yes	151-152, 183-184
PR	notPnotR		
$\mathbf{p}_Q$	$\mathbf{p}_Q$	yes	151-152, 183-184
$\mathbf{p}_Q$	$\mathbf{q}_Q$	no	none
$\mathbf{p}_Q$	$\mathbf{p}_{notQ}$	yes	190
$\mathbf{p}_Q$	$\mathbf{q}_{notQ}$	yes	all
PR	PnotR		
$\mathbf{p}_Q$	$\mathbf{p}_Q$	no	none
$\mathbf{p}_Q$	$\mathbf{q}_Q$	yes	151, 159, 183, 191
$\mathbf{p}_Q$	$\mathbf{p}_{notQ}$	yes	151-152, 159-160
$\mathbf{p}_Q$	$\mathbf{q}_{notQ}$	no	none
PR	notPR		
$\mathbf{p}_Q$	$\mathbf{p}_Q$	no	none
$\mathbf{p}_Q$	$\mathbf{q}_Q$	yes	151-152, 159-160
$\mathbf{p}_Q$	$\mathbf{p}_{notQ}$	yes	151, 159, 183, 191
$\mathbf{p}_Q$	$\mathbf{q}_{notQ}$	no	none

Similar tables can be constructed in relation to each partial or contingent form, till all conceivable combinations are exhausted, of course[90]. Some of these results are very significant. Look at each case and reflect on its practical meaning for causative reasoning.

For instance, that $\mathbf{p}_{PQR}$ and $\mathbf{p}_{PnotQR}$ are incompatible, since they have no moduses in common, means that something cannot be a partial cause of something else with both a certain complement (Q) and its negation (notQ) - if it is so with the one, it is certainly not so with the

[90] The results are all either explicit or implicit in the above table.

other; on the other hand, $\mathbf{p}_{PQR}$ is conjoinable with $\mathbf{p}_{notPQnotR}$ or $\mathbf{p}_{notPnotQnotR}$. Or again, causation of form $\mathbf{p}_{PQR}$ excludes prevention of form $\mathbf{p}_{PQnotR}$ or $\mathbf{p}_{notPQR}$, whereas it may well occur with prevention of form $\mathbf{p}_{PnotQnotR}$ or $\mathbf{p}_{notPnotQR}$. And so forth.

4. Categoricals and Conditionals.

Matricial analysis is applicable not only to causative propositions, but to their constituent conditional and categorical propositions. It is a universal method, as already stated. We initially, you will recall, defined causative propositions through specific combinations (conjunctions or disjunctions) of clauses, consisting of positive and negative conditionals and possible categoricals or conjunctions of categoricals.
Thus, for instances, complete causation was defined as the conjunction of "if P, then R", "if notP, not-then R" and "P is possible"; partial causation as that of "if (P + Q), then R", "if (notP + Q), not-then R", "if (P + notQ), not-then R" and "(P + Q) is possible"; and so forth. The negations of these conjunctions of clauses were then definable as inclusive disjunctions the negations of the clauses.
Eventually, we arrived at definitions of such causative propositions through lists of moduses. But each of their constituent clauses can themselves also be defined through moduses, i.e. microanalyzed; their conjunctions are then inferable by intersection and their disjunctions by merger. We could thus have begun our study by microanalyzing the constituent clauses, and then constructed the determinations with reference to their alternative moduses. By doing so, we shall close the circle, and demonstrate the completeness and consistency of the whole system.

Let us begin with **categorical propositions**.
An item P, whatever its form, can be considered as a categorical proposition in this context. If we construct a *one-item* grand matrix for it, we obtain the following table:

Table 13.9. Catalogue of moduses for a single item (P).

P	**1**	**2**	**3**	**4**
1	0	0	1	1
0	0	1	0	1

Column No. 1, which states that both P (first row) and notP (second row) are impossible, is an impossible modus, by the laws of logic. Columns 3-4 (in which the first row is coded '1', i.e. possible) represent the proposition "P is possible", while columns 2, 4 (in which the second row is coded '1', i.e. possible) represent the proposition "notP is possible". The common modus of these, No. 4, signifies that both P and notP are possible, i.e. that P is **contingent**[91];

[91] Note well that codes 1 and 0 in the moduses here signify possibility and impossibility, respectively. At a deeper level, that of 'radical' moduses, where they acquire the values of presence or absence (see Piecemeal Microanalysis, Section 1), the situation is of course different. In the latter case, modus 11 is impossible by the Law of Non-Contradiction (P and notP cannot be both present) and modus 00 is impossible by the Law of the Excluded Middle (P and notP cannot be both absent).

while modus 3 means that only P is possible (i.e. P is **necessary**) and modus 2 means that only notP is possible (i.e. P is **impossible**).
We thus see that all modalities are expressed in the grand matrix.
Note that "P is necessary" is equivalent to the proposition "P *but not* notP", i.e. it refers to P to the exclusion of notP, or more simply put to "P". Similarly, "P is impossible" can be written "notP". We may thus refer to the non-modal forms "P" or "notP" as exclusive categoricals, to distinguish them from the modal forms "P is possible" or "notP is possible"; note well the differences in moduses for them. "P" (modus 3) is included in "P is possible" (moduses 3-4), but more specific in scope.

Let us now consider the moduses of single items within a *two-item* framework, with reference to Table 12.1 of the previous chapter. They are:

Table 13.10. Enumeration of moduses of positive and negative categoricals in a two-item (PR) framework.

Proposition	Column number(s)	Comment
(necessarily) P	5, 9, 13	Three alternatives.
possibly P	5-16	All alternatives but those of notP; i.e. 12 cases.
(necessarily) notP	2-4	Three alternatives.
possibly notP	2-4, 6-8, 10-12, 14-16	All alternatives but those of P; i.e. 12 cases.
(necessarily) R	3, 9, 11	Three alternatives.
possibly R	3-4, 7-16	All alternatives but those of notR; i.e. 12 cases.
(necessarily) notR	2, 5-6	Three alternatives.
possibly notR	2, 4-8, 10, 12-16	All alternatives but those of R; i.e. 12 cases.

These results are obtained by reasoning in a similar manner. For instance, for the moduses of "P", select the columns where the two rows with P = 0 are both coded '0' (namely, Nos. 5, 9, 13); for the moduses of "P is possible", select the columns where one or both rows with P = 1 is/are coded '1' (namely, Nos. 5-16) or simply negate the three moduses corresponding to "notP". Similarly with regard to forms concerning item R.[92]
With regard to non-modal (i.e. necessary) conjunctions of (the positive or negative forms of) the items P and R, they may be obtained by appropriate intersections. Thus, for instance, "P and R" (or "PR"), being the conjunction of "P" (moduses 5, 9, 13) and "R" (moduses 3, 9, 11), yields a single common modus, viz. No. 9; and the negation of that conjunction, viz. "not(PR)", yields the leftover fourteen possible moduses. Similarly in the other cases; the following table lists results for all such cases, for the record[93]:

---

[92] It follows, incidentally, that the summary modus of P is '..00' (or, more precisely, 'μμ00') and that of R is '.0.0' (or, more precisely, 'μ0μ0'). Similarly for other cases.

[93] As regards summary moduses of the positive conjunctions, they are the same as the alternative moduses, since there is only one in each case. Thus, for instance, the summary of PR would be '1000'.

Table 13.11. Enumeration of moduses of positive and negative conjunctions in a two-item (PR) framework.

Proposition	Column number(s)	Comment
P + R	9	One common alternative.
P + notR	5	One common alternative.
notP + R	3	One common alternative.
notP + notR	2	One common alternative.
not(P + R)	2-8, 10-16	All alternatives but that of PR; i.e. 14 cases.
not(P + notR)	2-4, 6-16	All alternatives but that of PnotR; i.e. 14 cases.
not(notP + R)	2, 4-16	All alternatives but that of notPR; i.e. 14 cases.
not(notP + notR)	3-16	All alternatives but that of notPnotR; i.e. 14 cases.

Note that, since by "PR" we really mean "P is necessary and R is necessary" or "(P + R) is necessary", as already explained, the negation of such a conjunction, i.e. "not(PR)", is a modal proposition of the form "(P + R) is unnecessary".
Regarding modal conjunctions of the form "(P + R) is possible", they are equivalent to negative conditional propositions, which have the form "if P, not-then notQ". They will therefore make their appearance, implicitly, in the next table.

Let us now deal with **conditional propositions** (here logical conditionals, i.e. hypotheticals), whether positive (in the form if/then) or negative (in the form if/not-then). Their alternative moduses are listed in the following table, again with reference to a standard two-item grand matrix (i.e. Table 12.1 of the previous chapter):

Table 13.12. Enumeration of moduses of positive and negative conditionals in a two-item (PR) framework.

Proposition	Column number(s)	Comment
If P, then R	2-4, 9-12	Seven alternatives.
If P, then notR	2-8	Seven alternatives.
If notP, then R	3, 5, 7, 9, 11, 13, 15	Seven alternatives.
If notP, then notR	2, 5-6, 9-10, 13-14	Seven alternatives.
If P, not-then R	5-8, 13-16	All alternatives but those of "if P, then R", i.e. 8 cases.
If P, not-then notR	9-16	All alternatives but those of "if P, then notR", i.e. 8 cases.
If notP, not-then R	2, 4, 6, 8, 10, 12, 14, 16	All alternatives but those of "if notP, then R", i.e. 8 cases.
If notP, not-then notR	3-4, 7-8, 11-12, 15-16	All alternatives but those of "if notP, then notR", i.e. 8 cases.

The above information is obtained as follows. Take for instance "if P, then R"; it is understood to mean that the conjunction (P + notR) is impossible. Thus, referring to the said grand matrix, we must select the columns (alternative moduses) in which, for the PR sequence '10' (second row), this single condition is satisfied, i.e. the corresponding cells are coded '0' (impossible). This is true of the columns labeled 2-4, 9-12 (also of column 1, but that one is universally impossible, as we saw); so these are the applicable moduses, which we have listed in the table. The moduses of "if P, not-then R", meaning that (P + notR) is possible, follow by negation. Similarly in the other cases, *mutatis mutandis*.[94]

> Let us in this context look at the special cases of hypothetical form known as **paradoxical propositions.**
>
> First consider dilemmatic argument, to which paradoxical propositions may be assimilated. We can use the information in Table 13.12 to analyze it. For instance, if both "if P, then R" and "if notP, then R" are true, the common moduses are 3, 9, 11. The conclusion of such conjunction being "R", it is clear that "R" must include these three alternative moduses (at least). That is exactly what we found earlier (Table 13.10).
>
> Now look at Table 12.1, in the previous chapter. Rename R as P in this two-item grand matrix. Here, modus 1 is eliminated from the start because the PP sequences 11 and 00 cannot *both* be impossible (i.e. coded 0), by the law of contradiction. Moduses 3-8, 11-16 are all also eliminated because the PP sequences 10 or 01 cannot be possible (i.e. coded 1), by the law of contradiction. This leaves us only with the alternative moduses 2, 9 11. Given "if notP, then P" (i.e. 'notP and notP' is impossible), we can eliminate moduses 2 and 10, leaving modus 9 (= P). Similarly, given "if P, then notP" (i.e. 'P and P' is impossible), we can eliminate moduses 9 and 10, leaving modus 2 (= notP).
>
> In this way, paradoxical forms are made perfectly comprehensible under systematic microanalysis.

We can now interrelate the above forms with those of causative propositions, as follows.

Consider first the strong determinations, **m** and **n**. We may define **m** as the intersection of the moduses of "if P, then R" (namely, 2-4, 9-12), those of "if notP, not-then R" (2, 4, 6, 8, 10, 12, 14, 16) and those of "P is possible" (5-16) - which results in the common moduses 10, 12, as previously ascertained. Similarly, *mutatis mutandis*, for **n** (moduses 10, 14).

We see from the above table that **m** implies or is a species of "if P, then R" (which includes both its moduses 10 and 12)[95], is merely compatible with "if notP, then notR" (specifically, in modus 10), and is excluded from "if P, then notR" and from "if notP, then R" (which both lack moduses 10, 12). With regard to the negatives, **m** implies "if notP, not-then R" and "if P, not-then notR" (the latter implying that P is possible, note), is merely compatible with "if notP, not-then notR" (specifically, in modus 12), and is excluded from "if P, not-then R". We can similarly compare **n**.

Concerning now the absolute weak determinations, $\mathbf{p}_{abs}$ and $\mathbf{q}_{abs}$. Their moduses are respectively 14, 16 and 12, 16, so evidently neither of them implies a positive conditional proposition. Regarding $\mathbf{p}_{abs}$, it is excluded from three of them (which lack its two moduses) and is merely compatible with the fourth "if notP, then notR" (in modus 14, but not in modus

[94] The summary moduses can be worked out from the alternative moduses, here too.

[95] Clearly, though **m** is included in "if P, then R", it is not coextensive with it. The mere discovery of an implication does not signify causation; the other conditions have also to be fulfilled.

16). Accordingly, it implies three negative conditionals (which include both its moduses), while being merely compatible with the fourth "if notP, not-then notR" (in modus 16, but not in modus 14). We can similarly compare $\mathbf{q}_{abs}$.

We may therefore at last formally *define* absolute partial causation $\mathbf{p}_{abs}$ as the conjunction of the three negative conditionals (i) "**if P, not-then R**", (ii) "**if notP, not-then R**" and (iii) "**if P, not-then notR**", since their intersection results solely in its moduses 14, 16. **Similarly, we may *define* absolute contingent causation** $\mathbf{q}_{abs}$ as (i) "**if notP, not-then notR**", (ii) "**if P, not-then notR**" and (iii) "**if notP, not-then R**", whose common moduses are 12, 16. *Note well these are two interesting equations: we had not previously established or even guessed them.*[96]

If, by the way, we recall the summary moduses of $\mathbf{p}_{abs}$ and $\mathbf{q}_{abs}$, respectively "11.1" and "1.11", we realize that this is precisely what they mean, since every code "1" signifies that the PR sequence concerned cannot be "0". Thus, the first "1" means that the sequence PR = 11 is possible, and so that "if P, then notR" is false; the last "1" means that the sequence PR = 00 is possible, and so that "if notP, then R" is false; and similarly for the middle two positions (which differ in the two forms).

We can similarly treat, *mutatis mutandis*, the negative forms **not-m**, **not-n**, **not-$\mathbf{p}_{abs}$** and **not-$\mathbf{q}_{abs}$**. This is left to the reader as an exercise.

Note additionally that an exclusive categorical such as "P" (moduses 5, 9, 13) is incompatible with all forms of causation by P (**c**), since it has no common moduses with them (moduses 10, 12, 14, 16). Causation requires an underlying contingency for the items concerned (in the mode concerned), and is excluded at the outset where there is categorical necessity. Yet, "P" is compatible with "if P, then Q" (moduses 2-4, 9-12), for instance; taken together, they yield common modus 9, which means that R is also necessary.

All the above modus lists can easily be restated in terms of *three-item* moduses, by using Table 12.6 of the previous chapter. For examples, the latter moduses of "P + R" will be 33, 129, 161 (3 alternatives); those of "if P, then R" will be 2-16, 33-48, 129-144, 161-176 (63 alternatives); and so forth. We may skip indicating all correspondences; the reader is invited to work them out as an exercise.

We must, however, examine conjunctives or conditionals with three items in more detail, with reference to a three-item grand matrix. For this purpose, we need to know the alternative moduses of "P", "Q", "R", and their respective negations. With regard to "P" and "R", we need only expand the moduses given in Table 13.10 above, using Table 12.6 of the previous chapter. For "Q", we must in the usual manner refer directly to Table 12.3 of the previous chapter. The results are given in the following table:

---

96 Compare these definitions to those of **m**, **n**. Remember, too, that the negation of a conditional may be expressed as a possibility of conjunction. Thus (after reshuffling the three clauses), $\mathbf{p}_{abs}$ also means "(P + R) is possible, (notP + notR) is possible, and (P + notR) is possible"; and $\mathbf{q}_{abs}$ also means "(P + R) is possible, (notP + notR) is possible, and (notP + R) is possible". In each case, one conjunction remains open. The conjunction of these two forms, $\mathbf{p}_{abs}\mathbf{q}_{abs}$, therefore means that *all four* conjunctions of the items are possible.

Table 13.13. Enumeration of moduses of positive and negative categoricals in a three-item (PQR) framework.

Proposition	Column number(s)	Comment
(necessarily) P	17, 33, 49, 65, 81, 97, 113, 129, 145, 161, 177, 193, 209, 225, 241	15 alternatives.
(necessarily) notP	2-16	15 alternatives.
(necessarily) Q	5, 9, 13, 65, 69, 73, 77, 129, 133, 137, 141, 193, 197, 201, 205	15 alternatives.
(necessarily) notQ	2-4, 17-20, 33-36, 49-52	15 alternatives.
(necessarily) R	3, 9, 11, 33, 35, 41, 43, 129, 131, 137, 139, 161, 163, 169, 171	15 alternatives.
(necessarily) notR	2, 5-6, 17-18, 21-22, 65-66, 69-70, 81-82, 85-86	15 alternatives.

Propositions of the form "*possibly* P", etc., can be microanalyzed by negation[97]; they will have 255 - 15 = 240 alternative moduses, note. By combining the forms in the above table in every which way, we obtain the following results for conjunctions; and by negating the latter, for denials of conjunctions.

Table 13.14. Enumeration of moduses of three item positive and negative conjunctives in a three-item (PQR) framework.

Proposition	Column number(s)	Comment
P + Q + R	129	One alternative.
P + Q + notR	65	One alternative.
P + notQ + R	33	One alternative.
P + notQ + notR	17	One alternative.
notP + Q + R	9	One alternative.
notP + Q + notR	5	One alternative.
notP + notQ + R	3	One alternative.
notP + notQ + notR	2	One alternative.
not(P + Q + R)	2-128, 130-256	All alternatives but No. 129; i.e. 254 cases.
not(P + Q + notR)	2-64, 66-256	All alternatives but No. 65; i.e. 254 cases.
not(P + notQ + R)	2-32, 34-256	All alternatives but No. 33; i.e. 254 cases.
not(P + notQ + notR)	2-16, 18-256	All alternatives but No. 17; i.e. 254 cases.
not(notP + Q + R)	2-8, 10-256	All alternatives but No. 9; i.e. 254 cases.
not(notP + Q + notR)	2-4, 6-256	All alternatives but No. 5; i.e. 254 cases.

97 For instance, "possibly P" is the negation of "necessarily notP", and therefore has moduses 17-256.

Table 13.14 continued.

not(notP + notQ + R)	2, 4-256	All alternatives but No. 3; i.e. 254 cases.
not(notP + notQ + notR)	3-256	All alternatives but No. 2; i.e. 254 cases.

The following table, concerning conditionals and their negations, is constructed with reference to Table 12.3 of the previous chapter, in the usual manner. For instance, "if (P + Q), then R" means that (P + Q + notR) is impossible; therefore, we select the moduses which register a zero along the row for the PQR sequence 110. Similarly in other positive cases; then negatives are derived by listing the leftover moduses in each case.

Table 13.15. Enumeration of moduses of three item positive and negative conditionals in a three-item (PQR) framework.

Proposition	Column number(s)	Comment
If (P + Q), then R	2-64, 129-192	127 alternatives.
If (P + Q), then notR	2-128	127 alternatives.
If (P + notQ), then R	2-16, 33-48, 65-80, 97-112, 129-144, 161-176, 193-208, 225-240	127 alternatives.
If (P + notQ), then notR	2-32, 65-96, 129-160, 193-224	127 alternatives.
If (notP + Q), then R	2-4, 9-12, 17-20, 25-28, 33-36, 41-44, 49-52, 57-60, 65-68, 73-76, 81-84, 89-92, 97-100, 105-108, 113-116, 121-124, 129-132, 137-140, 145-148, 153-156, 161-164, 169-172, 177-180, 185-188, 193-196, 201-204, 209-212, 217-220, 225-228, 233-236, 241-244, 249-252	127 alternatives.
If (notP + Q), then notR	2-8, 17-24, 33-40, 49-56, 65-72, 81-88, 97-104, 113-120, 129-136, 145-152, 161-168, 177-184, 193-200, 209-216, 225-232, 241-248	127 alternatives.
If (notP + notQ), then R	3, 5, 7, 9, 11, 13, 15, 17, 19, 21, 23, 25, 27, 29, 31, 33, 35, 37, 39, 41, 43, 45, 47, 49, 51, 53, 55, 57, 59, 61, 63, 65, 67, 69, 71, 73, 75, 77, 79, 81, 83, 85, 87, 89, 91, 93, 95, 97, 99, 101, 103, 105, 107, 109, 111, 113, 115, 117, 119, 121, 123, 125, 127, 129, 131, 133, 135, 137, 139, 141, 143, 145, 147, 149, 151, 153, 155, 157, 159, 161, 163, 165, 167, 169, 171, 173, 175, 177, 179, 181, 183, 185, 187, 189, 191, 193, 195, 197, 199, 201, 203, 205, 207, 209, 211, 213, 215, 217, 219, 221, 223, 225, 227, 229, 231, 233, 235, 237, 239, 241, 243, 245, 247, 249, 251, 253, 255	127 alternatives.

Table 13.15 continued.

If (notP + notQ), then notR	2, 5-6, 9-10, 13-14, 17-18, 21-22, 25-26, 29-30, 33-34, 37-38, 41-42, 45-46, 49-50, 53-54, 57-58, 61-62, 65-66, 69-70, 73-74, 77-78, 81-82, 85-86, 89-90, 93-94, 97-98, 101-102, 105-106, 109-110, 113-114, 117-118, 121-122, 125-126, 129-130, 133-134, 137-138, 141-142, 145-146, 149-150, 153-154, 157-158, 161-162, 165-166, 169-170, 173-174, 177-178, 181-182, 185-186, 189-190, 193-194, 197-198, 201-202, 205-206, 209-210, 213-214, 217-218, 221-222, 225-226, 229-230, 233-234, 237-238, 241-242, 245-246, 249-250, 253-254	127 alternatives.
If (P + Q), not-then R	65-128, 193-256	The 128 remaining cases.
If (P + Q), not-then notR	129-256	The 128 remaining cases.
If (P + notQ), not-then R	17-32, 49-64, 81-96, 113-128, 145-160, 177-192, 209-224, 241-256	The 128 remaining cases.
If (P + notQ), not-then notR	33-64, 97-128, 161-192, 225-256	The 128 remaining cases.
If (notP + Q), not-then R	5-8, 13-16, 21-24, 29-32, 37-40, 45-48, 53-56, 61-64, 69-72, 77-80, 85-88, 93-96, 101-104, 109-112, 117-120, 125-128, 133-136, 141-144, 149-152, 157-160, 165-168, 173-176, 181-184, 189-192, 197-200, 205-208, 213-216, 221-224, 229-232, 237-240, 245-248, 253-256	The 128 remaining cases.
If (notP + Q), not-then notR	9-16, 25-32, 41-48, 57-64, 73-80, 89-96, 105-112, 121-128, 137-144, 153-160, 169-176, 185-192, 201-208, 217-224, 233-240, 249-256	The 128 remaining cases.
If (notP + notQ), not-then R	2, 4, 6, 8, 10, 12, 14, 16, 18, 20, 22, 24, 26, 28, 30, 32, 34, 36, 38, 40, 42, 44, 46, 48, 50, 52, 54, 56, 58, 60, 62, 64, 66, 68, 70, 72, 74, 76, 78, 80, 82, 84, 86, 88, 90, 92, 94, 96, 98, 100, 102, 104, 106, 108, 110, 112, 114, 116, 118, 120, 122, 124, 126, 128, 130, 132, 134, 136, 138, 140, 142, 144, 146, 148, 150, 152, 154, 156, 158, 160, 162, 164, 166, 168, 170, 172, 174, 176, 178, 180, 182, 184, 186, 188, 190, 192, 194, 196, 198, 200, 202, 204, 206, 208, 210, 212, 214, 216, 218, 220, 222, 224, 226, 228, 230, 232, 234, 236, 238, 240, 242, 244, 246, 248, 250, 252, 254, 256	The 128 remaining cases.

Table 13.15 continued.

If (notP + notQ), not-then notR	3-4, 7-8, 11-12, 15-16, 19-20, 23-24, 27-28, 31-32, 35-36, 39-40, 43-44, 47-48, 51-52, 55-56, 59-60, 63-64, 67-68, 71-72, 75-76, 79-80, 83-84, 87-88, 91-92, 95-96, 99-100, 103-104, 107-108, 111-112, 115-116, 119-120, 123-124, 127-128, 131-132, 135-136, 139-140, 143-144, 147-148, 151-152, 155-156, 159-160, 163-164, 167-168, 171-172, 175-176, 179-180, 183-184, 187-188, 191-192, 195-196, 199-200, 203-204, 207-208, 211-212, 215-216, 219-220, 223-224, 227-228, 231-232, 235-236, 239-240, 243-244, 247-248, 251-252, 255-256	The 128 remaining cases.

We can make similar comments here as before, elucidating the oppositions between causative and the less specific forms. This is left as an exercise for the reader.

In particular, the reader should compare the moduses of the relative weak determinations, given in Table 13.1 of the present chapter, with those derived from the two above tables and the original definitions of weak causation. For instance, note that "if (P + Q), then R" includes all 16 moduses of $\mathbf{p}_{PQR}$ (and so is a genus of it and serves in its definition); similarly for $\mathbf{q}_{PQR}$ in relation to "if (notP + notQ), then notR".

Additionally, observe that all the three-item moduses of "if P, then R" are included by "if (P + Q), then R" (but not vice-versa, of course), so that the former is a species of the latter. Note that the P-R form is more restrictive with only 63 moduses, while the P-Q-R form is broader in possibilities with 127 moduses. Similarly in other cases.

We have thus finished demonstrating that our grand matrices have universal utility, enabling us to express any form, whatever its breadth or polarity. We shall now move on to syllogistic applications, and show that all issues are resolvable by such matricial analysis.

Chapter 14. MAIN THREE-ITEM SYLLOGISMS.

1. Applying Microanalysis to Syllogism.

We shall now begin to use microanalytic methods for the understanding and solution of syllogistic problems. In this way, we shall develop an advanced and general theory of the syllogism, using a tool more powerful and universal than any previously used. Ciao Aristotle, hello future logic. We shall proceed in stages, from the simpler cases to the more complex.

As we are already aware, syllogism basically involves three items (terms or theses) in three propositions. The propositions (in this context, of causative form) are the major premise, the minor premise and the conclusion. The items involved are the minor item (which occurs somewhere in the minor premise, and as the cause in the conclusion), the major item (which occurs somewhere in the major premise, and as the effect in the conclusion), and the middle item (present in both premises, but ignored in the conclusion).

Syllogism, then, is argument aimed at discovering or establishing the relation (the conclusion) between two items (the minor and major), by way of their given relations (the premises) to a third item (the middle). If the premises are compatible, they will either imply a certain conclusion, in which case the syllogism is 'valid'; or fail to imply any conclusion, in which case the syllogism is 'invalid'.

And as we have seen, there are three main 'figures' for such argument, traditionally labeled Nos. 1, 2 and 3. In the first figure, the middle item has the roles of effect in the minor premise and cause in the major premise; in the second figure, it has the role of effect in both premises; and in the third figure, it has the role of cause in both figures.

Ordinarily, the three items occur only in the positions just said. But we may conceive of special cases where the items recur in other positions, as complements. This gives rise to arrangements we have called 'subfigures'. We need not concern ourselves with these complications at this stage, but will only focus our attention on the essential, Aristotelian arrangements.

Figures are, of course, abstractions regarding the positions of the items. Syllogism is concretized in individual formal 'moods', when the exact relations between these items are specified - i.e., in the present context, the determination and polarity of the causative relation in each premise. The latter are likewise useless without specification of the figure involved.

Thus, to end this brief review, we will here to begin with reexamine, using the microanalytic method, three-item causative syllogism in the three main figures (without consideration of the fourth figure or of subfigures). We shall start with positive moods, i.e. moods with both premises affirmative; after that, we shall deal with negative moods, i.e. those involving at least one negative premise.

Still later, we shall gradually treat all other conceivable situations, and thus demonstrate the universality of this technique.

Now, putting aside all the above mentioned details, syllogism is nothing more than a logical *conjunction* of two propositions (the premises), and all we seek through it is an evaluation or inference of their *intersection* (the conclusion). That is:

The major premise + the minor premise = the conclusion (if any).

It follows that, ***knowing the moduses of the premises, we can ascertain those of the conclusion***. The conclusion may thus be viewed as a summary the information of the premises taken together; i.e. as a statement of their combined value in knowledge. Each premise, taken separately, has a certain value - expressed in its modus list; in the conclusion, we find out their mutual impact - which is merely their common moduses, if any. We can refer to an image in mechanics, we have two vectors and we wish to calculate their resultant force.
The following simple example will illustrate how a valid conclusion arises. Mood 1/**m**/**m** (which, you will recall, we labeled 155) consists of the premises "Q is a complete cause of R" (the major, which we may label $\mathbf{m}_{QR}$) and "P is a complete cause of Q" (the minor, which we may label $\mathbf{m}_{PQ}$). We know by macroanalysis that its conclusion is "P is a complete cause of R" (which we may label $\mathbf{m}_{PR}$); as shown in the table below, microanalysis yields the same result.

Table 14.1. Microanalysis of a mood - the example of mood 1/**m**/**m** (No. 155).

Proposition	Moduses	Comments
The major premise, $\mathbf{m}_{QR}$	10, 12, 25-28, 42, 44, 57-60, 130, 132, 138, 140, 145-148, 153-156, 162, 164, 170, 172, 177-180, 185-188	36 alternatives, with reference to items Q and R.
The minor premise, $\mathbf{m}_{PQ}$	66-68, 70-72, 74-76, 78-80, 130-132, 134-136, 138-140, 142-144, 194-196, 198-200, 202-204, 206-208	36 alternatives, with reference to items P and Q.
The conclusion, $\mathbf{m}_{PR}$	130, 132, 138, 140	4 moduses in common to the above premises.

Note that the premises, being in the first figure, have respectively forms QR (major) and PQ (minor), while the conclusion must be of form PR. Once the modus or moduses of the premises are ascertained (as explained in the next section), we need only find our which modus or moduses, if any, they have in common. The form or forms which include *all* such common modus(es) is/are implied by them, and may be considered as constituting the conclusion. This intersection may be illustrated as follows

Diagram 14.1. Premises and Conclusion.

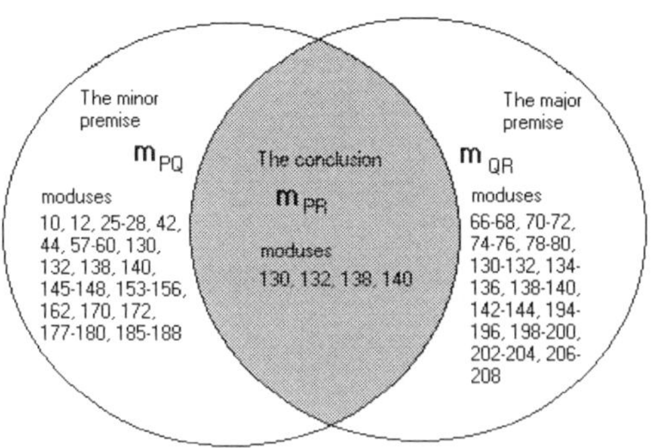

In our example, four moduses are found in common. With reference to Table 12.4 of the chapter on systematic microanalysis, which concerns causative propositions of form PR, we see that these four moduses are all *among* the 36 alternatives of **m**. It follows that $\mathbf{m}_{PR}$ is our conclusion or part thereof. Continuing our search, we see that *not all* of these four moduses are included in **mn** (only No. 132 is so) or $\mathbf{mq}_{abs}$ (only Nos. 132, 138, 140 are so), therefore we cannot obtain a more precise positive conclusion. Thus, $\mathbf{m}_{QR} + \mathbf{m}_{PQ}$ is $\mathbf{m}_{PR}$.

Note well that, to conclude $\mathbf{m}_{PR}$, we do not need to find all its moduses to be in common to the premises - just one would suffice; what is important is that all the moduses which are in common to the premises be included in the list of 36 for that putative conclusion.
Also note that *the above mentioned table*[98], *which enumerates three-item moduses for propositions of form PR, is appropriate for evaluating* the conclusion, *which has that same form in all figures.*
But the moduses of the premises *cannot be identified with reference to that table, since they have forms other than PR*, namely QR or PQ (in the first figure). In their case, we must revert to the grand matrix for three items, i.e. the preceding table in the same chapter, Table 12.3. How this is done is explained in detail in the next section, where the results for all conceivable strong or absolute weak premises in whatever figure are also tabulated.
We can, at the outset, formulate the following general **rules of inference**, based on our knowledge of the intersection process:

1. **If the premises have *no* modus in common, they are incompatible.** This signifies that the premises are already incoherent and cannot credibly occur together in knowledge. If we encounter such premises (without common modus, inconsistent) in practice, as often happens, not only can we draw no conclusion from them, but one or both of them must be reviewed and rejected or corrected.
2. **If the premises have *one or more* modus(es) in common and such modus(es) is/are *all* included by some causative propositional form(s), the latter constitute(s) our conclusion.** Note well again that, though all common moduses of the premises must be moduses of the putative conclusion, not all the moduses of the latter need be common to the former. For a form, remember, is implied by any one of its alternative moduses. Also note, a conclusion may consist of one causative form or a conjunction of such forms, of whatever polarities.
3. **But if the premises have *two or more* moduses in common and these moduses are *not all* included by some causative propositional form(s), though some of them are included by a form and others by its contradictory, we have no conclusion.** In this case, the premises are indeed compatible, since they have common moduses; but those common moduses do not give rise to a causative proposition, however vague, being too scattered[99]. The common moduses may of course still give rise to a proposition of form other than causative; but this is not of interest in the present context.

2. The Moduses of Premises.

In order to systematically apply the microanalytic method to the solution of syllogistic problems, we need first to identify the moduses of all conceivable *premises*. For we already have the moduses of eventual *conclusions*, which always have form PR: these are given in Table 12.4 of the chapter on systematic microanalysis.
It is marvelous that *we can refer to the same grand matrix for three items, with 255 possible moduses, which we used to interpret three-item propositions, to interpret three-item*

98 i.e. Table 12.4 of the chapter on systematic microanalysis.

99 Suppose, for instance, the premises have common moduses 147 and 148. Modus 147 implies **not-c**$_{abs}$, whereas modus 148 implies **c**$_{abs}$; therefore, neither of these, and indeed no other, causative propositions can be concluded. It is always best to first test the common moduses with reference to causation or non-causation; if they all fit into the one or the other, we can then check out whether they also fit some more precise (positive or negative) form(s).

syllogisms. For syllogisms with four items, we shall need to consider 65,535 possible moduses![100]
Regarding the premises in the three figures, we see that they involve four possible pairs and arrangements of items, namely QR or RQ for the major premise and PQ or QP for the minor premise. The summary moduses of the various generic forms of interest to us here are as follows. Note well that in the following table, the weak determinations **p**, **q** are intended as *absolute*, though not so specified.

Table 14.2. Three-item summary moduses for strong or absolute weak generic positive premises.

row labl	Items			**m**	**m**	**m**	**m**	**n**	**n**	**n**	**n**	**p**	**p**	**p**	**p**	**q**	**q**	**q**	**q**
	P	**Q**	**R**	QR	RQ	PQ	QP	QR	RQ	PQ	QP	QR	RQ	PQ	QP	QR	RQ	PQ	QP
a	**1**	**1**	**1**	ae	ae	ab	ab	ae	ae	ab	ab	ae	ae	ab	ab	ae	ae	ab	ab
b	**1**	**1**	**0**	0	·	ab	ab	·	0	ab	ab	bf	·	ab	ab	·	bf	ab	ab
c	**1**	**0**	**1**	·	0	0	·	0	·	·	0	·	cg	cd	·	cg	·	·	cd
d	**1**	**0**	**0**	dh	dh	0	·	dh	dh	·	0	dh	dh	cd	·	dh	dh	·	cd
e	**0**	**1**	**1**	ae	ae	·	0	ae	ae	0	·	ae	ae	·	ef	ae	ae	ef	·
f	**0**	**1**	**0**	0	·	·	0	·	0	0	·	bf	·	·	ef	·	bf	ef	·
g	**0**	**0**	**1**	·	0	gh	gh	0	·	gh	gh	·	cg	gh	gh	cg	·	gh	gh
h	**0**	**0**	**0**	dh	dh	gh	gh	dh	dh	gh	gh	dh	dh	gh	gh	dh	dh	gh	gh

Where two letters (among a-h) appear in two cells of a column, the intent is to indicate that this pair of cells can never both be zero, i.e. a 'μ' is intended[101]. Thus, for instance, instead of writing μ_{ae}, I write 'ae' in both rows a and e. This is done just to improve visibility of the information.
Notice the following details in the above table. First, note the similarity of sequence of modalities in two columns read in opposite directions (for instance the modus of $\mathbf{m}_{QR}$ read from top to bottom and that of $\mathbf{n}_{QR}$ read from bottom to top); this signifies mirroring. Second, note the identity of moduses of various pairs of propositions, like $\mathbf{m}_{QR}$ and $\mathbf{n}_{RQ}$, or $\mathbf{p}_{QR}$ and $\mathbf{q}_{RQ}$; this signifies convertibility from one to the other.
We can now, with reference to the rules implied by the above table, easily read the grand matrix for three items[102] in the usual manner, to obtain the alternative moduses of all the premises of concern to us here. For now, we need only deal with positive premises - first, the generics, based on the above summary moduses; then, the joint determinations, by recourse to intersections between the generics. The results are as follows:

100 For that, at a rate of 64 moduses to a page, we would need to construct a table of over 1,000 pages.
101 As in Table 11.7 of the chapter on piecemeal microanalysis.
102 See Table 12.3 of the chapter on systematic microanalysis.

Table 14.3. Enumeration of three-item alternative moduses for strong or absolute weak positive premises, generic or joint, for any figure of syllogism.

Determ.	Major QR	Major RQ	Minor PQ	Minor QP
m	10, 12, 25-28, 42, 44, 57-60, 130, 132, 138, 140, 145-148, 153-156, 162, 164, 170, 172, 177-180, 185-188	10, 14, 25-26, 29-30, 74, 78, 89-90, 93-94, 130, 134, 138, 142, 145-146, 149-150, 153-154, 157-158, 194, 198, 202, 206, 209-210, 213-214, 217-218, 221-222	66-68, 70-72, 74-76, 78-80, 130-132, 134-136, 138-140, 142-144, 194-196, 198-200, 202-204, 206-208	66-68, 82-84, 98-100, 114-116, 130-132, 146-148, 162-164, 178-180, 194-196, 210-212, 226-228, 242-244
n	10, 14, 25-26, 29-30, 74, 78, 89-90, 93-94, 130, 134, 138, 142, 145-146, 149-150, 153-154, 157-158, 194, 198, 202, 206, 209-210, 213-214, 217-218, 221-222	10, 12, 25-28, 42, 44, 57-60, 130, 132, 138, 140, 145-148, 153-156, 162, 164, 170, 172, 177-180, 185-188	66-68, 82-84, 98-100, 114-116, 130-132, 146-148, 162-164, 178-180, 194-196, 210-212, 226-228, 242-244	66-68, 70-72, 74-76, 78-80, 130-132, 134-136, 138-140, 142-144, 194-196, 198-200, 202-204, 206-208
$\mathbf{p}_{abs}$	14, 16, 29-32, 46, 48, 61-64, 74, 76, 78, 80, 89-96, 106, 108, 110, 112, 121-128, 134, 136, 142, 144, 149-152, 157-160, 166, 168, 174, 176, 181-184, 189-192, 194, 196, 198, 200, 202, 204, 206, 208-224, 226, 228, 230, 232, 234, 236, 238, 240-256	12, 16, 27-28, 31-32, 42, 44, 46, 48, 57-64, 76, 80, 91-92, 95-96, 106, 108, 110, 112, 121-128, 132, 136, 140, 144, 147-148, 151-152, 155-156, 159-160, 162, 164, 166, 168, 170, 172, 174, 176-192, 196, 200, 204, 208, 211-212, 215-216, 219-220, 223-224, 226, 228, 230, 232, 234, 236, 238, 240-256	82-84, 86-88, 90-92, 94-96, 98-100, 102-104, 106-108, 110-112, 114-116, 118-120, 122-124, 126-128, 146-148, 150-152, 154-156, 158-160, 162-164, 166-168, 170-172, 174-176, 178-180, 182-184, 186-188, 190-192, 210-212, 214-216, 218-220, 222-224, 226-228, 230-232, 234-236, 238-240, 242-244, 246-248, 250-252, 254-256	70-72, 74-76, 78-80, 86-88, 90-92, 94-96, 102-104, 106-108, 110-112, 118-120, 122-124, 126-128, 134-136, 138-140, 142-144, 150-152, 154-156, 158-160, 166-168, 170-172, 174-176, 182-184, 186-188, 190-192, 198-200, 202-204, 206-208, 214-216, 218-220, 222-224, 230-232, 234-236, 238-240, 246-248, 250-252, 254-256
$\mathbf{q}_{abs}$	12, 16, 27-28, 31-32, 42, 44, 46, 48, 57-64, 76, 80, 91-92, 95-96, 106, 108, 110, 112, 121-128, 132, 136, 140, 144, 147-148, 151-152, 155-156, 159-160, 162, 164, 166, 168, 170, 172, 174, 176-192, 196, 200, 204, 208, 211-212, 215-216, 219-220, 223-224, 226, 228, 230, 232, 234, 236, 238, 240-256	14, 16, 29-32, 46, 48, 61-64, 74, 76, 78, 80, 89-96, 106, 108, 110, 112, 121-128, 134, 136, 142, 144, 149-152, 157-160, 166, 168, 174, 176, 181-184, 189-192, 194, 196, 198, 200, 202, 204, 206, 208, 209-224, 226, 228, 230, 232, 234, 236, 238, 240, 241-256	70-72, 74-76, 78-80, 86-88, 90-92, 94-96, 102-104, 106-108, 110-112, 118-120, 122-124, 126-128, 134-136, 138-140, 142-144, 150-152, 154-156, 158-160, 166-168, 170-172, 174-176, 182-184, 186-188, 190-192, 198-200, 202-204, 206-208, 214-216, 218-220, 222-224, 230-232, 234-236, 238-240, 246-248, 250-252, 254-256	82-84, 86-88, 90-92, 94-96, 98-100, 102-104, 106-108, 110-112, 114-116, 118-120, 122-124, 126-128, 146-148, 150-152, 154-156, 158-160, 162-164, 166-168, 170-172, 174-176, 178-180, 182-184, 186-188, 190-192, 210-212, 214-216, 218-220, 222-224, 226-228, 230-232, 234-236, 238-240, 242-244, 246-248, 250-252, 254-256
mn	10, 25-26, 130, 138, 145-146, 153-154	10, 25-26, 130, 138, 145-146, 153-154	66-68, 130-132, 194-196	66-68, 130-132, 194-196
mqabs	12, 27-28, 42, 44, 57-60, 132, 140, 147-148, 155-156, 162, 164, 170, 172, 177-180, 185-188	14, 29-30, 74, 78, 89-90, 93-94, 134, 142, 149-150, 157-158, 194, 198, 202, 206, 209-210, 213-214, 217-218, 221-222	70-72, 74-76, 78-80, 134-136, 138-140, 142-144, 198-200, 202-204, 206-208	82-84, 98-100, 114-116, 146-148, 162-164, 178-180, 210-212, 226-228, 242-244

Table 14.3 continued.

np$_{abs}$	14, 29-30, 74, 78, 89-90, 93-94, 134, 142, 149-150, 157-158, 194, 198, 202, 206, 209-210, 213-214, 217-218, 221-222	12, 27-28, 42, 44, 57-60, 132, 140, 147-148, 155-156, 162, 164, 170, 172, 177-180, 185-188	82-84, 98-100, 114-116, 146-148, 162-164, 178-180, 210-212, 226-228, 242-244	70-72, 74-76, 78-80, 134-136, 138-140, 142-144, 198-200, 202-204, 206-208
p$_{abs}$**q**$_{abs}$	16, 31-32, 46, 48, 61-64, 76, 80, 91-92, 95-96, 106, 108, 110, 112, 121-128, 136, 144, 151-152, 159-160, 166, 168, 174, 176, 181-184, 189-192, 196, 200, 204, 208, 211-212, 215-216, 219-220, 223-224, 226, 228, 230, 232, 234, 236, 238, 240-256	16, 31-32, 46, 48, 61-64, 76, 80, 91-92, 95-96, 106, 108, 110, 112, 121-128, 136, 144, 151-152, 159-160, 166, 168, 174, 176, 181-184, 189-192, 196, 200, 204, 208, 211-212, 215-216, 219-220, 223-224, 226, 228, 230, 232, 234, 236, 238, 240-256	86-88, 90-92, 94-96, 102-104, 106-108, 110-112, 118-120, 122-124, 126-128, 150-152, 154-156, 158-160, 166-168, 170-172, 174-176, 182-184, 186-188, 190-192, 214-216, 218-220, 222-224, 230-232, 234-236, 238-240, 246-248, 250-252, 254-256	86-88, 90-92, 94-96, 102-104, 106-108, 110-112, 118-120, 122-124, 126-128, 150-152, 154-156, 158-160, 166-168, 170-172, 174-176, 182-184, 186-188, 190-192, 214-216, 218-220, 222-224, 230-232, 234-236, 238-240, 246-248, 250-252, 254-256

To repeat, the same information for causative propositions of form PR has already been tabulated, in a previous chapter; and those results are applicable to reading conclusions. The above table concerns positive premises that may arise in three-item syllogism. Though the same propositional form has always the same number of moduses, the modus numbers differ according to the items involved and their positions. Note this well, and compare and contrast.

3. The Moduses of Conclusions.

Putting the data in the above table together in various combinations, we can now ascertain the moduses of resulting conclusions. As already said, when the premises have no common modus, they are "*inconsistent*". When some resulting modus(es) fall under causation (**c**) and some other(s) under non-causation (**not-c**), we must admit that there is "*no conclusion*". In all other cases, there is a conclusion, namely the determination (whatever it be, at least causation or non-causation) which includes *all* the resulting moduses.

The following table lists all these results. Note well that all weak determinations mentioned in it are intended as *absolute*, though not so specified. Syllogism with relative weaks will be considered later.

Table 14.4. Moduses of conclusions of all syllogisms with strong or absolute weak positive premises, generic or joint.

	Major	Minor	Conclusion	Common moduses	No.
	First Figure				
Mood	QR	PQ	PR		
111	**mn**	**mn**	**mn**	130	1
112	**mn**	**mq**	**mq**	138	1
113	**mn**	**np**	**np**	146	1
114	**mn**	**pq**	**pq**	154	1
115	**mn**	**m**	**m**	130, 138	2
116	**mn**	**n**	**n**	130, 146	2
117	**mn**	**p**	**p**	146, 154	2
118	**mn**	**q**	**q**	138, 154	2
121	**mq**	**mn**	**mq**	132	1
122	**mq**	**mq**	**mq**	140	1
123	**mq**	**np**	*no conclusion*	147-148, 162, 164, 178-180	7
124	**mq**	**pq**	**not-n**	155-156, 170, 172, 186-188	7
125	**mq**	**m**	**mq**	132, 140	2
126	**mq**	**n**	*no conclusion*	132, 147-148, 162, 164, 178-180	8
127	**mq**	**p**	*no conclusion*	147-148, 155-156, 162, 164, 170, 172, 178-180, 186-188	14
128	**mq**	**q**	**not-n**	140, 155-156, 170, 172, 186-188	8
131	**np**	**mn**	**np**	194	1
132	**np**	**mq**	*no conclusion*	74, 78, 134, 142, 198, 202, 206	7
133	**np**	**np**	**np**	210	1
134	**np**	**pq**	**not-m**	90, 94, 150, 158, 214, 218, 222	7
135	**np**	**m**	*no conclusion*	74, 78, 134, 142, 194, 198, 202, 206	8
136	**np**	**n**	**np**	194, 210	2
137	**np**	**p**	**not-m**	90, 94, 150, 158, 210, 214, 218, 222	8
138	**np**	**q**	*no conclusion*	74, 78, 90, 94, 134, 142, 150, 158, 198, 202, 206, 214, 218, 222	14
141	**pq**	**mn**	**pq**	196	1
142	**pq**	**mq**	**not-n**	76, 80, 136, 144, 200, 204, 208	7
143	**pq**	**np**	**not-m**	211-212, 226, 228, 242-244	7
144	**pq**	**pq**	*no conclusion*	91-92, 95-96, 106, 108, 110, 112, 122-124, 126-128, 151-152, 159-160, 166, 168, 174, 176, 182-184, 190-192, 215-216, 219-220, 223-224, 230, 232, 234, 236, 238, 240, 246-248, 250-252, 254-256	49
145	**pq**	**m**	**not-n**	76, 80, 136, 144, 196, 200, 204, 208	8
146	**pq**	**n**	**not-m**	196, 211-212, 226, 228, 242-244	8

Table 14.4 continued.

147	**pq**	**p**	*no conclusion*	91-92, 95-96, 106, 108, 110, 112, 122-124, 126-128, 151-152, 159-160, 166, 168, 174, 176, 182-184, 190-192, 211-212, 215-216, 219-220, 223-224, 226, 228, 230, 232, 234, 236, 238, 240, 242-244, 246-248, 250-252, 254-256	56
148	**pq**	**q**	*no conclusion*	76, 80, 91-92, 95-96, 106, 108, 110, 112, 122-124, 126-128, 136, 144, 151-152, 159-160, 166, 168, 174, 176, 182-184, 190-192, 200, 204, 208, 215-216, 219-220, 223-224, 230, 232, 234, 236, 238, 240, 246-248, 250-252, 254-256	56
151	**m**	**mn**	**m**	130, 132	2
152	**m**	**mq**	**mq**	138, 140	2
153	**m**	**np**	*no conclusion*	146-148, 162, 164, 178-180	8
154	**m**	**pq**	**not-n**	154-156, 170, 172, 186-188	8
155	**m**	**m**	**m**	130, 132, 138, 140	4
156	**m**	**n**	*no conclusion*	130, 132, 146-148, 162, 164, 178-180	10
157	**m**	**p**	*no conclusion*	146-148, 154-156, 162, 164, 170, 172, 178-180, 186-188	16
158	**m**	**q**	**not-n**	138, 140, 154-156, 170, 172, 186-188	10
161	**n**	**mn**	**n**	130, 194	2
162	**n**	**mq**	*no conclusion*	74, 78, 134, 138, 142, 198, 202, 206	8
163	**n**	**np**	**np**	146, 210	2
164	**n**	**pq**	**not-m**	90, 94, 150, 154, 158, 214, 218, 222	8
165	**n**	**m**	*no conclusion*	74, 78, 130, 134, 138, 142, 194, 198, 202, 206	10
166	**n**	**n**	**n**	130, 146, 194, 210	4
167	**n**	**p**	**not-m**	90, 94, 146, 150, 154, 158, 210, 214, 218, 222	10
168	**n**	**q**	*no conclusion*	74, 78, 90, 94, 134, 138, 142, 150, 154, 158, 198, 202, 206, 214, 218, 222	16
171	**p**	**mn**	**p**	194, 196	2
172	**p**	**mq**	*no conclusion*	74, 76, 78, 80, 134, 136, 142, 144, 198, 200, 202, 204, 206, 208	14
173	**p**	**np**	**not-m**	210-212, 226, 228, 242-244	8
174	**p**	**pq**	*no conclusion*	90-92, 94-96, 106, 108, 110, 112, 122-124, 126-128, 150-152, 158-160, 166, 168, 174, 176, 182-184, 190-192, 214-216, 218-220, 222-224, 230, 232, 234, 236, 238, 240, 246-248, 250-252, 254-256	56
175	**p**	**m**	*no conclusion*	74, 76, 78, 80, 134, 136, 142, 144, 194, 196, 198, 200, 202, 204, 206, 208	16
176	**p**	**n**	**not-m**	194, 196, 210-212, 226, 228, 242-244	10
177	**p**	**p**	*no conclusion*	90-92, 94-96, 106, 108, 110, 112, 122-124, 126-128, 150-152, 158-160, 166, 168, 174, 176, 182-184, 190-192, 210-212, 214-216, 218-220, 222-224, 226, 228, 230, 232, 234, 236, 238, 240, 242-244, 246-248, 250-252, 254-256	64
178	**p**	**q**	*no conclusion*	74, 76, 78, 80, 90-92, 94-96, 106, 108, 110, 112, 122-124, 126-128, 134, 136, 142, 144, 150-152, 158-160, 166, 168, 174, 176, 182-184, 190-192, 198, 200, 202, 204, 206, 208, 214-216, 218-220, 222-224, 230, 232, 234, 236, 238, 240, 246-248, 250-252, 254-256	70

Table 14.4 continued.

181	**q**	**mn**	**q**	132, 196	2
182	**q**	**mq**	**not-n**	76, 80, 136, 140, 144, 200, 204, 208	8
183	**q**	**np**	*no conclusion*	147-148, 162, 164, 178-180, 211-212, 226, 228, 242-244	14
184	**q**	**pq**	*no conclusion*	91-92, 95-96, 106, 108, 110, 112, 122-124, 126-128, 151-152, 155-156, 159-160, 166, 168, 170, 172, 174, 176, 182-184, 186-188, 190-192, 215-216, 219-220, 223-224, 230, 232, 234, 236, 238, 240, 246-248, 250-252, 254-256	56
185	**q**	**m**	**not-n**	76, 80, 132, 136, 140, 144, 196, 200, 204, 208	10
186	**q**	**n**	*no conclusion*	132, 147-148, 162, 164, 178-180, 196, 211-212, 226, 228, 242-244	16
187	**q**	**p**	*no conclusion*	91-92, 95-96, 106, 108, 110, 112, 122-124, 126-128, 147-148, 151-152, 155-156, 159-160, 162, 164, 166, 168, 170, 172, 174, 176, 178-180, 182-184, 186-188, 190-192, 211-212, 215-216, 219-220, 223-224, 226, 228, 230, 232, 234, 236, 238, 240, 242-244, 246-248, 250-252, 254-256	70
188	**q**	**q**	*no conclusion*	76, 80, 91-92, 95-96, 106, 108, 110, 112, 122-124, 126-128, 136, 140, 144, 151-152, 155-156, 159-160, 166, 168, 170, 172, 174, 176, 182-184, 186-188, 190-192, 200, 204, 208, 215-216, 219-220, 223-224, 230, 232, 234, 236, 238, 240, 246-248, 250-252, 254-256	64
	Second Figure				
Mood	RQ	PQ	PR		
211	**mn**	**mn**	**mn**	130	1
212	**mn**	**mq**	**mq**	138	1
213	**mn**	**np**	**np**	146	1
214	**mn**	**pq**	**pq**	154	1
215	**mn**	**m**	**m**	130, 138	2
216	**mn**	**n**	**n**	130, 146	2
217	**mn**	**p**	**p**	146, 154	2
218	**mn**	**q**	**q**	138, 154	2
221	**mq**	**mn**	**np**	194	1
222	**mq**	**mq**	*no conclusion*	74, 78, 134, 142, 198, 202, 206	7
223	**mq**	**np**	**np**	210	1
224	**mq**	**pq**	**not-m**	90, 94, 150, 158, 214, 218, 222	7
225	**mq**	**m**	*no conclusion*	74, 78, 134, 142, 194, 198, 202, 206	8
226	**mq**	**n**	**np**	194, 210	2
227	**mq**	**p**	**not-m**	90, 94, 150, 158, 210, 214, 218, 222	8
228	**mq**	**q**	*no conclusion*	74, 78, 90, 94, 134, 142, 150, 158, 198, 202, 206, 214, 218, 222	14
231	**np**	**mn**	**mq**	132	1
232	**np**	**mq**	**mq**	140	1
233	**np**	**np**	*no conclusion*	147-148, 162, 164, 178-180	7
234	**np**	**pq**	**not-n**	155-156, 170, 172, 186-188	7
235	**np**	**m**	**mq**	132, 140	2

Table 14.4 continued.

236	**np**	**n**	*no conclusion*	132, 147-148, 162, 164, 178-180	8
237	**np**	**p**	*no conclusion*	147-148, 155-156, 162, 164, 170, 172, 178-180, 186-188	14
238	**np**	**q**	**not-n**	140, 155-156, 170, 172, 186-188	8
241	**pq**	**mn**	**pq**	196	1
242	**pq**	**mq**	**not-n**	76, 80, 136, 144, 200, 204, 208	7
243	**pq**	**np**	**not-m**	211-212, 226, 228, 242-244	7
244	**pq**	**pq**	*no conclusion*	91-92, 95-96, 106, 108, 110, 112, 122-124, 126-128, 151-152, 159-160, 166, 168, 174, 176, 182-184, 190-192, 215-216, 219-220, 223-224, 230, 232, 234, 236, 238, 240, 246-248, 250-252, 254-256	49
245	**pq**	**m**	**not-n**	76, 80, 136, 144, 196, 200, 204, 208	8
246	**pq**	**n**	**not-m**	196, 211-212, 226, 228, 242-244	8
247	**pq**	**p**	*no conclusion*	91-92, 95-96, 106, 108, 110, 112, 122-124, 126-128, 151-152, 159-160, 166, 168, 174, 176, 182-184, 190-192, 211-212, 215-216, 219-220, 223-224, 226, 228, 230, 232, 234, 236, 238, 240, 242-244, 246-248, 250-252, 254-256	56
248	**pq**	**q**	*no conclusion*	76, 80, 91-92, 95-96, 106, 108, 110, 112, 122-124, 126-128, 136, 144, 151-152, 159-160, 166, 168, 174, 176, 182-184, 190-192, 200, 204, 208, 215-216, 219-220, 223-224, 230, 232, 234, 236, 238, 240, 246-248, 250-252, 254-256	56
251	**m**	**mn**	**n**	130, 194	2
252	**m**	**mq**	*no conclusion*	74, 78, 134, 138, 142, 198, 202, 206	8
253	**m**	**np**	**np**	146, 210	2
254	**m**	**pq**	**not-m**	90, 94, 150, 154, 158, 214, 218, 222	8
255	**m**	**m**	*no conclusion*	74, 78, 130, 134, 138, 142, 194, 198, 202, 206	10
256	**m**	**n**	**n**	130, 146, 194, 210	4
257	**m**	**p**	**not-m**	90, 94, 146, 150, 154, 158, 210, 214, 218, 222	10
258	**m**	**q**	*no conclusion*	74, 78, 90, 94, 134, 138, 142, 150, 154, 158, 198, 202, 206, 214, 218, 222	16
261	**n**	**mn**	**m**	130, 132	2
262	**n**	**mq**	**mq**	138, 140	2
263	**n**	**np**	*no conclusion*	146-148, 162, 164, 178-180	8
264	**n**	**pq**	**not-n**	154-156, 170, 172, 186-188	8
265	**n**	**m**	**m**	130, 132, 138, 140	4
266	**n**	**n**	*no conclusion*	130, 132, 146-148, 162, 164, 178-180	10
267	**n**	**p**	*no conclusion*	146-148, 154-156, 162, 164, 170, 172, 178-180, 186-188	16
268	**n**	**q**	**not-n**	138, 140, 154-156, 170, 172, 186-188	10
271	**p**	**mn**	**q**	132, 196	2
272	**p**	**mq**	**not-n**	76, 80, 136, 140, 144, 200, 204, 208	8
273	**p**	**np**	*no conclusion*	147-148, 162, 164, 178-180, 211-212, 226, 228, 242-244	14

Table 14.4 continued.

274	**p**	**pq**	*no conclusion*	91-92, 95-96, 106, 108, 110, 112, 122-124, 126-128, 151-152, 155-156, 159-160, 166, 168, 170, 172, 174, 176, 182-184, 186-188, 190-192, 215-216, 219-220, 223-224, 230, 232, 234, 236, 238, 240, 246-248, 250-252, 254-256	56
275	**p**	**m**	**not-n**	76, 80, 132, 136, 140, 144, 196, 200, 204, 208	10
276	**p**	**n**	*no conclusion*	132, 147-148, 162, 164, 178-180, 196, 211-212, 226, 228, 242-244	16
277	**p**	**p**	*no conclusion*	91-92, 95-96, 106, 108, 110, 112, 122-124, 126-128, 147-148, 151-152, 155-156, 159-160, 162, 164, 166, 168, 170, 172, 174, 176, 178-180, 182-184, 186-188, 190-192, 211-212, 215-216, 219-220, 223-224, 226, 228, 230, 232, 234, 236, 238, 240, 242-244, 246-248, 250-252, 254-256	70
278	**p**	**q**	*no conclusion*	76, 80, 91-92, 95-96, 106, 108, 110, 112, 122-124, 126-128, 136, 140, 144, 151-152, 155-156, 159-160, 166, 168, 170, 172, 174, 176, 182-184, 186-188, 190-192, 200, 204, 208, 215-216, 219-220, 223-224, 230, 232, 234, 236, 238, 240, 246-248, 250-252, 254-256	64
281	**q**	**mn**	**p**	194, 196	2
282	**q**	**mq**	*no conclusion*	74, 76, 78, 80, 134, 136, 142, 144, 198, 200, 202, 204, 206, 208	14
283	**q**	**np**	**not-m**	210-212, 226, 228, 242-244	8
284	**q**	**pq**	*no conclusion*	90-92, 94-96, 106, 108, 110, 112, 122-124, 126-128, 150-152, 158-160, 166, 168, 174, 176, 182-184, 190-192, 214-216, 218-220, 222-224, 230, 232, 234, 236, 238, 240, 246-248, 250-252, 254-256	56
285	**q**	**m**	*no conclusion*	74, 76, 78, 80, 134, 136, 142, 144, 194, 196, 198, 200, 202, 204, 206, 208	16
286	**q**	**n**	**not-m**	194, 196, 210-212, 226, 228, 242-244	10
287	**q**	**p**	*no conclusion*	90-92, 94-96, 106, 108, 110, 112, 122-124, 126-128, 150-152, 158-160, 166, 168, 174, 176, 182-184, 190-192, 210-212, 214-216, 218-220, 222-224, 226, 228, 230, 232, 234, 236, 238, 240, 242-244, 246-248, 250-252, 254-256	64
288	**q**	**q**	*no conclusion*	74, 76, 78, 80, 90-92, 94-96, 106, 108, 110, 112, 122-124, 126-128, 134, 136, 142, 144, 150-152, 158-160, 166, 168, 174, 176, 182-184, 190-192, 198, 200, 202, 204, 206, 208, 214-216, 218-220, 222-224, 230, 232, 234, 236, 238, 240, 246-248, 250-252, 254-256	70
	Third Figure				
Mood	QR	QP	PR		
311	**mn**	**mn**	**mn**	130	1
312	**mn**	**mq**	**np**	146	1
313	**mn**	**np**	**mq**	138	1
314	**mn**	**pq**	**pq**	154	1
315	**mn**	**m**	**n**	130, 146	2
316	**mn**	**n**	**m**	130, 138	2
317	**mn**	**p**	**q**	138, 154	2
318	**mn**	**q**	**p**	146, 154	2

Table 14.4 continued.

321	**mq**	**mn**	**mq**	132	1
322	**mq**	**mq**	*no conclusion*	147-148, 162, 164, 178-180	7
323	**mq**	**np**	**mq**	140	1
324	**mq**	**pq**	**not-n**	155-156, 170, 172, 186-188	7
325	**mq**	**m**	*no conclusion*	132, 147-148, 162, 164, 178-180	8
326	**mq**	**n**	**mq**	132, 140	2
327	**mq**	**p**	**not-n**	140, 155-156, 170, 172, 186-188	8
328	**mq**	**q**	*no conclusion*	147-148, 155-156, 162, 164, 170, 172, 178-180, 186-188	14
331	**np**	**mn**	**np**	194	1
332	**np**	**mq**	**np**	210	1
333	**np**	**np**	*no conclusion*	74, 78, 134, 142, 198, 202, 206	7
334	**np**	**pq**	**not-m**	90, 94, 150, 158, 214, 218, 222	7
335	**np**	**m**	**np**	194, 210	2
336	**np**	**n**	*no conclusion*	74, 78, 134, 142, 194, 198, 202, 206	8
337	**np**	**p**	*no conclusion*	74, 78, 90, 94, 134, 142, 150, 158, 198, 202, 206, 214, 218, 222	14
338	**np**	**q**	**not-m**	90, 94, 150, 158, 210, 214, 218, 222	8
341	**pq**	**mn**	**pq**	196	1
342	**pq**	**mq**	**not-m**	211-212, 226, 228, 242-244	7
343	**pq**	**np**	**not-n**	76, 80, 136, 144, 200, 204, 208	7
344	**pq**	**pq**	*no conclusion*	91-92, 95-96, 106, 108, 110, 112, 122-124, 126-128, 151-152, 159-160, 166, 168, 174, 176, 182-184, 190-192, 215-216, 219-220, 223-224, 230, 232, 234, 236, 238, 240, 246-248, 250-252, 254-256	49
345	**pq**	**m**	**not-m**	196, 211-212, 226, 228, 242-244	8
346	**pq**	**n**	**not-n**	76, 80, 136, 144, 196, 200, 204, 208	8
347	**pq**	**p**	*no conclusion*	76, 80, 91-92, 95-96, 106, 108, 110, 112, 122-124, 126-128, 136, 144, 151-152, 159-160, 166, 168, 174, 176, 182-184, 190-192, 200, 204, 208, 215-216, 219-220, 223-224, 230, 232, 234, 236, 238, 240, 246-248, 250-252, 254-256	56
348	**pq**	**q**	*no conclusion*	91-92, 95-96, 106, 108, 110, 112, 122-124, 126-128, 151-152, 159-160, 166, 168, 174, 176, 182-184, 190-192, 211-212, 215-216, 219-220, 223-224, 226, 228, 230, 232, 234, 236, 238, 240, 242-244, 246-248, 250-252, 254-256	56
351	**m**	**mn**	**m**	130, 132	2
352	**m**	**mq**	*no conclusion*	146-148, 162, 164, 178-180	8
353	**m**	**np**	**mq**	138, 140	2
354	**m**	**pq**	**not-n**	154-156, 170, 172, 186-188	8
355	**m**	**m**	*no conclusion*	130, 132, 146-148, 162, 164, 178-180	10
356	**m**	**n**	**m**	130, 132, 138, 140	4
357	**m**	**p**	**not-n**	138, 140, 154-156, 170, 172, 186-188	10
358	**m**	**q**	*no conclusion*	146-148, 154-156, 162, 164, 170, 172, 178-180, 186-188	16
361	**n**	**mn**	*n*	130, 194	2
362	**n**	**mq**	*np*	146, 210	2

Table 14.4 continued.

363	**n**	**np**	*no conclusion*	74, 78, 134, 138, 142, 198, 202, 206	8
364	**n**	**pq**	**not-m**	90, 94, 150, 154, 158, 214, 218, 222	8
365	**n**	**m**	**n**	130, 146, 194, 210	4
366	**n**	**n**	*no conclusion*	74, 78, 130, 134, 138, 142, 194, 198, 202, 206	10
367	**n**	**p**	*no conclusion*	74, 78, 90, 94, 134, 138, 142, 150, 154, 158, 198, 202, 206, 214, 218, 222	16
368	**n**	**q**	**not-m**	90, 94, 146, 150, 154, 158, 210, 214, 218, 222	10
371	**p**	**mn**	**p**	194, 196	2
372	**p**	**mq**	**not-m**	210-212, 226, 228, 242-244	8
373	**p**	**np**	*no conclusion*	74, 76, 78, 80, 134, 136, 142, 144, 198, 200, 202, 204, 206, 208	14
374	**p**	**pq**	*no conclusion*	90-92, 94-96, 106, 108, 110, 112, 122-124, 126-128, 150-152, 158-160, 166, 168, 174, 176, 182-184, 190-192, 214-216, 218-220, 222-224, 230, 232, 234, 236, 238, 240, 246-248, 250-252, 254-256	56
375	**p**	**m**	**not-m**	194, 196, 210-212, 226, 228, 242-244	10
376	**p**	**n**	*no conclusion*	74, 76, 78, 80, 134, 136, 142, 144, 194, 196, 198, 200, 202, 204, 206, 208	16
377	**p**	**p**	*no conclusion*	74, 76, 78, 80, 90-92, 94-96, 106, 108, 110, 112, 122-124, 126-128, 134, 136, 142, 144, 150-152, 158-160, 166, 168, 174, 176, 182-184, 190-192, 198, 200, 202, 204, 206, 208, 214-216, 218-220, 222-224, 230, 232, 234, 236, 238, 240, 246-248, 250-252, 254-256	70
378	**p**	**q**	*no conclusion*	90-92, 94-96, 106, 108, 110, 112, 122-124, 126-128, 150-152, 158-160, 166, 168, 174, 176, 182-184, 190-192, 210-212, 214-216, 218-220, 222-224, 226, 228, 230, 232, 234, 236, 238, 240, 242-244, 246-248, 250-252, 254-256	64
381	**q**	**mn**	**q**	132, 196	2
382	**q**	**mq**	*no conclusion*	147-148, 162, 164, 178-180, 211-212, 226, 228, 242-244	14
383	**q**	**np**	**not-n**	76, 80, 136, 140, 144, 200, 204, 208	8
384	**q**	**pq**	*no conclusion*	91-92, 95-96, 106, 108, 110, 112, 122-124, 126-128, 151-152, 155-156, 159-160, 166, 168, 170, 172, 174, 176, 182-184, 186-188, 190-192, 215-216, 219-220, 223-224, 230, 232, 234, 236, 238, 240, 246-248, 250-252, 254-256	56
385	**q**	**m**	*no conclusion*	132, 147-148, 162, 164, 178-180, 196, 211-212, 226, 228, 242-244	16
386	**q**	**n**	**not-n**	76, 80, 132, 136, 140, 144, 196, 200, 204, 208	10
387	**q**	**p**	*no conclusion*	76, 80, 91-92, 95-96, 106, 108, 110, 112, 122-124, 126-128, 136, 140, 144, 151-152, 155-156, 159-160, 166, 168, 170, 172, 174, 176, 182-184, 186-188, 190-192, 200, 204, 208, 215-216, 219-220, 223-224, 230, 232, 234, 236, 238, 240, 246-248, 250-252, 254-256	64
388	**q**	**q**	*no conclusion*	91-92, 95-96, 106, 108, 110, 112, 122-124, 126-128, 147-148, 151-152, 155-156, 159-160, 162, 164, 166, 168, 170, 172, 174, 176, 178-180, 182-184, 186-188, 190-192, 211-212, 215-216, 219-220, 223-224, 226, 228, 230, 232, 234, 236, 238, 240, 242-244, 246-248, 250-252, 254-256	70

If we compare the results of microanalysis listed in the above table with those obtained by macroanalysis (those listed in chapter 6, and again in Tables 7.3, 7.4 and 7.5, and those listed in Tables 9.4 and 9.5), we discover that they are in most cases apparently inconsistent! However, before making any hasty judgments, it is well to become aware that *many apparent similarities between the cases treated are in fact superficial, so that differences in result should come as no surprise*.

We could at this stage create a table listing all positive moods in the three figures in one column, the collected positive and negative conclusions by macroanalysis in a second column, and the above conclusions obtained by microanalysis in a third column. Then, in a fourth column, we would want to explain, case by case, why any eventual divergences occurred. However, it is still too early for such a systematic comparison between our approaches, for reasons that will become clear in the following comments.

In the above table, we are dealing with only three items, P, Q and R. The weak determinations considered are always absolute (involving 108 alternative moduses in the PQR matrix). It follows that even when one or both premises are weak, the syllogism is always in subfigure 'a'[103].

We cannot, therefore, compare the conclusions obtained here to those previously found, when a subsidiary item (S) was specified and the minor item (P) was occasionally found as a complement in the major premise (as in subfigure 'd' and others). Here, when a weak determination (absolute) is affirmed or denied in premise(s) and/or in conclusion, the effective moduses might concern the subsidiary item *or* its negation, *or* the minor, middle or major item or its negation, *or again* some other item entirely!

We must thus tread very carefully before making comparisons. It is ultimately only by developing four-item microanalysis, and thereby four-item syllogism, that mechanical comparisons with macroanalysis can safely be made. We will consequently abstain from such consistency checking for now, while acknowledging its ultimate importance. Let us rather at this stage continue to develop three-item syllogistic theory. Some of these developments will clarify the issue raised here, as we shall see.

4. Dealing with Vaguer Propositions.

We shall now deal with three-item syllogism with vague premises; that is, strong causation (**s** = **m or n**), absolute weak causation ($\mathbf{w}_{abs}$ = $\mathbf{p}_{abs}$ **or** $\mathbf{q}_{abs}$), or absolute causation ($\mathbf{c}_{abs}$ = **s or** $\mathbf{w}_{abs}$ = **m or n or** $\mathbf{p}_{abs}$ **or** $\mathbf{q}_{abs}$). What we shall find is that *no conclusion* arises from any combination of such premises (with each other - we shall not bother to deal with combinations of vague premises with more precise premises, regarding them as practically unlikely to arise). This is an interesting result, though negative.

The following table merges the moduses of precise positive premises (given in Table 14.3, above), to obtain those of vague positive premises with forms **s**, $\mathbf{w}_{abs}$ or $\mathbf{c}_{abs}$. Note well that each of these forms is fully convertible, i.e. is the same for two given items whatever their positions (QR and RQ have the same moduses, PQ and QP have the same moduses).

---

103 See chapter 5 for the list of subfigures.

Table 14.5. Enumeration of three-item alternative moduses for vague positive premises, for any figure of syllogism.

Determination	Major QR	Major RQ	Minor PQ	Minor QP
s = m or n	10, 12, 14, 25-30, 42, 44, 57-60, 74, 78, 89-90, 93-94, 130, 132, 134, 138, 140, 142, 145-150, 153-158, 162, 164, 170, 172, 177-180, 185-188, 194, 198, 202, 206, 209-210, 213-214, 217-218, 221-222	10, 12, 14, 25-30, 42, 44, 57-60, 74, 78, 89-90, 93-94, 130, 132, 134, 138, 140, 142, 145-150, 153-158, 162, 164, 170, 172, 177-180, 185-188, 194, 198, 202, 206, 209-210, 213-214, 217-218, 221-222	66-68, 70-72, 74-76, 78-80, 82-84, 98-100, 114-116, 130-132, 134-136, 138-140, 142-144, 146-148, 162-164, 178-180, 194-196, 198-200, 202-204, 206-208, 210-212, 226-228, 242-244	66-68, 70-72, 74-76, 78-80, 82-84, 98-100, 114-116, 130-132, 134-136, 138-140, 142-144, 146-148, 162-164, 178-180, 194-196, 198-200, 202-204, 206-208, 210-212, 226-228, 242-244
$\mathbf{w}_{abs}$ = $\mathbf{p}_{abs}$ **or** $\mathbf{q}_{abs}$	12, 14, 16, 27-32, 42, 44, 46, 48, 57-64, 74, 76, 78, 80, 89-96, 106, 108, 110, 112, 121-128, 132, 134, 136, 140, 142, 144, 147-152, 155-160, 162, 164, 166, 168, 170, 172, 174, 176-192, 194, 196, 198, 200, 202, 204, 206, 208-224, 226, 228, 230, 232, 234, 236, 238, 240-256	12, 14, 16, 27-32, 42, 44, 46, 48, 57-64, 74, 76, 78, 80, 89-96, 106, 108, 110, 112, 121-128, 132, 134, 136, 140, 142, 144, 147-152, 155-160, 162, 164, 166, 168, 170, 172, 174, 176-192, 194, 196, 198, 200, 202, 204, 206, 208-224, 226, 228, 230, 232, 234, 236, 238, 240-256	70-72, 74-76, 78-80, 82-84, 86-88, 90-92, 94-96, 98-100, 102-104, 106-108, 110-112, 114-116, 118-120, 122-124, 126-128, 134-136, 138-140, 142-144, 146-148, 150-152, 154-156, 158-160, 162-164, 166-168, 170-172, 174-176, 178-180, 182-184, 186-188, 190-192, 198-200, 202-204, 206-208, 210-212, 214-216, 218-220, 222-224, 226-228, 230-232, 234-236, 238-240, 242-244, 246-248, 250-252, 254-256	70-72, 74-76, 78-80, 82-84, 86-88, 90-92, 94-96, 98-100, 102-104, 106-108, 110-112, 114-116, 118-120, 122-124, 126-128, 134-136, 138-140, 142-144, 146-148, 150-152, 154-156, 158-160, 162-164, 166-168, 170-172, 174-176, 178-180, 182-184, 186-188, 190-192, 198-200, 202-204, 206-208, 210-212, 214-216, 218-220, 222-224, 226-228, 230-232, 234-236, 238-240, 242-244, 246-248, 250-252, 254-256
$\mathbf{c}_{abs}$ = **m or n** **or** $\mathbf{p}_{abs}$ **or** $\mathbf{q}_{abs}$	10, 12, 14, 16, 25-32, 42, 44, 46, 48, 57-64, 74, 76, 78, 80, 89-96, 106, 108, 110, 112, 121-128, 130, 132, 134, 136, 138, 140, 142, 144-160, 162, 164, 166, 168, 170, 172, 174, 176-192, 194, 196, 198, 200, 202, 204, 206, 208-224, 226, 228, 230, 232, 234, 236, 238, 240-256	10, 12, 14, 16, 25-32, 42, 44, 46, 48, 57-64, 74, 76, 78, 80, 89-96, 106, 108, 110, 112, 121-128, 130, 132, 134, 136, 138, 140, 142, 144-160, 162, 164, 166, 168, 170, 172, 174, 176-192, 194, 196, 198, 200, 202, 204, 206, 208-224, 226, 228, 230, 232, 234, 236, 238, 240-256	66-68, 70-72, 74-76, 78-80, 82-84, 86-88, 90-92, 94-96, 98-100, 102-104, 106-108, 110-112, 114-116, 118-120, 122-124, 126-128, 130-132, 134-136, 138-140, 142-144, 146-148, 150-152, 154-156, 158-160, 162-164, 166-168, 170-172, 174-176, 178-180, 182-184, 186-188, 190-192, 194-196, 198-200, 202-204, 206-208, 210-212, 214-216, 218-220, 222-224, 226-228, 230-232, 234-236, 238-240, 242-244, 246-248, 250-252, 254-256	66-68, 70-72, 74-76, 78-80, 82-84, 86-88, 90-92, 94-96, 98-100, 102-104, 106-108, 110-112, 114-116, 118-120, 122-124, 126-128, 130-132, 134-136, 138-140, 142-144, 146-148, 150-152, 154-156, 158-160, 162-164, 166-168, 170-172, 174-176, 178-180, 182-184, 186-188, 190-192, 194-196, 198-200, 202-204, 206-208, 210-212, 214-216, 218-220, 222-224, 226-228, 230-232, 234-236, 238-240, 242-244, 246-248, 250-252, 254-256

As the table below shows, no positive or negative conclusion is obtainable from two vague positive premises (any combination of **s**, $\mathbf{w}_{abs}$, $\mathbf{c}_{abs}$). This means one has to go to a more precise level - of generics or joints - to obtain a conclusion in positive causative syllogism.

Note that the three figures yield the same moduses in each mood. This is because, as indicated above, the premises involved have the same moduses whatever the orientation of the items concerned (i.e. they are all fully convertible).

Table 14.6. Moduses of conclusions for selected vague positive premises.

Major	Minor	Conclusion
First Figure		
QR	PQ	PR
Second Figure		
RQ	PQ	PR
Third Figure		
QR	QP	PR

For all three figures			Common moduses	No.
s	**s**	*no conclusion*	74, 78, 130, 132, 134, 138, 140, 142, 146-148, 162, 164, 178-180, 194, 198, 202, 206, 210	21
s	$\mathbf{w}_{abs}$	*no conclusion*	74, 78, 90, 94, 134, 138, 140, 142, 146-148, 150, 154-156, 158, 162, 164, 170, 172, 178-180, 186-188, 198, 202, 206, 210, 214, 218, 222	33
$\mathbf{w}_{abs}$	**s**	*no conclusion*	74, 76, 78, 80, 132, 134, 136, 140, 142, 144, 147-148, 162, 164, 178-180, 194, 196, 198, 200, 202, 204, 206, 208, 210-212, 226, 228, 242-244	33
$\mathbf{w}_{abs}$	$\mathbf{w}_{abs}$	*no conclusion*	74, 76, 78, 80, 90-92, 94-96, 106, 108, 110, 112, 122-124, 126-128, 134, 136, 140, 142, 144, 147-148, 150-152, 155-156, 158-160, 162, 164, 166, 168, 170, 172, 174, 176, 178-180, 182-184, 186-188, 190-192, 198, 200, 202, 204, 206, 208, 210-212, 214-216, 218-220, 222-224, 226, 228, 230, 232, 234, 236, 238, 240, 242-244, 246-248, 250-252, 254-256	93
s	$\mathbf{c}_{abs}$	*no conclusion*	74, 78, 90, 94, 130, 132, 134, 138, 140, 142, 146-148, 150, 154-156, 158, 162, 164, 170, 172, 178-180, 186-188, 194, 198, 202, 206, 210, 214, 218, 222	36
$\mathbf{c}_{abs}$	**s**	*no conclusion*	74, 76, 78, 80, 130, 132, 134, 136, 138, 140, 142, 144, 146-148, 162, 164, 178-180, 194, 196, 198, 200, 202, 204, 206, 208, 210-212, 226, 228, 242-244	36
$\mathbf{w}_{abs}$	$\mathbf{c}_{abs}$	*no conclusion*	74, 76, 78, 80, 90-92, 94-96, 106, 108, 110, 112, 122-124, 126-128, 132, 134, 136, 140, 142, 144, 147-148, 150-152, 155-156, 158-160, 162, 164, 166, 168, 170, 172, 174, 176, 178-180, 182-184, 186-188, 190-192, 194, 196, 198, 200, 202, 204, 206, 208, 210-212, 214-216, 218-220, 222-224, 226, 228, 230, 232, 234, 236, 238, 240, 242-244, 246-248, 250-252, 254-256	96
$\mathbf{c}_{abs}$	$\mathbf{w}_{abs}$	*no conclusion*	74, 76, 78, 80, 90-92, 94-96, 106, 108, 110, 112, 122-124, 126-128, 134, 136, 138, 140, 142, 144, 146-148, 150-152, 154-156, 158-160, 162, 164, 166, 168, 170, 172, 174, 176, 178-180, 182-184, 186-188, 190-192, 198, 200, 202, 204, 206, 208, 210-212, 214-216, 218-220, 222-224, 226, 228, 230, 232, 234, 236, 238, 240, 242-244, 246-248, 250-252, 254-256	96

Table 14.6 continued.

$\mathbf{c}_{abs}$	$\mathbf{c}_{abs}$	*no conclusion*	74, 76, 78, 80, 90-92, 94-96, 106, 108, 110, 112, 122-124, 126-128, 130, 132, 134, 136, 138, 140, 142, 144, 146-148, 150-152, 154-156, 158-160, 162, 164, 166, 168, 170, 172, 174, 176, 178-180, 182-184, 186-188, 190-192, 194, 196, 198, 200, 202, 204, 206, 208, 210-212, 214-216, 218-220, 222-224, 226, 228, 230, 232, 234, 236, 238, 240, 242-244, 246-248, 250-252, 254-256	100

This table is interesting in that it shows that no causative conclusion can be drawn from vague causative premises in any of the three figures. In particular, it answers one of our first questions, teaching us that, generally speaking, if P causes Q and Q causes R, it does not follow that P causes R.

Chapter 15. SOME MORE THREE-ITEM SYLLOGISMS.

In the previous chapter, we examined absolute positive three-item syllogism. In the present chapter, we shall look for other syllogistic forms, involving relative or negative causative propositions. But keep in mind that this endeavor does not exhaust the matter: we shall eventually still be obliged to develop four-item syllogism.

1. Special Cases of Three-item Syllogism.

We shall now consider special cases of three-item syllogism, involving *weak determinations relative to the major and/or minor item*. That is, three-item syllogism with major premise of form Q(P)R or R(P)Q and/or minor premise of form P(R)Q or Q(R)P, with eventual conclusions of absolute form PR or relative form P(Q)R.

Note well that such syllogisms still involve *only three items*, even though one or more propositions in them are of *relative weak* determination. Such cases may conceivably arise in practice, though very rarely. For each of the three figures, we may conceive of seven such subfigures, in addition to the already dealt with standard Aristotelian arrangement. The eight subfigures are labeled 'a' and 'j' to 'p', for convenience, in the table below.

Table 15.1. Subfigures of syllogism with three items only.

	With PR conclusion				With P(Q)R conclusion			
Subfigures	a	j	k	l	m	n	o	p
Figure 1	QR PQ PR	QR P(R)Q PR	Q(P)R PQ PR	Q(P)R P(R)Q PR	QR PQ P(Q)R	QR P(R)Q P(Q)R	Q(P)R PQ P(Q)R	Q(P)R P(R)Q P(Q)R
Figure 2	RQ PQ PR	RQ P(R)Q PR	R(P)Q PQ PR	R(P)Q P(R)Q PR	RQ PQ P(Q)R	RQ P(R)Q P(Q)R	R(P)Q PQ P(Q)R	R(P)Q P(R)Q P(Q)R
Figure 3	QR QP PR	QR Q(R)P PR	Q(P)R QP PR	Q(P)R Q(R)P PR	QR QP P(Q)R	QR Q(R)P P(Q)R	Q(P)R QP P(Q)R	Q(P)R Q(R)P P(Q)R

Notice that in each figure there are two sets of four subfigures, with conclusions of form PR or P(Q)R. Only subfigure 'a' has been treated by macroanalysis, the other seven here were simply ignored in Phase One (being very special cases, not likely to often arise). They become relevant here only because they serve to clarify what we mean by three-item (as against four-item) syllogism.

Moods with major premise of form Q(P)R or R(P)Q involve a weak determination relative to the minor item P; and those with minor premise of form P(R)Q or Q(R)P involve a weak determination relative to the major item R. Their summary moduses are given in the following table. Note that the forms P(Q)R and Q(P)R yield the same moduses.

Table 15.2. Summary moduses for weak premises relative to the minor item P or to the major item R.

Row label	P	Q	R	p_P QR	p_P RQ	p_R PQ	p_R QP	q_P QR	q_P RQ	q_R PQ	q_R QP
a	**1**	**1**	**1**	1	1	1	1	·	·	·	·
b	**1**	**1**	**0**	0	·	·	·	·	1	1	1
c	**1**	**0**	**1**	·	0	0	·	1	·	·	1
d	**1**	**0**	**0**	1	1	1	·	·	·	·	0
e	**0**	**1**	**1**	·	·	·	0	1	1	1	·
f	**0**	**1**	**0**	1	·	·	1	·	0	0	·
g	**0**	**0**	**1**	·	1	1	1	0	·	·	·
h	**0**	**0**	**0**	·	·	·	·	1	1	1	1

The above table may be proved by appropriate reshuffling of columns and rows: in each case, we obtain the same summary modus of partial or contingent causation. For instance, for $\mathbf{p}_{QRP}$, move column P to the right of columns Q and R, then reorder the rows to 'aebfcgdh' (so that the sequences of QRP are 111, 110, 101, 100, etc.) – the result is summary modus 10.1.1.. as required to prove.

From the above summary moduses, we can derive the alternative moduses listed in the following table. The microanalyses of the strong forms (**m** and **n**) are carried over from Table 14.3. Notice the differences between the results below and those for relatives in Table 13.1. Here $\mathbf{mq}_{rel}$ = #s 42, 58, 170, 186, whereas there it = #s 42, 46, 170, 174; similarly, here $\mathbf{np}_{rel}$ = #s 149-150, 157-158, whereas there it = #s 149-150, 181-182. These differences (two out of four moduses in each case) are simply due to the forms of strong determination the weaks are combined with: there it was with PR, whereas here it is with QR.

Table 15.3. Enumeration of three-item alternative moduses weak positive premises relative to the minor item P or to the major item R.

Determ.	Major Q(P)R	Major R(P)Q	Minor P(R)Q	Minor Q(R)P
m	10, 12, 25-28, 42, 44, 57-60, 130, 132, 138, 140, 145-148, 153-156, 162, 164, 170, 172, 177-180, 185-188	10, 14, 25-26, 29-30, 74, 78, 89-90, 93-94, 130, 134, 138, 142, 145-146, 149-150, 153-154, 157-158, 194, 198, 202, 206, 209-210, 213-214, 217-218, 221-222	66-68, 70-72, 74-76, 78-80, 130-132, 134-136, 138-140, 142-144, 194-196, 198-200, 202-204, 206-208	66-68, 82-84, 98-100, 114-116, 130-132, 146-148, 162-164, 178-180, 194-196, 210-212, 226-228, 242-244
n	10, 14, 25-26, 29-30, 74, 78, 89-90, 93-94, 130, 134, 138, 142, 145-146, 149-150, 153-154, 157-158, 194, 198, 202, 206, 209-210, 213-214, 217-218, 221-222	10, 12, 25-28, 42, 44, 57-60, 130, 132, 138, 140, 145-148, 153-156, 162, 164, 170, 172, 177-180, 185-188	66-68, 82-84, 98-100, 114-116, 130-132, 146-148, 162-164, 178-180, 194-196, 210-212, 226-228, 242-244	66-68, 70-72, 74-76, 78-80, 130-132, 134-136, 138-140, 142-144, 194-196, 198-200, 202-204, 206-208

Table 15.3 continued.

$\mathbf{p}_{rel}$	149-152, 157-160, 181-184, 189-192	147-148, 151-152, 155-156, 159-160, 211-212, 215-216, 219-220, 223-224	147-148, 151-152, 155-156, 159-160, 211-212, 215-216, 219-220, 223-224	135-136, 151-152, 167-168, 183-184, 199-200, 215-216, 231-232, 247-248
$\mathbf{q}_{rel}$	42, 46, 58, 62, 106, 110, 122, 126, 170, 174, 186, 190, 234, 238, 250, 254	74, 76, 90, 92, 106, 108, 122, 124, 202, 204, 218, 220, 234, 236, 250, 252	74, 76, 90, 92, 106, 108, 122, 124, 202, 204, 218, 220, 234, 236, 250, 252	98, 100, 102, 104, 106, 108, 110, 112, 226, 228, 230, 232, 234, 236, 238, 240
$\mathbf{mq}_{rel}$	42, 58, 170, 186	74, 90, 202, 218	74, 76, 202, 204	98, 100, 226, 228
$\mathbf{np}_{rel}$	149-150, 157-158	147-148, 155-156	147-148, 211-212	135-136, 199-200
$\mathbf{p}_{rel}\mathbf{q}_{rel}$	190	220	220	232
$\mathbf{w}_{rel}$	42, 46, 58, 62, 106, 110, 122, 126, 149-152, 157-160, 170, 174, 181-184, 186, 189-192, 234, 238, 250, 254	74, 76, 90, 92, 106, 108, 122, 124, 147-148, 151-152, 155-156, 159-160, 202, 204, 211-212, 215-216, 218-220, 223-224, 234, 236, 250, 252	74, 76, 90, 92, 106, 108, 122, 124, 147-148, 151-152, 155-156, 159-160, 202, 204, 211-212, 215-216, 218-220, 223-224, 234, 236, 250, 252	98, 100, 102, 104, 106, 108, 110, 112, 135-136, 151-152, 167-168, 183-184, 199-200, 215-216, 226, 228, 230-232, 234, 236, 238, 240, 247-248
$\mathbf{c}_{rel}$	10, 12, 14, 25-30, 42, 44, 46, 57-60, 62, 74, 78, 89-90, 93-94, 106, 110, 122, 126, 130, 132, 134, 138, 140, 142, 145-160, 162, 164, 170, 172, 174, 177-192, 194, 198, 202, 206, 209-210, 213-214, 217-218, 221-222, 234, 238, 250, 254	10, 12, 14, 25-30, 42, 44, 57-60, 74, 76, 78, 89-90, 92-94, 106, 108, 122, 124, 130, 132, 134, 138, 140, 142, 145-150, 151-160, 162, 164, 170, 172, 177-180, 185-188, 194, 198, 202, 204, 206, 209-224, 234, 236, 250, 252	66-68, 70-72, 74-76, 78-80, 82-84, 90, 92, 98-100, 106, 108, 114-116, 122, 124, 130-132, 134-136, 138-140, 142-144, 146-148, 151-152, 155-156, 159-160, 162-164, 178-180, 194-196, 198-200, 202-204, 206-208, 210-212, 215-216, 218-220, 223-224, 226-228, 234, 236, 242-244, 250, 252	66-68, 70-72, 74-76, 78-80, 82-84, 98-100, 102, 104, 106, 108, 110, 112, 114-116, 130-132, 134-136, 138-140, 142-144, 146-148, 151-152, 162-164, 167-168, 178-180, 183-184, 194-196, 198-200, 202-204, 206-208, 210-212, 215-216, 226-228, 230-232, 234, 236, 238, 240, 242-244, 247-248

Note that for $\mathbf{p}_{rel}$, $\mathbf{q}_{rel}$, $\mathbf{p}_{rel}\mathbf{q}_{rel}$ and $\mathbf{w}_{rel}$, a major premise of form R(P)Q and a minor premise of form P(R)Q, of the same weak determination, have the same alternative moduses. This is not surprising, since these weak forms involve the same items as causes (P, R) and effect (Q). A similar equation is not obtained where a strong determination is involved because in such case the items involved are in fact only RQ and PQ (without complements).

The following tables list a selection of syllogisms one or both of whose premises involve all three items. They ignore syllogisms with an exclusively strong premise (such as moods 111-118, 121, 125, 126, etc.), not because such syllogisms are impossible or uninteresting, but simply for brevity's sake (the reader is invited to look into such cases as an exercise).

The conclusions are implicit in the common moduses of the premises. Once these common moduses, if any, are identified, the conclusion has to be sought in Table 12.4, if of absolute form PR, or in Table 13.1, if of relative form P(Q)R, these forms being those predicted in Table 15.1 above.

Forms without subscript are intended as absolute, those with a subscript as relative to the mentioned complement (P, Q or R). A conclusion of the form **p** or **q** or **pq** has absolute form PR; a similar conclusion of relative form P(Q)R is not implied by it. A conclusion of the form $\mathbf{not\text{-}c}_{Q}$ means **not-m** + **not-n** + $\mathbf{not\text{-}p}_{Q}$ + $\mathbf{not\text{-}q}_{Q}$, which does not imply $\mathbf{not\text{-}c}_{abs}$. Lone determinations are sometimes concluded, but these are of course relative and not absolute. Note well that the subsidiary item S is never mentioned here, since we have not performed four-item microanalysis yet.

Table 15.4. Moduses of conclusions for selected relative weak positive minor premises (subfigures j, n).

Mood	Major	Minor	Conclusion	Common moduses	No.
1st	QR	PQ	PR		
122	**mq**	**mq**$_R$	*inconsistent*	None	0
123	**mq**	**np**$_R$	**not-c**$_Q$	147-148	2
124	**mq**	**p**$_R$**q**$_R$	*inconsistent*	None	0
127	**mq**	**p**$_R$	**not-c**$_Q$	147-148, 155-156	4
128	**mq**	**q**$_R$	*inconsistent*	None	0
132	**np**	**mq**$_R$	**not-c**$_Q$	74, 202	2
133	**np**	**np**$_R$	*inconsistent*	None	0
134	**np**	**p**$_R$**q**$_R$	*inconsistent*	None	0
137	**np**	**p**$_R$	*inconsistent*	None	0
138	**np**	**q**$_R$	**not-c**$_Q$	74, 90, 202, 218	4
142	**pq**	**mq**$_R$	**not-c**$_Q$	76, 204	2
143	**pq**	**np**$_R$	**not-c**$_Q$	211-212	2
144	**pq**	**p**$_R$**q**$_R$	**not-c**$_Q$	220	1
147	**pq**	**p**$_R$	**not-s**	151-152, 159-160, 211-212, 215-216, 219-220, 223-224	12
148	**pq**	**q**$_R$	**not-s**	76, 92, 106, 108, 122, 124, 204, 220, 234, 236, 250, 252	12
172	**p**	**mq**$_R$	**not-c**$_Q$	74, 76, 202, 204	4
173	**p**	**np**$_R$	**not-c**$_Q$	211-212	2
174	**p**	**p**$_R$**q**$_R$	**not-c**$_Q$	220	1
177	**p**	**p**$_R$	**not-s**	151-152, 159-160, 211-212, 215-216, 219-220, 223-224	12
178	**p**	**q**$_R$	**not-s**	74, 76, 90, 92, 106, 108, 122, 124, 202, 204, 218, 220, 234, 236, 250, 252	16
182	**q**	**mq**$_R$	**not-c**$_Q$	76, 204	2
183	**q**	**np**$_R$	**not-c**$_Q$	147-148, 211-212	4
184	**q**	**p**$_R$**q**$_R$	**not-c**$_Q$	220	1
187	**q**	**p**$_R$	**not-s**	147-148, 151-152, 155-156, 159-160, 211-212, 215-216, 219-220, 223-224	16
188	**q**	**q**$_R$	**not-s**	76, 92, 106, 108, 122, 124, 204, 220, 234, 236, 250, 252	12
2nd	**RQ**	**PQ**	**PR**		
222	**mq**	**mqR**	**not-cQ**	74, 202	2
223	**mq**	**npR**	**inconsistent**	None	0
224	**mq**	**pRqR**	**inconsistent**	None	0
227	**mq**	**pR**	**inconsistent**	None	0
228	**mq**	**qR**	**not-cQ**	74, 90, 202, 218	4
232	**np**	**mqR**	**inconsistent**	None	0
233	**np**	**npR**	**not-cQ**	147-148	2
234	**np**	**pRqR**	**inconsistent**	None	2

Table 15.4 continued.

237	**np**	p_R	**not-c_Q**	147-148, 155-156	4
238	**np**	q_R	*inconsistent*	None	0
242	**pq**	**m**q_R	**not-c_Q**	76, 204	2
243	**pq**	**n**p_R	**not-c_Q**	211-212	2
244	**pq**	p_Rq_R	**not-c_Q**	220	1
247	**pq**	p_R	**not-s**	151-152, 159-160, 211-212, 215-216, 219-220, 223-224	12
248	**pq**	q_R	**not-s**	76, 92, 106, 108, 122, 124, 204, 220, 234, 236, 250, 252	12
272	**p**	**m**q_R	**not-c_Q**	76, 204	2
273	**p**	**n**p_R	**not-c_Q**	147-148, 211-212	4
274	**p**	p_Rq_R	**not-c_Q**	220	1
277	**p**	p_R	**not-s**	147-148, 151-152, 155-156, 159-160, 211-212, 215-216, 219-220, 223-224	16
278	**p**	q_R	**not-s**	76, 92, 106, 108, 122, 124, 204, 220, 234, 236, 250, 252	12
282	**q**	**m**q_R	**not-c_Q**	74, 76, 202, 204	4
283	**q**	**n**p_R	**not-c_Q**	211-212	2
284	**q**	p_Rq_R	**not-c_Q**	220	1
287	**q**	p_R	**not-s**	151-152, 159-160, 211-212, 215-216, 219-220, 223-224	12
288	**q**	q_R	**not-s**	74, 76, 90, 92, 106, 108, 122, 124, 202, 204, 218, 220, 234, 236, 250, 252	16
3rd	QR	QP	PR		
322	**mq**	**m**q_R	*inconsistent*	None	0
323	**mq**	**n**p_R	*inconsistent*	None	0
324	**mq**	p_Rq_R	*inconsistent*	None	0
327	**mq**	p_R	*inconsistent*	None	0
328	**mq**	q_R	*inconsistent*	None	0
332	**np**	**m**q_R	*inconsistent*	None	0
333	**np**	**n**p_R	*inconsistent*	None	0
334	**np**	p_Rq_R	*inconsistent*	None	0
337	**np**	p_R	*inconsistent*	None	0
338	**np**	q_R	*inconsistent*	None	0
342	**pq**	**m**q_R	**p + not-p_Q**	226, 228	2
343	**pq**	**n**p_R	**q + not-q_Q**	136, 200	2
344	**pq**	**pRqR**	**not-cQ**	232	1
347	**pq**	**pR**	**q + not-qQ**	136, 151-152, 168, 183-184, 200, 215-216, 232, 247-248	12
348	**pq**	**qR**	**p + not-pQ**	106, 108, 110, 112, 226, 228, 230, 232, 234, 236, 238, 240	12
372	**p**	**mqR**	**p + not-pQ**	226, 228	2
373	**p**	**npR**	**q + not-qQ**	136, 200	2
374	**p**	**pRqR**	**not-cQ**	232	1
377	**p**	**pR**	**q + not-qQ**	136, 151-152, 168, 183-184, 200, 215-216, 232, 247-248	12
378	**p**	**qR**	**p + not-pQ**	106, 108, 110, 112, 226, 228, 230, 232, 234, 236, 238, 240	12
382	**q**	**mqR**	**p + not-pQ**	226, 228	2

Table 15.4 continued.

383	$\mathbf{q}$	$\mathbf{np}_R$	$\mathbf{q + not\text{-}q}_Q$	136, 200	2
384	$\mathbf{q}$	$\mathbf{p}_R\mathbf{q}_R$	$\mathbf{not\text{-}c}_Q$	232	1
387	$\mathbf{q}$	$\mathbf{p}_R$	$\mathbf{q + not\text{-}q}_Q$	136, 151-152, 168, 183-184, 200, 215-216, 232, 247-248	12
388	$\mathbf{q}$	$\mathbf{q}_R$	$\mathbf{p + not\text{-}p}_Q$	106, 108, 110, 112, 226, 228, 230, 232, 234, 236, 238, 240	12

Table 15.5. Moduses of conclusions for selected relative weak positive major premises (subfigures k, o).

Mood	Major	Minor	Conclusion	Common moduses	No.
1st	QR	PQ	PR		
122	$\mathbf{mq}_P$	$\mathbf{mq}$	*inconsistent*	None	0
123	$\mathbf{mq}_P$	$\mathbf{np}$	*inconsistent*	None	0
124	$\mathbf{mq}_P$	$\mathbf{pq}$	$\mathbf{q}_Q$	170, 186	2
127	$\mathbf{mq}_P$	$\mathbf{p}$	$\mathbf{q}_Q$	170, 186	2
128	$\mathbf{mq}_P$	$\mathbf{q}$	$\mathbf{q}_Q$	170, 186	2
132	$\mathbf{np}_P$	$\mathbf{mq}$	*inconsistent*	None	0
133	$\mathbf{np}_P$	$\mathbf{np}$	*inconsistent*	None	0
134	$\mathbf{np}_P$	$\mathbf{pq}$	$\mathbf{p}_Q$	150, 158	2
137	$\mathbf{np}_P$	$\mathbf{p}$	$\mathbf{p}_Q$	150, 158	2
138	$\mathbf{np}_P$	$\mathbf{q}$	$\mathbf{p}_Q$	150, 158	2
142	$\mathbf{p}_P\mathbf{q}_P$	$\mathbf{mq}$	*inconsistent*	None	0
143	$\mathbf{p}_P\mathbf{q}_P$	$\mathbf{np}$	*inconsistent*	None	0
144	$\mathbf{p}_P\mathbf{q}_P$	$\mathbf{pq}$	$\mathbf{p}_Q\mathbf{q}_Q$	190	1
147	$\mathbf{p}_P\mathbf{q}_P$	$\mathbf{p}$	$\mathbf{p}_Q\mathbf{q}_Q$	190	1
148	$\mathbf{p}_P\mathbf{q}_P$	$\mathbf{q}$	$\mathbf{p}_Q\mathbf{q}_Q$	190	1
172	$\mathbf{p}_P$	$\mathbf{mq}$	*inconsistent*	None	0
173	$\mathbf{p}_P$	$\mathbf{np}$	*inconsistent*	None	0
174	$\mathbf{p}_P$	$\mathbf{pq}$	$\mathbf{p}_Q$	150-152, 158-160, 182-184, 190-192	12
177	$\mathbf{p}_P$	$\mathbf{p}$	$\mathbf{p}_Q$	150-152, 158-160, 182-184, 190-192	12
178	$\mathbf{p}_P$	$\mathbf{q}$	$\mathbf{p}_Q$	150-152, 158-160, 182-184, 190-192	12
182	$\mathbf{q}_P$	$\mathbf{mq}$	*inconsistent*	None	0
183	$\mathbf{q}_P$	$\mathbf{np}$	*inconsistent*	None	0
184	$\mathbf{q}_P$	$\mathbf{pq}$	$\mathbf{q}_Q$	106, 110, 122, 126, 170, 174, 186, 190, 234, 238, 250, 254	12
187	$\mathbf{q}_P$	$\mathbf{p}$	$\mathbf{q}_Q$	106, 110, 122, 126, 170, 174, 186, 190, 234, 238, 250, 254	12
188	$\mathbf{q}_P$	$\mathbf{q}$	$\mathbf{q}_Q$	106, 110, 122, 126, 170, 174, 186, 190, 234, 238, 250, 254	12
2nd	RQ	PQ	PR		
222	$\mathbf{mq}_P$	$\mathbf{mq}$	$\mathbf{not\text{-}c}_Q$	74, 202	2
223	$\mathbf{mq}_P$	$\mathbf{np}$	*inconsistent*	None	0
224	$\mathbf{mq}_P$	$\mathbf{pq}$	$\mathbf{not\text{-}c}_Q$	90, 218	2

Table 15.5 continued.

227	$\mathbf{mq}_{P}$	**p**	$\mathbf{not\text{-}c}_{Q}$	90, 218	2
228	$\mathbf{mq}_{P}$	**q**	$\mathbf{not\text{-}c}_{Q}$	74, 90, 202, 218	4
232	$\mathbf{np}_{P}$	**mq**	*inconsistent*	None	0
233	$\mathbf{np}_{P}$	**np**	$\mathbf{not\text{-}c}_{Q}$	147-148	2
234	$\mathbf{np}_{P}$	**pq**	$\mathbf{not\text{-}c}_{Q}$	155-156	2
237	$\mathbf{np}_{P}$	**p**	$\mathbf{not\text{-}c}_{Q}$	147-148, 155-156	4
238	$\mathbf{np}_{P}$	**q**	$\mathbf{not\text{-}c}_{Q}$	155-156	2
242	$\mathbf{p}_{P}\mathbf{q}_{P}$	**mq**	*inconsistent*	None	0
243	$\mathbf{p}_{P}\mathbf{q}_{P}$	**np**	*inconsistent*	None	0
244	$\mathbf{p}_{P}\mathbf{q}_{P}$	**pq**	$\mathbf{pq + not\text{-}w}_{Q}$	220	1
247	$\mathbf{p}_{P}\mathbf{q}_{P}$	**p**	$\mathbf{pq + not\text{-}w}_{Q}$	220	1
248	$\mathbf{p}_{P}\mathbf{q}_{P}$	**q**	$\mathbf{pq + not\text{-}w}_{Q}$	220	1
272	$\mathbf{p}_{P}$	**mq**	*inconsistent*	None	0
273	$\mathbf{p}_{P}$	**np**	$\mathbf{not\text{-}c}_{Q}$	147-148, 211-212	4
274	$\mathbf{p}_{P}$	**pq**	$\mathbf{not\text{-}s + not\text{-}q}_{Q}$	151-152, 155-156, 159-160, 215-216, 219-220, 223-224	12
277	$\mathbf{p}_{P}$	**p**	$\mathbf{not\text{-}s + not\text{-}q}_{Q}$	147-148, 151-152, 155-156, 159-160, 211-212, 215-216, 219-220, 223-224	16
278	$\mathbf{p}_{P}$	**q**	$\mathbf{not\text{-}s + not\text{-}q}_{Q}$	151-152, 155-156, 159-160, 215-216, 219-220, 223-224	12
282	$\mathbf{q}_{P}$	**mq**	$\mathbf{not\text{-}s + not\text{-}c}_{Q}$	74, 76, 202, 204	4
283	$\mathbf{q}_{P}$	**np**	*inconsistent*	None	0
284	$\mathbf{q}_{P}$	**pq**	$\mathbf{not\text{-}s + not\text{-}p}_{Q}$	90, 92, 106, 108, 122, 124, 218, 220, 234, 236, 250, 252	12
287	$\mathbf{q}_{P}$	**p**	$\mathbf{not\text{-}s + not\text{-}p}_{Q}$	90, 92, 106, 108, 122, 124, 218, 220, 234, 236, 250, 252	12
288	$\mathbf{q}_{P}$	**q**	$\mathbf{not\text{-}s + not\text{-}p}_{Q}$	74, 76, 90, 92, 106, 108, 122, 124, 202, 204, 218, 220, 234, 236, 250, 252	16
3rd	QR	QP	PR		
322	$\mathbf{mq}_{P}$	**mq**	*inconsistent*	None	0
323	$\mathbf{mq}_{P}$	**np**	*inconsistent*	None	0
324	$\mathbf{mq}_{P}$	**pq**	$\mathbf{q}_{Q}$	170, 186	2
327	$\mathbf{mq}_{P}$	**p**	$\mathbf{q}_{Q}$	170, 186	2
328	$\mathbf{mq}_{P}$	**q**	$\mathbf{q}_{Q}$	170, 186	2
332	$\mathbf{np}_{P}$	**mq**	*inconsistent*	None	0
333	$\mathbf{np}_{P}$	**np**	*inconsistent*	None	0
334	$\mathbf{np}_{P}$	**pq**	$\mathbf{p}_{Q}$	150, 158	2
337	$\mathbf{np}_{P}$	**p**	$\mathbf{p}_{Q}$	150, 158	2
338	$\mathbf{np}_{P}$	**q**	$\mathbf{p}_{Q}$	150, 158	2
342	$\mathbf{p}_{P}\mathbf{q}_{P}$	**mq**	*inconsistent*	None	0
343	$\mathbf{p}_{P}\mathbf{q}_{P}$	**np**	*inconsistent*	None	0
344	$\mathbf{p}_{P}\mathbf{q}_{P}$	**pq**	$\mathbf{p}_{Q}\mathbf{q}_{Q}$	190	1
347	$\mathbf{p}_{P}\mathbf{q}_{P}$	**p**	$\mathbf{p}_{Q}\mathbf{q}_{Q}$	190	1
348	$\mathbf{p}_{P}\mathbf{q}_{P}$	**q**	$\mathbf{p}_{Q}\mathbf{q}_{Q}$	190	1
372	$\mathbf{p}_{P}$	**mq**	*inconsistent*	None	0

Table 15.5 continued.

373	$\mathbf{p}_{P}$	**np**	*inconsistent*	None	0
374	$\mathbf{p}_{P}$	**pq**	$\mathbf{p}_{Q}$	150-152, 158-160, 182-184, 190-192	12
377	$\mathbf{p}_{P}$	**p**	$\mathbf{p}_{Q}$	150-152, 158-160, 182-184, 190-192	12
378	$\mathbf{p}_{P}$	**q**	$\mathbf{p}_{Q}$	150-152, 158-160, 182-184, 190-192	12
382	$\mathbf{q}_{P}$	**mq**	*inconsistent*	None	0
383	$\mathbf{q}_{P}$	**np**	*inconsistent*	None	0
384	$\mathbf{q}_{P}$	**pq**	$\mathbf{q}_{Q}$	106, 110, 122, 126, 170, 174, 186, 190, 234, 238, 250, 254	12
387	$\mathbf{q}_{P}$	**p**	$\mathbf{q}_{Q}$	106, 110, 122, 126, 170, 174, 186, 190, 234, 238, 250, 254	12
388	$\mathbf{q}_{P}$	**q**	$\mathbf{q}_{Q}$	106, 110, 122, 126, 170, 174, 186, 190, 234, 238, 250, 254	12

Table 15.6. Moduses of conclusions for selected relative weak positive premises (subfigures l, p).

Mood	Major	Minor	Conclusion	Common moduses	No.
1st	QR	PQ	PR		
122	$\mathbf{mq}_{P}$	$\mathbf{mq}_{R}$	*inconsistent*	None	0
123	$\mathbf{mq}_{P}$	$\mathbf{np}_{R}$	*inconsistent*	None	0
124	$\mathbf{mq}_{P}$	$\mathbf{p}_{R}\mathbf{q}_{R}$	*inconsistent*	None	0
127	$\mathbf{mq}_{P}$	$\mathbf{p}_{R}$	*inconsistent*	None	0
128	$\mathbf{mq}_{P}$	$\mathbf{q}_{R}$	*inconsistent*	None	0
132	$\mathbf{np}_{P}$	$\mathbf{mq}_{R}$	*inconsistent*	None	0
133	$\mathbf{np}_{P}$	$\mathbf{np}_{R}$	*inconsistent*	None	0
134	$\mathbf{np}_{P}$	$\mathbf{p}_{R}\mathbf{q}_{R}$	*inconsistent*	None	0
137	$\mathbf{np}_{P}$	$\mathbf{p}_{R}$	*inconsistent*	None	0
138	$\mathbf{np}_{P}$	$\mathbf{q}_{R}$	*inconsistent*	None	0
142	$\mathbf{p}_{P}\mathbf{q}_{P}$	$\mathbf{mq}_{R}$	*inconsistent*	None	0
143	$\mathbf{p}_{P}\mathbf{q}_{P}$	$\mathbf{np}_{R}$	*inconsistent*	None	0
144	$\mathbf{p}_{P}\mathbf{q}_{P}$	$\mathbf{p}_{R}\mathbf{q}_{R}$	*inconsistent*	None	0
147	$\mathbf{p}_{P}\mathbf{q}_{P}$	$\mathbf{p}_{R}$	*inconsistent*	None	0
148	$\mathbf{p}_{P}\mathbf{q}_{P}$	$\mathbf{q}_{R}$	*inconsistent*	None	0
172	**pP**	**mqR**	*inconsistent*	None	0
173	**pP**	**npR**	*inconsistent*	None	0
174	**pP**	**pRqR**	*inconsistent*	None	0
177	**pP**	**pR**	*p-aloneQ*	151-152, 159-160	4
178	**pP**	**qR**	*inconsistent*	None	0
182	**qP**	**mqR**	*inconsistent*	None	0
183	**qP**	**npR**	*inconsistent*	None	0
184	**qP**	**pRqR**	*inconsistent*	None	0
187	**qP**	**pR**	*inconsistent*	None	0
188	**qP**	**qR**	*q-aloneQ*	106, 122, 234, 250	4

Table 15.6 continued.

2nd	RQ	PQ	PR		
222	**mq**$_P$	**mq**$_R$	**not-c**$_Q$	74, 202	2
223	**mq**$_P$	**np**$_R$	*inconsistent*	None	0
224	**mq**$_P$	**p**$_R$**q**$_R$	*inconsistent*	None	0
227	**mq**$_P$	**p**$_R$	*inconsistent*	None	0
228	**mq**$_P$	**q**$_R$	**not-c**$_Q$	74, 90, 202, 218	4
232	**np**$_P$	**mq**$_R$	*inconsistent*	None	0
233	**np**$_P$	**np**$_R$	**not-c**$_Q$	147-148	2
234	**np**$_P$	**p**$_R$**q**$_R$	*inconsistent*	None	0
237	**np**$_P$	**p**$_R$	**not-c**$_Q$	147-148, 155-156	4
238	**np**$_P$	**q**$_R$	*inconsistent*	None	0
242	**p**$_P$**q**$_P$	**mq**$_R$	*inconsistent*	None	0
243	**p**$_P$**q**$_P$	**np**$_R$	*inconsistent*	None	0
244	**p**$_P$**q**$_P$	**p**$_R$**q**$_R$	**not-c**$_Q$	220	1
247	**p**$_P$**q**$_P$	**p**$_R$	**not-c**$_Q$	220	1
248	**p**$_P$**q**$_P$	**q**$_R$	**not-c**$_Q$	220	1
272	**p**$_P$	**mq**$_R$	*inconsistent*	None	0
273	**p**$_P$	**np**$_R$	**not-c**$_Q$	147-148, 211-212	4
274	**p**$_P$	**p**$_R$**q**$_R$	**not-c**$_Q$	220	1
277	**p**$_P$	**p**$_R$	**not-s**	147-148, 151-152, 155-156, 159-160, 211-212, 215-216, 219-220, 223-224	16
278	**p**$_P$	**q**$_R$	**not-c**$_Q$	220	1
282	**q**$_P$	**mq**$_R$	**not-c**$_Q$	74, 76, 202, 204	4
283	**q**$_P$	**np**$_R$	*inconsistent*	None	0
284	**q**$_P$	**p**$_R$**q**$_R$	**not-c**$_Q$	220	1
287	**q**$_P$	**p**$_R$	**not-c**$_Q$	220	1
288	**q**$_P$	**q**$_R$	**not-s**	74, 76, 90, 92, 106, 108, 122, 124, 202, 204, 218, 220, 234, 236, 250, 252	16
3rd	QR	QP	PR		
322	**mq**$_P$	**mq**$_R$	*inconsistent*	None	0
323	**mq**$_P$	**np**$_R$	*inconsistent*	None	0
324	**mq**$_P$	**p**$_R$**q**$_R$	*inconsistent*	None	0
327	**mq**$_P$	**p**$_R$	*inconsistent*	None	0
328	**mqP**	**qR**	*inconsistent*	None	0
332	**npP**	**mqR**	*inconsistent*	None	0
333	**npP**	**npR**	*inconsistent*	None	0
334	**npP**	**pRqR**	*inconsistent*	None	0
337	**npP**	**pR**	*inconsistent*	None	0
338	**npP**	**qR**	*inconsistent*	None	0
342	**pPqP**	**mqR**	*inconsistent*	None	0
343	**pPqP**	**npR**	*inconsistent*	None	0
344	**pPqP**	**pRqR**	*inconsistent*	None	0

Table 15.6 continued.

347	$\mathbf{p}_P\mathbf{q}_P$	$\mathbf{p}_R$	*inconsistent*	None	0
348	$\mathbf{p}_P\mathbf{q}_P$	$\mathbf{q}_R$	*inconsistent*	None	0
372	$\mathbf{p}_P$	$\mathbf{mq}_R$	*inconsistent*	None	0
373	$\mathbf{p}_P$	$\mathbf{np}_R$	*inconsistent*	None	0
374	$\mathbf{p}_P$	$\mathbf{p}_R\mathbf{q}_R$	*inconsistent*	None	0
377	$\mathbf{p}_P$	$\mathbf{p}_R$	**p-alone**$_Q$	151-152, 183-184	4
378	$\mathbf{p}_P$	$\mathbf{q}_R$	*inconsistent*	None	0
382	$\mathbf{q}_P$	$\mathbf{mq}_R$	*inconsistent*	None	0
383	$\mathbf{q}_P$	$\mathbf{np}_R$	*inconsistent*	None	0
384	$\mathbf{q}_P$	$\mathbf{p}_R\mathbf{q}_R$	*inconsistent*	None	0
387	$\mathbf{q}_P$	$\mathbf{p}_R$	*inconsistent*	None	0
388	$\mathbf{q}_P$	$\mathbf{q}_R$	**q-alone**$_Q$	106, 110, 234, 238	4

We can similarly look into syllogisms involving vague positive premises, absolute or relative weak. The following table gives some examples.

Table 15.7. Moduses of conclusions for selected combinations of relative weak and absolute vague positive premises (various subfigures).

Fig.	Major	Minor	Conclusion	Common moduses	No.
1st	QR	PQ	PR		
	s	$\mathbf{w}_R$	**not-s**	74, 90, 147-148, 155-156, 202, 218	8
	w	$\mathbf{w}_R$	**not-s**	74, 76, 90, 92, 106, 108, 122, 124, 147-148, 151-152, 155-156, 159-160, 202, 204, 211-212, 215-216, 218-220, 223-224, 234, 236, 250, 252	31
	c	$\mathbf{w}_R$	**not-s**	74, 76, 90, 92, 106, 108, 122, 124, 147-148, 151-152, 155-156, 159-160, 202, 204, 211-212, 215-216, 218-220, 223-224, 234, 236, 250, 252	31
	$\mathbf{w}_P$	**s**	*inconsistent*	None	0
	$\mathbf{w}_P$	**w**	$\mathbf{w}_Q$	106, 110, 122, 126, 150-152, 158-160, 170, 174, 182-184, 186, 190-192, 234, 238, 250, 254	23
	$\mathbf{w}_P$	**c**	$\mathbf{w}_Q$	106, 110, 122, 126, 150-152, 158-160, 170, 174, 182-184, 186, 190-192, 234, 238, 250, 254	23
	$\mathbf{w}_P$	$\mathbf{w}_R$	$\mathbf{w}_Q$	106, 122, 151-152, 159-160, 234, 250	8
2nd	RQ	PQ	PR		
	s	$\mathbf{w}_R$	**not-s**	74, 90, 147-148, 155-156, 202, 218	8
	w	$\mathbf{w}_R$	**not-s**	74, 76, 90, 92, 106, 108, 122, 124, 147-148, 151-152, 155-156, 159-160, 202, 204, 211-212, 215-216, 218-220, 223-224, 234, 236, 250, 252	31
	c	$\mathbf{w}_R$	**not-s**	74, 76, 90, 92, 106, 108, 122, 124, 147-148, 151-152, 155-156, 159-160, 202, 204, 211-212, 215-216, 218-220, 223-224, 234, 236, 250, 252	31
	$\mathbf{w}_P$	**s**	**not-s**	74, 76, 147-148, 202, 204, 211-212	8
	$\mathbf{w}_P$	**w**	**not-s**	74, 76, 90, 92, 106, 108, 122, 124, 147-148, 151-152, 155-156, 159-160, 202, 204, 211-212, 215-216, 218-220, 223-224, 234, 236, 250, 252	31

Table 15.7 continued.

	$\mathbf{w}_P$	**c**	**not-s**	74, 76, 90, 92, 106, 108, 122, 124, 147-148, 151-152, 155-156, 159-160, 202, 204, 211-212, 215-216, 218-220, 223-224, 234, 236, 250, 252	31
	$\mathbf{w}_P$	$\mathbf{w}_R$	**not-s**	74, 76, 90, 92, 106, 108, 122, 124, 147-148, 151-152, 155-156, 159-160, 202, 204, 211-212, 215-216, 218-220, 223-224, 234, 236, 250, 252	31
3rd	QR	QP	PR		
	s	$\mathbf{w}_R$	*inconsistent*	None	0
	w	$\mathbf{w}_R$	**w**	106, 108, 110, 112, 136, 151-152, 168, 183-184, 200, 215-216, 226, 228, 230, 232, 234, 236, 238, 240, 247-248	23
	c	$\mathbf{w}_R$	**w**	106, 108, 110, 112, 136, 151-152, 168, 183-184, 200, 215-216, 226, 228, 230, 232, 234, 236, 238, 240, 247-248	23
	$\mathbf{w}_P$	**s**	*inconsistent*	None	0
	$\mathbf{w}_P$	**w**	$\mathbf{w}_Q$	106, 110, 122, 126, 150-152, 158-160, 170, 174, 182-184, 186, 190-192, 234, 238, 250, 254	23
	$\mathbf{w}_P$	**c**	$\mathbf{w}_Q$	106, 110, 122, 126, 150-152, 158-160, 170, 174, 182-184, 186, 190-192, 234, 238, 250, 254	23
	$\mathbf{w}_P$	$\mathbf{w}_R$	$\mathbf{w}_Q$	106, 110, 151-152, 183-184, 234, 238	8

Note that some combinations are inconsistent. Some moods yield conclusion **w** in relative (which implies absolute) form, and some only in absolute form. In other cases, though the number of common moduses may vary considerably, there is no conclusion other than the form **not-s**, which (remember) means 'not strong causation', implying both **not-m** + **not-n**; it can also be read as the disjunction $\mathbf{w}_{abs}$ **or not-**$\mathbf{c}_{abs}$.

2. Dealing with Negatives.

Let us now sketch the application of our methods to the solution of syllogisms involving a negative premise or two. We start as always by identifying the alternative moduses of the various possible premises; then we list their conjunctions in syllogisms of various figures, and with reference to the common alternative moduses, if any, evaluate their possible conclusions.

> We have learned in the preceding chapters and the above section the lessons that: the number of common moduses in the premises may be small, yet yield a conclusion; or it may be large, yet yield no conclusion. The less number of alternative moduses there are in the premises, the less likely are they to intersect, and be compatible and yield a conclusion. On the other hand, the more number of alternative moduses in the premises, the more often they will have some in common, but also the more likely will there be a vague conclusion, or no conclusion at all by virtue of including moduses of both causation and non-causation. These generalities are equally applicable to the findings below.

In the following table, the alternative moduses of the strong and absolute weak determinations are obtained by negation from Table 14.3. Those of the relative weaks are obtained by negation from Table 15.3 above. All the rest follows by intersection.

Table 15.8. Enumeration of three-item alternative moduses for negative premises, for any figure of syllogism (generic forms).

Determination	Major QR	Major RQ	Minor PQ	Minor QP
not-m	2-9, 11, 13-24, 29-41, 43, 45-56, 61-129, 131, 133-137, 139, 141-144, 149-152, 157-161, 163, 165-169, 171, 173-176, 181-184, 189-256	2-9, 11-13, 15-24, 27-28, 31-73, 75-77, 79-88, 91-92, 95-129, 131-133, 135-137, 139-141, 143-144, 147-148, 151-152, 155-156, 159-193, 195-197, 199-201, 203-205, 207-208, 211-212, 215-216, 219-220, 223-256	2-65, 69, 73, 77, 81-129, 133, 137, 141, 145-193, 197, 201, 205, 209-256	2-65, 69-81, 85-97, 101-113, 117-129, 133-145, 149-161, 165-177, 181-193, 197-209, 213-225, 229-241, 245-256
not-n	2-9, 11-13, 15-24, 27-28, 31-73, 75-77, 79-88, 91-92, 95-129, 131-133, 135-137, 139-141, 143-144, 147-148, 151-152, 155-156, 159-193, 195-197, 199-201, 203-205, 207-208, 211-212, 215-216, 219-220, 223-256	2-9, 11, 13-24, 29-41, 43, 45-56, 61-129, 131, 133-137, 139, 141-144, 149-152, 157-161, 163, 165-169, 171, 173-176, 181-184, 189-256	2-65, 69-81, 85-97, 101-113, 117-129, 133-145, 149-161, 165-177, 181-193, 197-209, 213-225, 229-241, 245-256	2-65, 69, 73, 77, 81-129, 133, 137, 141, 145-193, 197, 201, 205, 209-256
not-p_{abs}	2-13, 15, 17-28, 33-45, 47, 49-60, 65-73, 75, 77, 79, 81-88, 97-105, 107, 109, 111, 113-120, 129-133, 135, 137-141, 143, 145-148, 153-156, 161-165, 167, 169-173, 175, 177-180, 185-188, 193, 195, 197, 199, 201, 203, 205, 207, 225, 227, 229, 231, 233, 235, 237, 239	2-11, 13-15, 17-26, 29-30, 33-41, 43, 45, 47, 49-56, 65-75, 77-79, 81-90, 93-94, 97-105, 107, 109, 111, 113-120, 129-131, 133-135, 137-139, 141-143, 145-146, 149-150, 153-154, 157-158, 161, 163, 165, 167, 169, 171, 173, 175, 193-195, 197-199, 201-203, 205-207, 209-210, 213-214, 217-218, 221-222, 225, 227, 229, 231, 233, 235, 237, 239	2-81, 85, 89, 93, 97, 101, 105, 109, 113, 117, 121, 125, 129-145, 149, 153, 157, 161, 165, 169, 173, 177, 181, 185, 189, 193-209, 213, 217, 221, 225, 229, 233, 237, 241, 245, 249, 253	2-69, 73, 77, 81-85, 89, 93, 97-101, 105, 109, 113-117, 121, 125, 129-133, 137, 141, 145-149, 153, 157, 161-165, 169, 173, 177-181, 185, 189, 193-197, 201, 205, 209-213, 217, 221, 225-229, 233, 237, 241-245, 249, 253
not-q_{abs}	2-11, 13-15, 17-26, 29-30, 33-41, 43, 45, 47, 49-56, 65-75, 77-79, 81-90, 93-94, 97-105, 107, 109, 111, 113-120, 129-131, 133-135, 137-139, 141-143, 145-146, 149-150, 153-154, 157-158, 161, 163, 165, 167, 169, 171, 173, 175, 193-195, 197-199, 201-203, 205-207, 209-210, 213-214, 217-218, 221-222, 225, 227, 229, 231, 233, 235, 237, 239	2-13, 15, 17-28, 33-45, 47, 49-60, 65-73, 75, 77, 79, 81-88, 97-105, 107, 109, 111, 113-120, 129-133, 135, 137-141, 143, 145-148, 153-156, 161-165, 167, 169-173, 175, 177-180, 185-188, 193, 195, 197, 199, 201, 203, 205, 207, 225, 227, 229, 231, 233, 235, 237, 239	2-69, 73, 77, 81-85, 89, 93, 97-101, 105, 109, 113-117, 121, 125, 129-133, 137, 141, 145-149, 153, 157, 161-165, 169, 173, 177-181, 185, 189, 193-197, 201, 205, 209-213, 217, 221, 225-229, 233, 237, 241-245, 249, 253	2-81, 85, 89, 93, 97, 101, 105, 109, 113, 117, 121, 125, 129-145, 149, 153, 157, 161, 165, 169, 173, 177, 181, 185, 189, 193-209, 213, 217, 221, 225, 229, 233, 237, 241, 245, 249, 253
not-p_{rel}	2-148, 153-156, 161-180, 185-188, 193-256	2-146, 149-150, 153-154, 157-158, 161-210, 213-214, 217-218, 221-222, 225-256	2-146, 149-150, 153-154, 157-158, 161-210, 213-214, 217-218, 221-222, 225-256	2-134, 137-150, 153-166, 169-182, 185-198, 201-214, 217-230, 233-246, 249-256
not-q_{rel}	2-41, 43-45, 47-57, 59-61, 63-105, 107-109, 111-121, 123-125, 127-169, 171-173, 175-185, 187-189, 191-233, 235-237, 239-249, 251-253, 255-256	2-73, 75, 77-89, 91, 93-105, 107, 109-121, 123, 125-201, 203, 205-217, 219, 221-233, 235, 237-249, 251, 253-256	2-73, 75, 77-89, 91, 93-105, 107, 109-121, 123, 125-201, 203, 205-217, 219, 221-233, 235, 237-249, 251, 253-256	2-97, 99, 101, 103, 105, 107, 109, 111, 113-225, 227, 229, 231, 233, 235, 237, 239, 241-256

Readers are encouraged to continue this research as an exercise, by filling in a table like that below for both absolute and relative weak determinations. The negations of the joint determinations can be worked out with reference to Table 14.3.

Table 15.9. Enumeration of three-item alternative moduses for negative premises, for any figure of syllogism (specific forms).

Determination	Major QR	Major RQ	Minor PQ	Minor QP
not(mn)				
not(mq)				
not(np)				
not(pq)				

Table 15.9 continued.

Determination	Major QR	Major RQ	Minor PQ	Minor QP
m + not-n				
not-m + n				
not-m + not-n = not(m or n)				
m + not-p				
not-m + p				
not-m + not-p = not(m or p)				
n + not-q				
not-n + q				
not-n + not-q = not(n or q)				
m + not-q				
not-m + q				
not-m + not-q = not(m or q)				
n + not-p				
not-n + p				
not-n + not-p = not(n or p)				
p + not-q				
not-p + q				
not-p + not-q = not(p or q)				
no-causation				

Once the researcher has filled in the above table, he or she may investigate all interesting combinations of premises, and see whether they are compatible (i.e. have common alternative moduses), and if so what conclusions, if any, may be drawn (i.e. whether the common moduses all fall under some form of causation or non-causation). I will do the job in the following table for some selected negative premises, and leave it to the reader to finish the table with a more exhaustive treatment, involving mixtures of positive and negative premises.

Table 15.10. Moduses of conclusions for selected (generic, absolute) negative premises, in Figure 1.

Fig.	Major	Minor	Conclusion	Common moduses	No.
1st	QR	PQ	PR		
	not-m	**not-m**	*no conclusion*	2-9, 11, 13-24, 29-41, 43, 45-56, 61-65, 69, 73, 77, 81-129, 133, 137, 141, 149-152, 157-161, 163, 165-169, 171, 173-176, 181-184, 189-193, 197, 201, 205, 209-256	187
	not-n	**not-n**	*no conclusion*	2-9, 11-13, 15-24, 27-28, 31-65, 69-73, 75-77, 79-81, 85-88, 91-92, 95-97, 101-113, 117-129, 133, 135-137, 139-141, 143-144, 151-152, 155-156, 159-161, 165-177, 181-193, 197, 199-201, 203-205, 207-208, 215-216, 219-220, 223-225, 229-241, 245-256	187
	not-m	**not-n**	*no conclusion*	2-9, 11, 13-24, 29-41, 43, 45-56, 61-65, 69-81, 85-97, 101-113, 117-129, 133-137, 139, 141-144, 149-152, 157-161, 165-169, 171, 173-176, 181-184, 189-193, 197-209, 213-225, 229-241, 245-256	193
	not-n	**not-m**	*no conclusion*	2-9, 11-13, 15-24, 27-28, 31-65, 69, 73, 77, 81-88, 91-92, 95-129, 133, 137, 141, 147-148, 151-152, 155-156, 159-193, 197, 201, 205, 211-212, 215-216, 219-220, 223-256	193
	not-m	**not-p**	*no conclusion*	2-9, 11, 13-24, 29-41, 43, 45-56, 61-81, 85, 89, 93, 97, 101, 105, 109, 113, 117, 121, 125, 129, 131, 133-137, 139, 141-144, 149, 157, 161, 165, 169, 173, 181, 189, 193-209, 213, 217, 221, 225, 229, 233, 237, 241, 245, 249, 253	127
	not-n	**not-q**	*no conclusion*	2-9, 11-13, 15-24, 27-28, 31-69, 73, 77, 81-85, 97-101, 105, 109, 113-117, 121, 125, 129, 131-133, 137, 141, 147-148, 161-165, 169, 173, 177-181, 185, 189, 193, 195-197, 201, 205, 211-212, 225-229, 233, 237, 241-245, 249, 253	127
	not-m	**not-q**	*no conclusion*	2-9, 11, 13-24, 29-41, 43, 45-56, 61-69, 73, 77, 81-85, 89, 93, 97-101, 105, 109, 113-117, 121, 125, 129, 131, 133, 137, 141, 149, 157, 161, 163, 165, 169, 173, 181, 189, 193-197, 201, 205, 209-213, 217, 221, 225-229, 233, 237, 241-245, 249, 253	121
	not-n	**not-p**	*no conclusion*	2-9, 11-13, 15-24, 27-28, 31-73, 75-77, 79-81, 85, 97, 101, 105, 109, 113, 117, 121, 125, 129, 131-133, 135-137, 139-141, 143-144, 161, 165, 169, 173, 177, 181, 185, 189, 193, 195-197, 199-201, 203-205, 207-208, 225, 229, 233, 237, 241, 245, 249, 253	121
	not-p	**not-m**	*no conclusion*	2-13, 15, 17-28, 33-45, 47, 49-60, 65, 69, 73, 77, 81-88, 97-105, 107, 109, 111, 113-120, 129, 133, 137, 141, 145-148, 153-156, 161-165, 167, 169-173, 175, 177-180, 185-188, 193, 197, 201, 205, 225, 227, 229, 231, 233, 235, 237, 239	127
	not-q	**not-n**	*no conclusion*	2-11, 13-15, 17-26, 29-30, 33-41, 43, 45, 47, 49-56, 65, 69-75, 77-79, 81, 85-90, 93-94, 97, 101-105, 107, 109, 111, 113, 117-120, 129, 133-135, 137-139, 141-143, 145, 149-150, 153-154, 157-158, 161, 165, 167, 169, 171, 173, 175, 193, 197-199, 201-203, 205-207, 209, 213-214, 217-218, 221-222, 225, 229, 231, 233, 235, 237, 239	127

Table 15.10 continued.

	not-p	**not-n**	*no conclusion*	2-13, 15, 17-28, 33-45, 47, 49-60, 65, 69-73, 75, 77, 79, 81, 85-88, 97, 101-105, 107, 109, 111, 113, 117-120, 129, 133, 135, 137-141, 143, 145, 153-156, 161, 165, 167, 169-173, 175, 177, 185-188, 193, 197, 199, 201, 203, 205, 207, 225, 229, 231, 233, 235, 237, 239	121
	not-q	**not-m**	*no conclusion*	2-11, 13-15, 17-26, 29-30, 33-41, 43, 45, 47, 49-56, 65, 69, 73, 77, 81-90, 93-94, 97-105, 107, 109, 111, 113-120, 129, 133, 137, 141, 145-146, 149-150, 153-154, 157-158, 161, 163, 165, 167, 169, 171, 173, 175, 193, 197, 201, 205, 209-210, 213-214, 217-218, 221-222, 225, 227, 229, 231, 233, 235, 237, 239	121
	not-p	**not-p**	*no conclusion*	2-13, 15, 17-28, 33-45, 47, 49-60, 65-73, 75, 77, 79, 81, 85, 97, 101, 105, 109, 113, 117, 129-133, 135, 137-141, 143, 145, 153, 161, 165, 169, 173, 177, 185, 193, 195, 197, 199, 201, 203, 205, 207, 225, 229, 233, 237	103
	not-q	**not-q**	*no conclusion*	2-11, 13-15, 17-26, 29-30, 33-41, 43, 45, 47, 49-56, 65-69, 73, 77, 81-85, 89, 93, 97-101, 105, 109, 113-117, 129-131, 133, 137, 141, 145-146, 149, 153, 157, 161, 163, 165, 169, 173, 193-195, 197, 201, 205, 209-210, 213, 217, 221, 225, 227, 229, 233, 237	103
	not-p	**not-q**	*no conclusion*	2-13, 15, 17-28, 33-45, 47, 49-60, 65-69, 73, 77, 81-85, 97-101, 105, 109, 113-117, 129-133, 137, 141, 145-148, 153, 161-165, 169, 173, 177-180, 185, 193, 195, 197, 201, 205, 225, 227, 229, 233, 237	109
	not-q	**not-p**	*no conclusion*	2-11, 13-15, 17-26, 29-30, 33-41, 43, 45, 47, 49-56, 65-75, 77-79, 81, 85, 89, 93, 97, 101, 105, 109, 113, 117, 129-131, 133-135, 137-139, 141-143, 145, 149, 153, 157, 161, 165, 169, 173, 193-195, 197-199, 201-203, 205-207, 209, 213, 217, 221, 225, 229, 233, 237	109
2nd	RQ	PQ	PR		
3rd	QR	QP	PR		

We can similarly proceed for the same combinations of generic, absolute, negative premises in Figures 2 and 3. But there is no need to, for it is easy to predict that the conclusion will be "*no conclusion (of the form PR)*" in all similar cases. If you look at all the cells of Table 15.8, you will observe that they have certain alternative moduses in common to all of them – for instances Nos. 2 and 237. Since modus 2 is an alternative of non-causation (**not-c**$_{abs}$) and modus 237 is an alternative of causation (**c**$_{abs}$), as shown in Table 12.4, no pair of the premises listed in the above table can yield any causative conclusion.

Rather, our next step would be to develop syllogisms with negative premises from Table 15.9, as well as 15.8a, and then – and probably much more fruitfully – syllogisms with mixtures of positive and negative premises. This can and should be carried out till all possible combinations of all conceivable forms are carefully exhausted. I leave the job to other researchers (any reader willing to do it), so as to move on and deal with other issues in causative logic.

Chapter 16. OUTSTANDING ISSUES.

In this closing chapter, my purpose is to break some additional ground, discussing certain outstanding issues in causation without attempting to exhaust them at this time.

1. Four-Item Syllogism.

Our attempt, in the preceding five chapters, to solve by microanalysis all the problems of causative logic with reference to only three items has evidently failed. Some problems have indeed been solved – in particular, *the refutation of absolute lone determinations* should be cited as an important breakthrough. However, many syllogistic problems have been left without solution – and worse still, we are faced with *apparent contradictions between the results of macro- and microanalysis*, at least superficially, though these fears may[104] turn out to be unfounded after further scrutiny.

Evidently, we will not succeed in definitive solution of the technical issues relative to causation until[105] we *develop microanalysis with four items*. This essential and daunting task remains to be done at the present time. Only through microanalysis including the subsidiary item (S), in addition to that involving the major, middle and minor items (P, Q, R), will be have full control of syllogistic argument involving relative weak determinations.

To develop such level of detail, i.e. for syllogisms with four items, we would have to build a much larger grand matrix – a table with **16 rows and 2^{16} = 65,536 columns**. This could be done easily enough through a computer spreadsheet. We would prepare the said number of cells, filling in 0s and 1s in a systematic pattern and labeling the rows a-p and the columns 1-65536.

But the difficulty comes after that, in view of the inhumanly large quantity of data involved. We would need to write a search and flag program identifying the alternative moduses for each determination (based on conditions and rules like those exemplified when developing three-item microanalysis). Then we would infer the moduses for derivatives, including absolute vs. relative weak determinations, joint determinations, negative determinations, and preventive vs. causative propositions. Finally, we would consider all conceivable combinations of premises and identify their common modus(es) if any (if they have no common moduses, they are inconsistent); and find out what conclusion(s) if any these results allow in each case. I am personally not a good enough computer programmer to even try that.

---

[104] I say 'may' – I mean 'hopefully will'. If it turns out that there are indeed contradictions, then they are due to human errors (errors of inattention on my part). I say this with certainty, because I do not believe that any of the postulates adopted in developing macro- and microanalysis are faulty; and they cannot be in mutual contradiction, being essentially the same in both instances.

[105] Other logicians might prefer to resolve the issues of causative logic through symbolic formulas. Although such an effort would not be without value, I personally eschew such means, which to my mind are obscure, lacking in transparency. Rather than express things in such esoteric language, I have preferred to make the message explicit, so that everyone can see for him or her self with a minimum of mental effort just what is meant and why it is true.

Instead, my intention is to try and develop a relational database, through some existing software. The ultimate goal pursued here is to develop a quick method for identifying applicable determinations and for computing fusions of determinations in a syllogistic context. If the software is made to classify the determination relating certain items, from an analysis of their observed relative presences and absences (alternative moduses), the job is as good as done. It should then be possible to input the syllogism under investigation (specifying its figure and mood) and have the software output all the conclusion(s) if any that can be validly drawn.

Such a program would have not only theoretical value, but also practical value. In this day and age, when computers are readily available, this would give us a universal tool for causative reasoning, inductive (formation of causative propositions from alternative moduses) and deductive (immediate or syllogistic inferences). I am currently developing this tool.

> Note also that neither in Phase One nor so far in Phase Two of the present research have I made any systematic attempt to test for syllogistic conclusions other than the *regular* ones, involving the major and minor items (form PR) and where appropriate also a subsidiary item (form PSR). When I say that there is "no valid conclusion" to some syllogism, I mean that there is none of such form. And when I find and list a valid conclusion, it is only one of such form (or the inverse causative forms, notP.notR or notP.notS.notR).
>
> But in many cases, there may be a valid conclusion or additional valid conclusions of *irregular* form, i.e. involving the negation of the subsidiary item (notS), or the negation of the minor (notP) or the major (notR); that is to say, of causative form P.notS.R or notP.S.R, or of preventive form P.notR or notP.R or P.S.notR or notP.S.R or P.notS.notR or notP.notS.R. Any program written to spew out conclusions from regular premises (PQRS) should also test for such irregular conclusions, i.e. should be made to tell us the *full* conclusion all forms considered.

2. On Laws of Causation.

The expression 'law of causation' can also be applied to each and every theorem we have proved concerning causation. All our reductions of causative propositions to simpler conjunctive or conditional propositions, or to specified alternative moduses, all the immediate or syllogistic inferences from causative propositions that we have established, constitute so many 'laws' about causation.

The 'grand matrices', of 15 possible moduses for any two items, or 255 possible moduses for three items, or 65,535 possible moduses for four items, and so forth, may be viewed as the nearest thing to *a universal law of causation* that we can formally guarantee:

> ***Any* two or more items *must* be related by some modus(es) within these frameworks, although the modus(es) by which they are related are *not* necessarily those of causation (or prevention).**

The only alternative modus that is formally impossible is the one in each framework (labeled No. 1) consisting entirely of zeros: this the laws of thought interdict in advance for all items. Two (or more) items are always 'tied together' by one or more of moduses (each of which can be visually imagined as a sort of ticker-tape in which zeros and ones are punched), but we cannot predict how many and precisely which moduses are effective for that particular pair of items (or more).

A grand matrix represents all the ways that any two (or more) items might *in principle*, i.e. from an epistemological perspective, at first sight, be found to co-exist or not co-exist. But *in practice*, from an ontological point of view, after thorough research, not all these ways are applicable in every case: in each given case, only some alternative moduses are likely to be applicable.

As previously discussed[106], we can group the alternative moduses in various ways, according to what sort of relationship they signify between the items concerned. We can thus distinguish between 'connective' relationships (causation or prevention) and 'non-connective' relationships (one or more items incontingent), as shown in the table below for two items.

In one case, **the last modus of any grand matrix (that involving only '1' codes)**, i.e. modus #16 in a two-item framework (conventionally classified as absolute partial and contingent causation or prevention, i.e. the form $\mathbf{p}_{abs}\mathbf{q}_{abs}$), **we cannot strictly say whether connection or nonconnection is ultimately involved** (i.e. when more items are eventually taken into account, in a larger grand matrix). So this modus might be placed under either heading, or under neither of them[107].

Table 16.1. Possible relations between any two items P and R.

Relationship		**Modus Nos.**
Connection between P and R		*7-8, 10, 12, 14-15, (16)*
Causation by P of R		10, 12, 14, (16)
Prevention by P of R		7, 8, 15, (16)
Non-connection between P and R		*2-6, 9, 11, 13, (16)*
Both P and R are incontingent		2, 3, 5, 9
P impossible	R impossible	2
P impossible	R necessary	3
P necessary	R impossible	5
P necessary	R necessary	9
Only one of P or R is incontingent		4, 6, 11, 13
P incontingent	R contingent	4, 13
P contingent	R incontingent	6, 11
Indefinite regarding connection or nonconnection		***16***

Some groups of alternative moduses signify incontingency (necessity or impossibility) of one (or more) of the items concerned, while the others signify contingency of the two (or more) items concerned. An incontingent item is independent of all others. Only where all items involved are contingent can causation or prevention (i.e. some connection) occur between them. Different combinations of moduses have been identified as different determinations of causation or prevention. These determinations have been classified in various hierarchies and

106 See Chapter 13.2.

107 Note well that this is a relatively late realization of mine, in Chapter 13.2: that the last modus is not necessarily always to be interpreted as signifying causation; it is only indicative of possible causation. Consequently, my classification of the 2-item modus #16, or the 3-item modus #256, etc., under the heading of causation was not accurate and could be misleading. It should more precisely be classified as 'indefinite'.

polarities: strongs/weaks, absolutes/relatives, generics/joints/lones, positives/negatives, causative/preventive, each of which is signified by a certain group of moduses. But contingency of all items does not signify their connection.

Having thus put matricial analysis in perspective, it is easier for us to evaluate on purely formal grounds certain philosophical claims for or against causation that have arisen over the centuries. We shall here use the word 'cause' in the specific sense of causative connection, including in it both causation and prevention, but excluding other causal relations (such as volition). As we shall see, *none* of these claims can be *formally* established from our definitions of causation and all the properties of causation emerging from them.

1. Some philosophies have claimed that ***everything has a cause***. This is commonly referred to as 'the law of universal causation' and is the position most widely adhered to. It is a claim that causation is to be found everywhere, that all things are ruled by it – i.e. that every thing is caused by some other things, themselves in turn caused by others, and so forth *ad infinitum*. There are different versions of this proposed law, "nothing is without cause".
   a) Oriental philosophies would opt for a *radical* interpretation, based on the belief that all things in the empirical world (*dharmas*) are impermanent, so that nothing exists that is independent. Clearly, this viewpoint eliminates a certain number of alternative moduses (those signifying incontingency) from consideration at the outset: but no formal grounds for such a narrowing of scope have been proposed. Indeed, if one reflects, the claim in question is *self-contradictory*, since it is itself put forward as a permanent fact, a necessity. So we can on *formal* grounds reject it.
   b) Most Western advocates of universal causation would more *moderately* understand it as "nothing *contingent* is without cause". They would allow that some things are necessary or impossible, but consider that those things which are possible but not necessary have to have a cause. It is important thing to realize that, contrary to what many of its advocates believe, this alleged law *cannot be formally deduced* from the definitions of causation. It is only conceivable on *inductive* grounds, by *generalization* from previously encountered cases.

   Note that, unlike the radical version, the moderate version of universal causation is not inconsistent and does not prejudicially exclude any alternative moduses. What it does exclude at the outset, in advance of empirical research and without formal proof, is that some contingent item may exist that has no causative (or preventive) relation to at least one other item in the universe.
2. Some philosophies have claimed that ***nothing has a cause***. We may cite as proponents: in the East, the Indian philosopher Nagarjuna (2nd Cent. CE), and in the West, the Scottish philosopher David Hume (18th Cent.). This viewpoint is essentially a denial that there is any such relation as causation; it is a claim that the concept is meaningless, a human invention without corresponding reality, an error of reasoning. Here again, we could distinguish a *radical* version, which excludes incontingency in principle and so is internally inconsistent, and a *moderate* version, which reserves indeterminacy to contingents.

   Either way, the negative thesis that 'nothing is causatively related to anything else' *arbitrarily eliminates for any two (or more) items the vast majority of alternative moduses*: all those signifying causation or prevention, and does so *for all items* past, present or future, everywhere in the universe. It gives no formal ground for such a sweeping measure, but bases it on denial of the possibility of conceptualization or generalization. This may be claimed as an empiricist posture, or may be coupled with skepticism about perceptual evidence. But, since any such claims themselves use concepts and appeal to generalities that could only be admitted by granting generalization, they are *self-contradictory* and therefore logically untenable.

The antagonists of causation attempt to mitigate this paradox by claiming that causative propositions are "conventional" (Nagarjuna) or "habitual" (Hume), ignoring that such *explanations* themselves rely on admission of the causative relation. Some instead argue that though causation may be theoretically meaningful, it is impossible to establish in practice. **But as shown in the present work, a concept of causation *can* readily be constructed, using indubitable simpler concepts of presence, absence, conjunction and disjunction, possibility and impossibility.** Moreover, the concept would have to be convincingly defined, before it could be declared empty! So it cannot be meaningless. As for the fear that causative relations have no actual instances or are in practice unknowable, we shall now explain it.

The deep reason for such antagonisms is the failure to understand causative propositions as simply *records of conjunctions of presences and/or absences* of two (or more) items. Such summaries are generalized from observation, subject to corrective particularization if new observation belies them. The antagonists have not emotionally reconciled themselves to the *tentative, inductive* nature of knowledge, and so set up and cling to badly defined and impossible deductive ideals of knowledge (without noticing that they themselves cannot possibly satisfy them).

Causative judgment is indeed based not only on empirical evidence, but also on ordering of information by the rational faculty, since it concerns not only presence but also absence, and all negation involves rational projection. This however only means that reason provides an 'overlay' (the grand matrix) through which to order (summarize and predict) events, but the evidence this overlay is laid over (i.e. the 1s and 0s exhibited by the items, the moduses applicable in their case) is empirical (ultimately, though we may thereafter get to know some by immediate or syllogistic inference from previous experiences).

3. Some philosophies have claimed that ***some things have a cause and some things do not have any cause***. This position gained acceptance among physicists and consequent popularity in the 20th Century, after the advent of quantum mechanics (as interpreted by Niels Bohr) and Big-Bang astronomy (Stephen Hawking seems to advocate that the apparition of matter and its primeval explosion were simultaneous and causeless). This position, note well, is a compromise between the preceding two. It admits of non-causation[108] in specific areas (the beginning of time) or at certain levels (the subatomic), together with causation in all other cases or situations. This position is formally neither provable nor disprovable: it is consistent and does no violence to matricial analysis.

   It is neither more nor less conceivable than the (moderate) 'universal causation' thesis. They are simply – equally conceivable contrary hypotheses or predictions. Each of these two theses must be viewed as an epistemological postulate or an ontological generalization. 'Universal causation' is a generalization from cases where causation was apparently (after certain generalizations) found, to cases where it is *not yet* found; whereas 'particular causation and particular noncausation' emphasizes the cases where we have not

---

108 This refers to *absolute* noncausation, which is not to be confused with *relative* noncausation: it means *no cause whatsoever*, and not merely *not this specified cause.*

yet found a credible cause, and suggests that we generalize this failure to 'no cause *will ever be* found, because none exists'.

A grand matrix, remember, foresees every conceivable way two or more items *might* appear together or apart (refer Table 16.1 above). To establish that a given item has ***some*** cause, it suffices that we find ***one other*** item that has a relationship of connection with it (two-item moduses #s 7-8, 10, 12, 14-15, and some cases of #16). But to establish that a given item has ***no*** cause is much more difficult! It is not sufficient to show that one other item is not its cause (two-item moduses 2-6, 9, 11, 13, and some cases of 16): one has to show that this is true of ***all other*** items.

Obviously, for those of us who make no claim to omniscience, this is an impossible task. We can only – either appeal to a law of universal causation or accept the possibility that some things are causeless. In any case, generalization is doubly involved: first, in the inductive proof that any modus is applicable to the set of items observed; second, in the inductive passage from those items to items not observed. Denying *both* these principles is not a viable third alternative, as already explained.

To repeat, neither of the coherent doctrines can be proved or disproved deductively; they are neither self-evident nor self-contradictory. They may only conceivably be established inductively, through generalization from respectively "we have found causes for everything encountered so far" (which is far from the case) or "there are things for which no causes have so far been found" (which is true, but since "there are things for which causes have eventually been found" is also true, we are inhibited from quick generalization). There is a standoff.

Since no formal ground for either position is evident, the science of Logic must make formal allowance for *both* positions. Its task is to provide the formal means for open-minded debate of this topic (as of all others): it cannot prejudicially exclude the one or the other from language and block discourse in advance.

It is important to be clearly aware where in a grand matrix causelessness or spontaneity is allowed for. Refer to the last three rows of Table 16.1 above, where one or both items are contingent, yet the moduses of causation or prevention do not apply to them. If item R is the contingency under scrutiny, our table implies that R *might* be without cause if its relation to P falls under modus 4 or 13 or under modus 16. For an item to certainly *be* causeless, it would have to have one of these relations not only to P, but also to all other items in the universe – P1, P2, P3, etc.

Absolute noncausation of R can be expressed in the form "**nothing causes R**", which collects together innumerable statements of relative noncausation, of the form "P does not cause R; P1 does not cause R; P2 does not cause R;...etc.", where P, P1, P2,... are *all existents other than* R.

Now, to formally deny that there exists anything such that nothing causes it, one would have to find an inconsistency in the said mass of statements, or in their summing up in one sentence. No such inconsistency arises. Therefore, we have to admit absolute non-causation as a formal possibility, i.e. as at least *conceivable*. It may still be factually false, i.e. there may indeed be no such animal. The issue must therefore remain open; that is, formal logic may and must proceed without resolving it. Epistemology and ontology may still, nevertheless, postulate the one or other position with reference to wider considerations.

With regard to the question: what would the relationship be between the causation apparent at the level of our ordinary sensory experiences and the spontaneity assumed by physicists to be operative at a deeper, subatomic level (known indirectly, by postulates and experimental observations) – the answer is simple enough. It is the relationship implied by a dilemmatic argument like "**whether X or Y occurs, nevertheless Z is bound to occur**" – that is: whether X or Y blossoms spontaneously at the subatomic level, *they both have the same effect Z, or equally fail to affect Z,* at the commonplace level. Here, X and Y may refer to events underlying Z, or to magnitudes or degrees of certain events (namely, velocity and position), whose average result is equally Z or of which Z remains independent. Thus, the deeper level may be open to spontaneity while the more superficial level remains governed by causation, without any incoherence being implied.

3. Interdependence.

The universal causation doctrine predicts that every existent has *at least some causative relation(s)* to some other existents. This is usually understood in a moderate sense as *only some other things* cause each thing, but Buddhism understands it more extremely as *all other things* cause each thing. This 'universal universal causation' is referred to as the **interdependence** (or codependence) of all things.

We normally suppose that only the past and present can cause the present or future; and indeed this principle should primarily be read that way. But some might go further and claim that time is transcended by causation, and that literally everything causes everything; I am not sure Buddhism goes to that extreme. Note also that, in truth, Buddhism intends its interdependence principle restrictively, as applicable only to *dharmas*, i.e. the transient phenomena constituting the world of appearances; in the higher or deeper realm of the quiescent and undifferentiated "original ground" there is no causation.

Be it said in passing, this version of "karmic law" must be distinguished from the more narrow statement, which most of us agree with, that *actions have consequences*. The latter does not imply the former! More deeply, I think what the Buddhists really meant by their law of karma was that each human (or other living) being is somewhat locked within recurring behavior patterns, very difficult (or impossible) to get out of. This is another issue, concerning not causation but volition.

That is the sense of "the wheel": our cultural and personal habits as well as our physical limitations, keep influencing our behavior and are reinforced by repetition. Much meditation and long-term corrective action are required to change them; they cannot be overcome by immediate measures, by a sheer act of will. We are thus burdened by a "baggage" of karma, which we carry out through our lives with usually little change; it may be lightened with sustained effort, but is more likely to be made heavier as time passes.

If we logically examine the claim that "everything causes everything", we see that if everything is causatively connected to everything else, then nothing is without such connection to any other thing, let alone without causative connection to anything whatsoever. That is, this doctrine is effectively *a denial that relative as well as absolute noncausation ever occurs*, which no one in Western culture would admit. To evaluate it objectively, let us look back on the findings in the present volume.

First, in defense of the idea of interdependence, it should be recalled that when we discussed the significance of the "last modus" in any grand matrix (modus #16 for two items, or #256 for three, etc.), which declares any combination of the items concerned or their negations as *possible* (code 1 in every cell of the modus), we saw that there was an uncertainty as to whether this indicated causation (or more broadly, connection) or its absence. If the last modus could be shown on formal grounds to indicate causation in *all* cases, then all contingents in the universe would have to be considered as causatively related to all others (i.e. *any two contingents taken at random* could be affirmed as causatively related, specifically in the way of the partial contingent determination, **pq**).

However, since such formal demonstration is lacking, and the idea is anyway disagreeable to common sense (at least that of non-Buddhists), we estimated that the science of Logic had to keep an open mind and grant the possibility of the alternative interpretation, namely that two items may or may not be causatively related to each other (i.e. relative noncausation is possible), and moreover that spontaneity (i.e. absolute noncausation) is at least conceivable in some cases. However, in this context, the Buddhist thesis of interdependence, remains a legitimate formal postulate. But note well, only a possible alternative hypothesis; and not a very probable one for most observers (those of us who believe in freewill, for example; as well as physicists who reify the Heisenberg Uncertainty Principle).

An important formal criticism we can level against the notion of interdependence is to ask what manner or degree of causation is meant by it. The term 'causes' in 'everything causes everything' is used very vaguely. Is only causation intended, to the exclusion of volition? And if causation is intended, surely this is meant broadly to include prevention? And are the different determinations of causation admitted, i.e. strong (complete and/or necessary) as well as weak (partial and/or contingent)? The definition of causation traditionally attributed to the Buddha is:

> *When this is, that is; this arising, that arises. When this is not, that is not; this ceasing, that ceases.*

This definition would suggest that only complete necessary causation is intended. But other discussions within Buddhism suggest that this definition is only intended as a paradigm, as the most obvious case, and partial and contingent causation is also in practice admitted, as use of the plural in the expression "causes and conditions" testifies. We may regard prevention as formally subsumed by all these concepts, by negation of an item. Some discourses also seem to accept volition, but this need not concern us here. Focusing, then, on causation in a broad sense, we may make the following criticism.

If everything is causatively related to everything else, then *the only conceivable kind of causation would be weak (both partial and contingent)*. For strong causation (complete and/or necessary) surely implies a certain exclusiveness of relationship between the items. If all items are involved to some degree in the existence of a given item, then none of those causes can be claimed to predominate. So finally, it seems to me, this Buddhist doctrine of multilateral causation requires all bilateral causative relations to be weak, and ultimately abandons strong determinations (including mixtures), and all the more so the strongest determination (which it originally rightly claimed as the definition of causation).

> One way to show that the interdependence theory implies specifically a 'universal weak link' is as follows. If we claim interdependence to apply indiscriminately to *all* 'things', i.e. not only to experiential things (*dharmas*), but also to abstract things, we fall into *formal* difficulties as soon as we suppose some causative relations to be strong. For then such abstract relations (i.e. causations) also count as 'things', and are therefore subject to interdependence. We might thus ask how a cause can be complete or necessary when that relationship is itself dependent on some yet other cause: we are forced to contradict our premise and conclude that the cause is not as complete or necessary as it seemed.

I suppose the proposed state of affairs (universal interdependence) is formally conceivable, although I do not see on what grounds we could possibly allow such rejection in one fell swoop of a large number of moduses (i.e. all alternative moduses concerning the strong determinations). Unless a reasonable formal or empirical ground is provided, there is no justification in such a radical measure: it would constitute prejudice. The Buddhist claim is of course based on a meditative experience; but since this is esoteric, not readily available to all observers at will, we must remain critical and view it as speculative. We cannot categorically eliminate it on firm rational grounds, but we cannot just take it on faith.

It should be realized that causation is a conceptual object, not a percept. Before we can discern a causative relation between two or more percepts (and all the more so between concepts) we have to distinguish the percepts from each other (and conceptualize them by comparison and contrast of many percepts, in the case of concepts). Also, causation refers to negation, which is a product of rational as well as empirical factors. Thus, if we approach the issue of causation with respect to the phenomenological order of things, we must recognize that it is a rather high-level *abstract*, although of basic importance in the organization of knowledge. It is not something we just directly see or otherwise sense. For this reason, we may remain skeptical that there is some flash of insight that would instantly reveal the causal relations of all things in the universe.

Thus, while the interdependence doctrine apparently does not give rise to formal inconsistency, we have good reason to doubt it with reference to normal human knowledge development. Causation is ordinarily known only gradually, through painstaking observation and analysis of particular data, always subject to review and revision as new data makes its appearance and possible contradictions are encountered. Our minds are not omniscient or rigidly deductive, but cumulative and flexibly inductive: we proceed by trial and error, constantly adjusting our positions to match up with new input and logical insight. Therefore, we cannot rely on sweeping statements, like that about interdependence, without being very careful.

Of course, some philosophers would argue back that causation as such is a man-made illusion, since pure experience only reveals *undifferentiated presence*. Differentiation into 'distinct' percepts, and finding that some sought things are 'absent', and conceptualization on the basis of 'similarities and differences', are all acts of reason. Indeed, if all perceived appearances are regarded as mere wave motions in a single, otherwise uniform substrate of existence (the 'original ground' of Buddhists or the Unified Field of physicists), then the boundaries we think we perceive or conceive for individuated things are in fact mere fictions, and all things (including even our fantasies about causation) are ultimately One in a very real sense.

So let us keep an open mind either way, and cheerfully move on. I just want to add one more small set of reflections, which the Buddhist idea of interdependence generated in me. This idea is often justified with reference to causal chains[109]. I tried therefore to imagine the world as a large body of water, like Lake Geneva say. According to this theory, supposedly, a disturbance anywhere in the lake eventually ripples through *the whole* lake, to an ever-diminishing degree but never dampening to zero. I then translated this image into the language of causal chains, for purposes of formal evaluation.

Looking at the results of macroanalysis, one would immediately answer that the Buddhist expectation is wrong. As we have seen, ***a cause of a cause of something is not necessarily***

[109] See for instance Thich Naht Hanh, *The Heart of Understanding* (Berkeley, Ca.: Parallax, 1988).

itself a cause of that thing; and even if it is a cause, it may be so to a lesser degree. Many first figure syllogisms yield no causative conclusion, although their premises are compatible. Some do yield a conclusion, but that conclusion is often weaker in determination than the premises. Thus, we have formal reasons to doubt the idea of interdependence, if it is taken to imply that 'a cause of cause of something is itself in turn a cause of that thing'.

All the same, I thought, thinking of the movement of disturbances in the lake, there is some truth in the contention. I then thought that maybe we should conceive of '*orders of causation*' – and postulate that even "if A causes B and B causes C, but nevertheless A does not syllogistically cause C" is true in a given case in terms of first-order causation, it can still be said that A causes C in second-order causation. And we could perhaps continue, and declare that if the latter (meaning, causes a cause of) is not applicable in a given case, we could appeal to a third order of causation, etc. We might thus, in an attempt to give credence to all theories, explain the Buddhist notion as involving a diluted sense of 'causation'.

This idea seemed plausible for a while, until I got into microanalysis. In the latter approach, conclusions are given in terms of alternative moduses. There is no room for a fanciful, more abstract, additional order of causation: the result would be identical, still the same number (one or more) of legitimate alternative moduses. No useful purpose would be served in inventing new (narrower or broader) sets of alternative moduses, and giving such groups new names. We could only at best regard all moduses in a grand matrix (other than the first, composed of all zeros) as indicative of some 'causation' (in a maximal sense), and so say that any alternative modus found at the conclusion of a syllogistic intersection is '*residual causation*'.

But having reached this bottom line, we see how trite the suggestion is.

4. Other Features of Causation Worthy of Study.

Before closing the present chapter, I would like to add some brief comments on some features of causation that should be further highlighted.

a) **Parallel Causation**. This concept was presented in some detail in our initial discussion of the generic and specific determinations, and thereafter no longer mentioned. I here just wish to remind the reader of the possibility that different causes, which are not necessarily causatively related to each other, may nevertheless have a causative relation to the same effect. That is, two things, say A and B may separately (strongly or weakly) cause some third thing C, and yet A does not cause B and B does not cause A. As the proverb says, many roads lead to Rome. If this is forgotten, one may easily get confused and think of 'pluralities of causes' as only possible within a single weak causation or in a chain of (weak or strong) causations.

This feature of causation is implicit in the microanalytic approach, insofar the possibility of *several grand matrices having common items* is not formally excluded.

b) **Degrees of Causation**. We have developed the concept of weak causation without distinction between the different possible degrees of such weak causation. That is, we have to also ask: what is more effective, what plays a larger part in producing the effect, the item (or collection of items) called partial and/or contingent cause or the item (or collection of items) called the 'complement'? We did set up a gross hierarchy between the joint determinations, **mn** being the strongest, **mq** and **np** being middling, and **pq** being the weakest. But we also

mentioned that in weak causation, the participant items may have unequal shares in the causation.
This feature of causation has not been made apparent in matricial analysis so far, and therefore needs to be accounted for in some way. I would suggest offhand that the way to include it may be to consider *the degrees of probability underlying each possibility* mentioned in the alternative moduses concerned. Thus, instead of a code '1' in each cell of an alternative modus, we might have some as worth 20%, others 40%, etc., with all non-zeros adding up to 100% probability. For example, if P and Q are complementary partial causes of R, P without Q may be more likely to be followed by R than Q without P.
In some cases, the issue may be dealt with by considering concomitant variations (see below). In any case, this topic requires further attention.

c) **Reciprocity and Direction**.
A cause and effect may (in some cases, not all) be interchangeable. For example, if we refer to the 'ideal gas equation' PV/T = constant, and consider a gas at constant temperature, we know that if the pressure is varied (increased or decreased), then the volume varies accordingly (decreased or increased). It is also true that if the volume is varied, the pressure is proportionately affected. This is *mutual causation*. Some things in a cause-effect relation do not have similar reciprocity. For example, no matter what we do, entropy further increases: our relation to entropy is one-way.
It should be stressed that even if we acknowledge that the *direction of causation* may only go in the direction of time, cause and effect are often simultaneous events (this is especially common in the extensional mode of causation, but also occurs in the natural mode). Cases of mutual causation, as well as cases of non-reciprocity, may occur either way, i.e. with cause temporally before effect or with both at the same time.
The essence of causation is certain possibilities and impossibilities of conjunctions – it does not concern questions of reciprocity or direction. These issues are left implicit in matricial analysis, acknowledged as formally possible by virtue of being ignored.

d) **Concomitant Variation**. We analyzed J. S. Mill's fifth method, that of Concomitant Variations, in some detail in Appendix 1. Although I mentioned this there, I want to here stress that this method concerns not only strong causation, but also weak causation. The above mentioned 'ideal gas equation' is an excellent example[110]. In strong causation, the concomitant variation between cause and effect is one-to-one, although not necessarily proportional. In cases of weak causation, where two or more causes together produce the effect, the part played by each factor is clarified by (if possible) holding any other factor in check (i.e. constant) while varying the one examined. This is of course not always possible.
When it is possible, the standard technique is to tabulate or graphically represent the results of experiment and then try and express them in a mathematical formula, like PV/T = constant, which summarizes a mass of if-then statements as already explained. Epistemologically, this constitutes generalization from observation. When such simple approach is not possible, because we cannot directly control the situation (for instance, in some sociological or medical researches), we resort to adductive methodology. We posit certain postulates, construct a formula out of them, and then test that formula with reference to empirical data.

---

[110] I always feel a certain affection for that example, which I learned in my teens. It shows how education has an impact on us.

It should be seen that concomitant variation deals essentially in concepts, rather than percepts. A percept is only what it is: if change occurs, another percept has replaced it. A concept, on the other hand, is an abstraction, which may well have different particular values in different cases or situations. Our formulae are algebra, not arithmetic.
We shall have to analyze concomitant variation further with reference to matricial analysis. Can the latter method be enlarged or clarified to include consideration of the former within it?

5. To Be Continued.

The Logic of Causation is a research and book project that I started several years ago, and which will no doubt take a few more years to complete. It is itself just a stage within my larger *Causal Logic* research and book project.

I published, on a small scale, an "unedited and unfinished draft" of *The Logic of Causation* back in 1999 (Phase One). The present "revised and expanded edition", published on a small scale in 2003, corrects some errors found in the 1999 version relating to the issue of lone determinations, and adds new developments of 2000 (Phase Two), as well as some recently written material such as chapters 10 and 16.

The reason why I "pre-publish" like this is that I am periodically forced to leave this research work to earn my living by other means. I do not know when I will get another chance to continue it, and wish to share with other people the results already obtained, if only through my Internet site, www.TheLogician.net. Furthermore, knowing that life is unpredictable and often short, I want to make sure the work already done is not lost to humanity, if my days happen to come to an end prematurely. I pray, however, that G-d allows I finish this work (and more still) long before!

Phase Two is in truth far from over at this time. We have here introduced the basic principles and formulas of microanalysis, but only listed most of the significant *three-item* syllogisms. But a very important development still in process is *four-item* syllogism. For this, because of the enormous matrices involved, I have to work with complex relational databases. Only after this work is completed can we compare Phases One and Two, and make sure that all previous work is consistent and error-free.

After all these technicalities are finished, and the facts of the case are settled, I will be able to devote my full attention to remaining philosophical issues relating to causation. Thereafter, I shall turn to volition and other issues in causality.

Avi Sion
Anières (GE), Switzerland, 2003.

AVI SION

THE LOGIC OF CAUSATION:

PHASE THREE: SOFTWARE ASSISTED ANALYSIS

Phase III: Software Assisted Analysis. The approach in the second phase was very 'manual' and time consuming; the third phase is intended to 'mechanize' much of the work involved by means of spreadsheets (to begin with). This increases reliability of calculations (though no errors were found, in fact) – but also allows for a wider scope. Indeed, we are now able to produce a larger, 4-item grand matrix, and on its basis find the moduses of causative and other forms needed to investigate 4-item syllogism. As well, now each modus can be interpreted with greater precision and causation can be more precisely defined and treated.

In this latest phase, ***the research is brought to a successful finish!*** Its main ambition, to obtain a complete and reliable listing of all 3-item and 4-item causative syllogisms, being truly fulfilled. This was made technically feasible, in spite of limitations in computer software and hardware, by cutting up problems into smaller pieces. For every mood of the syllogism, it was thus possible to scan for conclusions 'mechanically' (using spreadsheets), testing all forms of causative and preventive conclusions. Until now, this job could only be done 'manually', and therefore not exhaustively and with certainty. It took over 72'000 pages of spreadsheets to generate the sought for conclusions.

This is a historic breakthrough for causal logic and logic in general. Of course, not all conceivable issues are resolved. There is still some work that needs doing, notably with regard to 5-item causative syllogism. But what has been achieved solves the core problem. The method for the resolution of all outstanding issues has definitely now been found and proven. The only obstacle to solving most of them is the amount of labor needed to produce the remaining (less important) tables. As for 5-item syllogism, bigger computer resources are also needed.

Note: Phase III of the research resulted in so many and so large tables (some of them hundreds and even thousands of pages long) that it was impossible to include them all in the present printed edition. For this reason all are published on the Internet, in my website www.TheLogician.net, for your scrutiny.

Chapter 17. RESUMING THE RESEARCH.

1. History of the Research.

I have been dreaming of systematizing causal logic since my teens, I think, when I first studied works on logic and philosophy.

My first book, *Future Logic* (1990), mentions the manifest modal foundations of causality and indeed the tacit causal foundations of modality, stressing that different types (or modes) of causality exist reflecting the different types of modality (see chapters 11-12 there). And of course, knowing approximately the basic definitions of causation in terms of conditional propositions, the work done on the latter in *Future Logic* was incidentally work on the logic of causation (see parts III and IV there).

Moreover, having understood the formal continuity between categorical and de re conditional propositions, and indeed between the different modes of modality (including the logical), the work done with regard to factorial induction of categoricals was also significant in the long run to induction of conditionals – and thence to that of causative propositions (which are, after all, just conjunctions of selected conditional and categorical ones).

I made some general remarks relating to causal logic in my book *Judaic Logic*, first published in 1995 (see there chapter 10.2, *binyan av*), showing my continuing interest.

My research efforts into the logic of causation per se started in earnest in the late 1990s, with a macroanalytic approach to the problem. My purpose then was, simply put, to clearly define all the varieties of causation (its determinations, indicative of degrees of causation), then correlate them all (oppositions, eductions) and work out all syllogistic reasoning possible between them (which necessitated the development of matricial analysis). It was, I believe the first time anyone had ever tried so ambitious a project in this field of logic. This first phase of the research was published in October 1999 as *The Logic of Causation*. However, I soon realized that there were some problems in these initial results, and tried to improve on them in a second edition published in July 2000.

But it was by then clear that I needed to develop a much deeper and more systematic approach to obtain reliable results. It was, I think, not until the later half of 2002 that I found the time to proceed with microanalysis of causation, the second phase of my research. The massive amount of work involved was completed rather quickly, because I devoted all my time and concentration to it. By about March 2003, I was able to publish the results. This work involved very many painstaking manual 'calculations', and produced a very profound understanding of causation, which allowed me to *formally* settle some age-old difficult issues concerning it.

Thus, various "laws of causation" traditionally proposed were examined and evaluated. Criticisms of causation as such, such as those of Nagarjuna or David Hume, were rebutted. The notion of natural spontaneity used in modern quantum physics, as well as the Buddhist notion of interdependence were scrutinized and judged. And a critical analysis of J. S. Mill's proposed methodology for identifying causation was made possible. (See chapter 16 and appendix 1, here.) In same period, I wrote two other works which had some bearing on the understanding of causality, namely *Buddhist Illogic* (2002 – see chapters 7 & 8) and *Phenomenology* (2003 – see chapter 2.5).

However, even as I was completing this new phase of the research, it was clear to me that some uncertainties remained, due to the manual method of calculation used (subject in principle to human error, though all results were double-checked) and because some problems could not be solved by considering only three items. It was clear to me that a third phase of research, involving a more mechanical approach (using spreadsheets, database software or ad hoc programming) to increase reliability, and a larger scope (i.e. at least four items) to increase reliability, were needed. There and then, I started doing some work in that direction; but ran out of time, having to deal with many mundane matters.

In 2004, I devoted my time to writing *Volition and Allied Causal Concepts*, a study relevant to causation by implication. I continued thinking about causation in 2005, writing down my insights in *Ruminations* (see part I chapter 8 there), and even made some effort to advance phase III causation research. In 2006, my time was taken up writing *Meditations*, and in 2007-8 writing *Logical and Spiritual Reflections*. The latter work including some insights relating to causality (notably in book 1, chapters 3 & 6, and book 3, chapter 11). I also made some more effort in 2008 to advance phase III research, but was soon stopped by other concerns. The year 2009 was devoted to improving my website and to creating an online bookshop to sell my books.

I first posted some phase III results in my website, TheLogician.net, in October 2009, partly to encourage myself to pursue the matter further. In January 2010, I decided to try and complete phase III – and the work done is described in the following pages.

My initial idea with regard to phase III research was to develop a computer program capable of 'calculating' the value of causative propositions and syllogisms directly from the matrix relating the items concerned[111]. Realizing that in the absence of professional help such programming was beyond my immediate capabilities, I thought instead of using database software, such as Access. I began indeed doing so, but soon realized I had difficulty visualizing the interrelationships involved, not having made use of such software for many years. I therefore decided that the best way for me to proceed was through the use of spreadsheets, namely Excel software; and this is what I did.

2. Matrices of the Frameworks.

As explained in phase II, a 'matrix' is a condensed statement or catalogue of *all logically conceivable ontological situations* relating any number of items; note that ignorance or uncertainty is not counted as a situation, being merely epistemological. Each such completely defined situation is called a 'modus'.

The first major task of our resumed research was to develop a matrix for four items similar to the matrices for two and three items developed in chapter 12. It was soon clear that to achieve that, I had to *transpose* the earlier tables so that the modus numbers henceforth appeared as a column instead of as a row as heretofore. It is much easier to view and work with 65,536 rows than so many columns. This simple change of perspective makes all processes so much easier, I wish I had thought of doing it from the start – it would certainly have saved me much time and trouble! Sometimes by hurrying blindly we slow ourselves down.

[111] This is why I have called this phase that of Software Assisted Analysis, although of course its ultimate motive is to investigate the 4-item framework.

Anyway, in this manner Table 12.1, cataloguing the 16 moduses for 2 items (PR), became Table 17.1 shown below. As can be seen, the modus numbers 1 to 16 are in the first column, labeled ID. The next four column headings signify the different possible combinations of the two items concerned, P and R (0 and 1 here meaning present and absent, note well) – '11' meaning (P + R), '10' meaning (P + notR), '01' meaning (notP + R), and '00' meaning (notP + notR). Note that 11>10>01>00.

The 'summary' column merely summarizes the information in the preceding four (ignoring leading zeros); notice that the numbers in it range from 0 (i.e. '0000') to 1111 in an orderly manner (i.e. each number is greater than the one above it). It is evident from the summary that the moduses are not numbered randomly, but in increasing magnitude. The last column counts the zeros in each of the moduses[112]; this number is an indication of the degree of freedom or lack of it in the situation signified by the modus concerned (since here, 0 means 'impossible' while 1 means 'possible' combination). Using a spreadsheet program, it is easy to generate the modus numbers, the summaries and count of zeros; and to verify that the summaries are indeed in order.

Table 17.1 List of 16 Moduses for 2 Items PR

ID	**11**	**10**	**01**	**00**	**summary**	**number of zeros**
1	0	0	0	0	0000	4
2	0	0	0	1	0001	3
3	0	0	1	0	0010	3
4	0	0	1	1	0011	2
5	0	1	0	0	0100	3
6	0	1	0	1	0101	2
7	0	1	1	0	0110	2
8	0	1	1	1	0111	1
9	1	0	0	0	1000	3
10	1	0	0	1	1001	2
11	1	0	1	0	1010	2
12	1	0	1	1	1011	1
13	1	1	0	0	1100	2
14	1	1	0	1	1101	1
15	1	1	1	0	1110	1
16	1	1	1	1	1111	0

Similarly, Table 12.3, cataloguing the 256 moduses for 3 items (PR), became Table 17.2, consisting of rows labeled 1 to 256 and a matrix of 8 columns labeled 111, 110, 101, 100, 011, 010, 001, 000, followed by a summary column, with numbers ranging from 0 (i.e. '00000000') to 11111111, and a count of zeros.

Finally, a new table cataloguing the 65,536 moduses for 4 items (PQRS), Table 17.3, was generated. Note that the latter table, having a matrix of 16 columns, implied some high summary numbers to be 16 digits long; this was a technical problem in that Excel software

112 This information was previously given in the second column of Table 12.6.

cannot handle more than 15 digits, so that the last digit is made 0 instead 1 in certain cases (e.g. 1111111111111111). Here two a final column was added on, showing the number of zeros in each modus.

For the record: the matrix of a single item P comprises 2*4=8 cells (see Table 13.9). The matrix for two items, say P and R, comprises 4*16=48 cells (Table 17.1). That for three items, PQR has 8*256=2,048 cells (Table 17.2). The corresponding table for four items has 16*65,536=1,048,576 cells (Table 17.3). *Each cell represents a bit of information (about a possible or impossible combination of the items concerned) that needs to be induced or deduced to determine the modus applicable; only when all the cells adding up to a modus are so determined can we claim to know that there is or is not a relation of causation or whatever between the items.* Note this well, because it shows how far true causal logic is from the simplistic claims of the likes of David Hume.

Thus, our first three tables display the matrices at the root of the three frameworks, which we will in the present phase of our study: ***These matrices contain the basic data from which all other tables will be constructed*** – i.e. they house the information needed to develop the definitions of all conjunctive, conditional and causative forms, their oppositions, eductions from them, and syllogisms involving them. These three tables are:

- Table 17.1 – 2-Item Matrix: 16 Moduses. (1 page in pdf file).
- Table 17.2 – 3-Item Matrix: 256 Moduses. (4 pages in pdf file).
- Table 17.3 – 4-Item Matrix: 65,536 Moduses. (565 pages in pdf file).

The first of these tables, being brief, is reproduced above, though most of its content is already given in Table 12.1, in order to show how the original was transposed. The second, could have been reprinted here, but most of its information is already given in Table 12.3, so there is no point. The third table is of course far too long to include in the present printed report. However, this table and the two preceding it, and indeed all subsequent tables relating to phase III work, are made available as .pdf files in my website[113], at the following address (there click on 'Phase III'):

http://www.thelogician.net/4_logic_of_causation/4_lc_tables.htm

Please do carefully examine the phase III tables there, for they are the purpose of the whole research! Note that the first three tables were developed successively, in order of complexity, starting with two items, then three, then four. The advantage of this method is that the formulae used in each table to generate data or calculate results can be passed on to the next table, i.e. to the larger framework, with appropriate modifications to adjust to the increased complexity at hand. Though I may not point it out repeatedly, keep in mind that this pattern of development is used throughout the present research. It has made things rather easy for me!

---

113 Note that some tables were produced in 'landscape' (instead of 'portrait') orientation. Adobe pdf Reader shows such tables sideways – you have to click on View then on Rotate View Clockwise to redress them. Needless to say, if the image seems too small, you can increase its size as much as you want using the appropriate button on the Reader tool bar.

3. Comparing Frameworks.

The next challenge was to compare moduses in the different frameworks. This is not only done out of curiosity, but to understand in detail just how each larger framework passes on information previously found and provides new information.

We had already answered this question in part in Table 12.6, listing the correspondences between 2-item moduses and 3-item moduses – but the work was then performed manually, and now it needed to be done mechanically, i.e. using formulae in a spreadsheet. This new work resulted in the following two tables, displayed in the website: Table 17.4 (equivalent to Table 12.6), comparing the 2- and 3- item frameworks, and Tables 17.5 and 17.6 (two new ones), comparing the 3- and 4- item frameworks.

These tables were produced in stages, briefly put, as follows. I started with a 3-item matrix with 8 columns, and used it to produce another table with only 4 columns. Note well – rather than move from 2 to 3 items, I worked backwards from 3 to 2 items. Each of the cells in the latter had an appropriate formula deriving it from relevant cells in the same row of the former. A summary column was then added to this derivative table, from which – using the vertical lookup function of Excel – each 3-item row was given a 2-item modus number (ranging from 1 to 16), and the job was done.

Thereafter, it was easy to derive a further table listing and then counting the 3-item moduses corresponding to each 2-item modus. See the additional notes at the bottom of these tables. All information obtained was checked with reference to the relevant tables produced in phase II and found consistent.

The same method was used to identify the 3-item modus number corresponding to each 4-item modus number. However, whereas for the '2 to 3 items' comparison the results are lumped together in one pdf (Table 17.4), for the '3 to 4 items' comparison the results are split into two pdf files (Tables 17.5 & 17-6), in view of the mass of data involved. The first part lists all the moduses for 4 items and next to each of them the corresponding the 3-item modus, and it also lists the count of 4-item moduses corresponding to each 3-item modus. The second part specifies the 4-item modus numbers corresponding to each 3-item modus number, and counts them.

Needless to say, the latter three tables all provide us with some valuable new information not previously generated. Thus, regarding comparison of frameworks, the following three tables are made available for your scrutiny on The Logician website:

- Table 17.4 – From 2 to 3 Items Moduses. (6 pages in pdf file).
- Table 17.5 – From 3 to 4 Items Moduses – 1st part. (1192 pages in pdf file).
- Table 17.6 – From 3 to 4 Items Moduses – 2nd part. (2792 pages in pdf file).

We can now look into the moduses applicable to each of the forms of causation and various other forms.

Chapter 18. MODUSES OF THE FORMS.

1. 2-Item Framework Moduses.

The next set of tables was produced with several purposes in mind. As we have established in the past, all propositions relating to causation – and indeed all their underlying categorical (individual or conjunctive) and conditional bases and connections – can be expressed entirely with reference to the moduses in the relevant matricial framework, depending on the number of items involved.

Thus, the main task of causative logic is to systematically identify the moduses of all the forms of causation. This work has of course already been largely done in phase II, but here our task was to do it mechanically instead of manually and to develop it from the 2- and 3- item frameworks to the 4-item framework. I first produced the following two tables, which are on display at the The Logician website.

- Table 18.1 – 2-item PR Moduses of Forms. (6 pages in pdf file).
- Table 18.2 – 2-item PR Moduses of Forms – Formulae Used. (1 page in pdf file).

The first table comprises a mass of information previously scattered in several tables (notably, 12.1, 12.2, 13.3, 13.10, 13.11, 13.12, and 16.1). The second table is merely an auxiliary one, showing (for the record) the formulae used to generate the first table. See the notes at the bottom of each of these tables for further information. All the results obtained in Table 18.1 were compared to corresponding results obtained in phase II, and they were found in agreement – showing that no errors were made in either research.

Notice that Table 18.1, which totals 79 columns, can be divided into distinct segments: the first segment shows the 2-item matrix from which all subsequent values are calculated through transparent formulae. The next segment lists the moduses for each of the individual propositions involved (P is possible, P is impossible, etc.). The following segment similarly treats conjunctions of the two items (P+R is possible, P+R is impossible, etc.). Then comes the moduses of the generic forms of causation and their negations (**m**, **n**, **p**, **q**, etc.). This is followed by the moduses of the specific forms of causation and their negations (**mn**, **mq**, **np**, **pq**, etc.). The segment after that deals with absolute **lone** forms (they are confirmed to be non-existent) and vaguer forms of causation and their negations (**s**, **w**, **c**, etc.).

The next segments deal with prevention, inverse causation and inverse prevention, connection and their negations. Prevention by P of R, remember, means causation by P of notR. Inverse causation by P of R refers to causation by notP of notR; and inverse prevention by P of R refers to causation by notP of R. Although these are all derivable from causation, they needed to be shown here to obtain the very last segment of the table, which gives us *a full verbal interpretation of each and every modus*. The interpretations in this table will be discussed at length a bit further on (section 4).

A valuable insight I had while preparing Table 18.1, which I must mention here, is that when in it I equate, say, "if P, then R" to "P + notR is impossible", I have in mind *de dicta* (logical) conditioning, which in accordance with common practice only requires the 'connection' to be specified. But it is clear that if we wish to deal with *de re* (extensional, natural or temporal) conditioning, we cannot make this equation, for here the 'base' that "P is possible" (whence, "R is possible" too) must be specified as well[114]. This is important only insofar as we are still dealing with conditional propositions as such; for as soon as we get into causative propositions, the 'bases' are always tacitly implied anyway, so the logic for de dicta and de re is the same.

2. 3-Item Framework Moduses.

Although Table 18.1 (and its auxiliary 18.2) contains little new information on the 2-item moduses of forms, its production was very useful to producing an equivalent table for the 3-item framework, because I could copy the formulae used in the former and paste them in the latter, and then expand them to apply to the enlarged framework. This saved a lot of time and trouble. Of course, the 3-item framework table involved many new forms, but even these could be derived from the preceding using appropriate equivalence tables. The result was a very large spreadsheet of 415 columns (including 10 for the matrix).

Rather than present all this data in one massive table, which readers would get lost in and so miss out on important information, I split it up for publication into smaller tables. These include Tables 18.3-18.6, which respectively deal with categorical and conditional forms, causation, prevention, and interpretation of the moduses. To which must be added two auxiliary tables, Tables 18.7-18.8, which spell out the formulae used and the equivalences exploited in producing the original big table. We thus have six tables[115] for the 3-item framework, which are as usual posted at the The Logician website for your scrutiny:

- Table 18.3 – 3-item PQR Moduses of Forms – Categoricals and Conditionals. (12 pages in pdf file).
- Table 18.4 – 3-item PQR Moduses of Forms – Causation. (18 pages in pdf file).
- Table 18.5 – 3-item PQR Moduses of Forms – Prevention. (18 pages in pdf file).
- Table 18.6 – 3-item PQR Moduses of Forms – Interpretation. (4 pages in pdf file).
- Table 18.7 – 3-item PQR Moduses of Forms – Formulae Used. (8 pages in pdf file).
- Table 18.8 – 3-item PQR Moduses of Forms – Equivalences. (1 page in pdf file).

Notice the subdivisions into segments within these tables. Some of the data in these tables has already been generated in phase II, notably in Tables 11.3, 11.4, 12.3, 12.4, 13.1, 13.3, 13.4, 13.7, 13.14, 13.15, 14.3, 14.5. See the notes at the bottom of the segments; notice that wherever information was already given in phase II, results were compared and found consistent. However, much data in these phase III tables is new, generated in pursuit of enlarged scope, more symmetry of treatment, and thorough interpretation of the moduses. The more mechanical nature of data generation in phase III enabled such increased ambition.

Thus, Table 18.3 (96 columns, plus the matrix) shows the 3-item moduses of all possible categorical propositions (P is possible, etc.), then of all 2-item conjunctions (P+R is possible, etc.), and then of all 3-item conjunctions (P+Q+R is possible, etc.) – many of which

---

114 I do not think I realized that during phase II – see for instance Table 13.12. Needless to say, this insight applies not only to the 2-item framework but equally to the 3-item framework and on.

115 More tables are introduced in later chapters.

conjunctions of course signify conditional propositions[116]. Note that every combination and permutation of the three items P, Q, R are treated here, and in the subsequent tables.
Table 18.4 (130 columns, plus the matrix) shows the 3-item moduses of the generic and specific, absolute and relative, forms of causation, including lones and vaguer forms, and their negations (and, for the record, of inverse causation). Note the equal treatment here of forms relative to notQ; the motives of this and similar expansions of scope being, not mere curiosity, but (a) to make possible interpretation of all the moduses in Table 18.6 and (b) to enable us to draw as much conclusion as possible when we get to the syllogistic stage (whereas in phase II, we deliberately limited our possibilities of conclusion).
Table 18.5 (130 columns, plus the matrix) repeats the work of 18.4 with regard to prevention (i.e. causation by P of notR). This is done, again, both to facilitate interpretation of the moduses and to ensure maximization of conclusions at the later syllogistic stage.
Table 18.6 (43 columns) constitutes the crucial interpretation of the results obtained in the preceding two tables. It verbalizes and makes sense of all the information collected in them. The need to develop such a table for the 3-item framework propelled most of the work preceding it. Notice here the symmetry of the results for causation and prevention (as can be expected). Note the use of certain summaries of information on causation, lone causation, prevention and lone prevention. Note also the last columns, concerning modus 16. The interpretations in this table will be discussed at length a bit further on (section 4).
Table 18.7, to repeat, lists the formulae used in producing the original table comprising Tables 18.3-18.6. It did not seem necessary or useful to split this table in four. What is noteworthy here is that most formulae are written in terms of the initial matrix. Table 18.8 reveals how some of the formulae in Table 18.7 were derived from others, simply by reordering and/or changing the polarities of the terms involved. I include it here for the sake of transparency.

3. 4-Item Moduses of the Forms.

Clearly, the 2-item framework table is of value only to begin with, to teach us how to analyze the forms – but this information is not enough to produce all conceivable syllogisms. On the other hand, the 3-item framework does give rise to systematic syllogistic work, so that many forms have to be analyzed in their many guises, i.e. with respect to the various combinations and permutations of the three items P, Q, R, and their negations.
This is of course all the more true in the 4-item framework – but in the latter case we have to be more restrained, otherwise the tables would be far too large for comfort. With this reasoning in mind, I only analyzed in the 4-item framework a selection of forms, the minimum needed to answer some previously unanswered syllogistic questions. The resulting Table 18.9 and its auxiliary 18.10 can be viewed at the The Logician website, as usual:

- Table 18.9 – 4-item PQR Moduses of Forms. (2408 pages in pdf file).
- Table 18.10 – 4-item PQR Moduses of Forms – Formulae Used. (3 pages in pdf file).

[116] Or more precisely, the 'connection' of such propositions, without their 'bases', as mentioned in the previous section. This is consistent with the usual formulation of logical conditioning; but for de re forms of conditioning, we would have to include consideration of the underlying possibilities before identifying the conjunctives with conditionals.

Table 18.9 (173 columns, including 17 for the matrix) shows the 4-item moduses of *selected* generic and specific, absolute and relative, forms of causation, including some lones, and their negations. This is already a big mass of information to have to deal with. Table 18.10 lists the formulae used to produce Table 18.9.

Notice that the applicable moduses for any of the forms examined are in here signaled by a '1' instead of by the modus number (as in similar tables for 2-items and for 3-items). The reasons for this are simply to avoid overly wide columns and to make the file as a whole more manageable in size. The 1s in the columns of the 4-item table are just meant to indicate that yes, the modus number opposite (to the left of) the cell concerned is a possible modus number for the form concerned. Accordingly, '0' means the adjacent modus is not applicable.

It is important to understand that the 3-item framework is in principle sufficient to fulfill the task of causative logic. That is because two items suffice to define the genera of strong causation and three items suffice to define those of weak causation. From the start of our research, remember (see chapter 2.3), we conceived of partial or contingent causation as consisting of two causes, say P and Q, and one effect R. We arbitrarily viewed P as the main cause and Q as its complement, so as to conjoin the weak forms of causation with strong forms expressed in terms of P and R. Just as the effect R could be a mass of phenomena lumped together under this name, so could P and Q respectively be far from unitary. Thus, by definition, the complement Q was designed to accommodate any number of phenomena – of which Q would be the effective single resultant in the PQR causative formula.

In other words, when there are more than two partial and/or contingent causes, or more to the point when in addition to P we have several complements, Q1, Q2, etc. – we are called upon to first determine a *resultant* complement Q whose behavior within the causative proposition concerned would correspond to the behavior of the several narrower complements Q1, Q2, etc. By this artifice, we were able to reduce the problem of relative causation to only three items, P, Q, R.

The need for more items than these three arises only at the syllogistic stage of the study of causation, when we need to investigate how relative causation is transmitted from either premise to the conclusion (if any), and what perhaps happens (if anything) when both premises are relative causations each involving a different complement. Thus, conceivably, we might need matrices of four, and maybe even five, items to find all possible syllogistic conclusions. This issue will be further discussed later on.

4. Interpretation of the Moduses.

We shall now interpret and discuss the individual moduses in the 2-item and 3-item frameworks. It is important to understand at the outset that each modus represents one complete situation – meaning that the two or three items whose relations are found to fit into the pattern symbolized by a certain modus may be said to be causatively or otherwise related as that modus signifies. For this reason, it is important to clearly identify the significance of each modus; such identification has enduring, universal value.

a. Regarding the interpretations of the 2-item moduses, please refer to Table 18.1 (page 6), an extract from which is printed here, below. This table is not new, since it corresponds to Table 16.1 presented in phase II.

This table teaches us that of the 16 moduses in a 2-item framework: one modus is logically impossible anyway (#1); eight moduses have one or both items incontingent (i.e. necessary or impossible) and so cannot signify any causative connection between them (since an item that is incontingent, in the mode of modality concerned, is independent of all else); three moduses signify a strong causation (**mn**, **mq**, **np**) and three more a strong prevention (ditto); and the last modus (#16) refers to both weak (**pq**) causation and prevention. The weak causations mentioned in this framework are of course all absolute, since we do not know the complement they concern, though it may be assumed that they do concern *some* complement(s).

For connection (i.e. causation or prevention) to occur and be claimed, the two items concerned have to both be contingent. This occurs only in seven of the moduses, namely numbers 7, 8, 10, 12, 14, 15, 16 (the remaining eight being either impossible or incontingent, as already pointed out). This result is very surprising, for it means that apart from incontingency, logic has found no place from non-connection! That is, this tabulation of possibilities being exhaustive, we are left with no way to rationally express a situation of non-connection between individually contingent items. This seems to imply that any two contingent items in the universe, taken at random, are somewhat connected together, by causation and/or by prevention, to whatever degree (i.e. as **mn**, **mq**, **np or pq** – the latter three referring partly or wholly to absolute weak determinations).

Though I am again alarmed upon encountering this result, it must be stressed that I had already noticed it and tried hard to explain it in phase II (see Chapters 13.2 and 16.2). I will try to propose new insights regarding it, further on, armed with a similar analysis for the 3-item framework.

Detail from Table 18.1 – Interpretation of the 2-Item Moduses[117].

ID	summary	
1	impossible	
2	incontingency	
3	incontingency	
4	incontingency	
5	incontingency	
6	incontingency	
7	only strong prevention	**mn**
8	joint s-w prevention	**mq abs**
9	incontingency	
10	only strong causation	**mn**
11	incontingency	
12	joint s-w causation	**mq abs**
13	incontingency	
14	joint s-w causation	**np abs**
15	joint s-w prevention	**np abs**
16	both causation and prevention	**pq abs + pq abs**

[117] Note my use here and elsewhere of '**mq** abs' instead of $\mathbf{mq}_{abs}$, '**np** abs' instead of $\mathbf{np}_{abs}$., '**pq** abs' instead of $\mathbf{p}_{abs}\mathbf{q}_{abs}$ – no new meaning is intended in such cases; I just find it more convenient. Similarly of course with regard to '**p** rel' and '**q** rel', and their compounds, later on.

For now, note one thing that I did not clearly realize before – it is that the last modus (#16) refers to *both* weak absolute causation and weak absolute prevention, and not as I previously wrote or implied to either the one or the other. This new observation is significant, in that it teaches us that causation and prevention at this low degree of determination are not mutually exclusive, but rather apparently occur in tandem (this is later confirmed in the 3-item framework).

Moreover, it is well to remember in this context, before moving on, that causation here includes both causation by P of R and inverse causation by P of R, i.e. causation by notP of notR, since these have the same moduses, though **mq** becomes **np** and vice versa. Similarly, prevention, here includes both prevention by P of R, i.e. causation by P of notR, and inverse prevention by P of R, i.e. causation by notP of R, since these have the same moduses, though **mq** becomes **np** and vice versa. That is to say, the above table of 16 moduses covers every logical possibility.

b. Let us now look at the similar interpretations of the 3-item moduses. These may be examined in detail in Table 18.6 given online. The following is an extract from that table:

Table 18.6 (detail) – Interpretations of the individual moduses

ID	summary	interpretations of the individual moduses
1	00000000	impossible modus
2	00000001	non-connection due to incontingency of P and/or R, and of Q
3	00000010	non-connection due to incontingency of P and/or R, and of Q
4	00000011	non-connection due to incontingency of P and/or R, and of Q
5	00000100	non-connection due to incontingency of P and/or R, and of Q
6	00000101	non-connection due to incontingency of P and/or R
7	00000110	non-connection due to incontingency of P and/or R
8	00000111	non-connection due to incontingency of P and/or R
9	00001000	non-connection due to incontingency of P and/or R, and of Q
10	00001001	non-connection due to incontingency of P and/or R
11	00001010	non-connection due to incontingency of P and/or R
12	00001011	non-connection due to incontingency of P and/or R
13	00001100	non-connection due to incontingency of P and/or R, and of Q
14	00001101	non-connection due to incontingency of P and/or R
15	00001110	non-connection due to incontingency of P and/or R
16	00001111	non-connection due to incontingency of P and/or R
17	00010000	non-connection due to incontingency of P and/or R, and of Q
18	00010001	non-connection due to incontingency of P and/or R, and of Q
19	00010010	strong prevention, with Q incontingent
20	00010011	joint prevention (absolute), with Q incontingent
21	00010100	non-connection due to incontingency of P and/or R
22	00010101	non-connection due to incontingency of P and/or R
23	00010110	joint prevention (relative to Q)
24	00010111	joint prevention (absolute)
25	00011000	strong prevention
26	00011001	joint prevention (absolute)
27	00011010	strong prevention
28	00011011	joint prevention (absolute)

Table 18.6 (detail) – Interpretations of the individual moduses (continued)

ID	summary	interpretations of the individual moduses
29	00011100	joint prevention (absolute)
30	00011101	joint prevention (absolute)
31	00011110	joint prevention (relative to Q)
32	00011111	joint prevention (absolute)
33	00100000	non-connection due to incontingency of P and/or R, and of Q
34	00100001	strong causation, with Q incontingent
35	00100010	non-connection due to incontingency of P and/or R, and of Q
36	00100011	joint causation (absolute), with Q incontingent
37	00100100	strong causation
38	00100101	strong causation
39	00100110	joint causation (absolute)
40	00100111	joint causation (absolute)
41	00101000	non-connection due to incontingency of P and/or R
42	00101001	joint causation (relative to Q)
43	00101010	non-connection due to incontingency of P and/or R
44	00101011	joint causation (absolute)
45	00101100	joint causation (absolute)
46	00101101	joint causation (relative to Q)
47	00101110	joint causation (absolute)
48	00101111	joint causation (absolute)
49	00110000	non-connection due to incontingency of P and/or R, and of Q
50	00110001	joint causation (absolute), with Q incontingent
51	00110010	joint prevention (absolute), with Q incontingent
52	00110011	weak causation (absolute) and weak prevention (absolute), Q incontingent
53	00110100	joint causation (absolute)
54	00110101	joint causation (absolute)
55	00110110	weak causation (absolute) and weak prevention (relative to Q)
56	00110111	weak causation (absolute) and weak prevention (absolute), Q incontingent
57	00111000	joint prevention (absolute)
58	00111001	weak causation (relative to Q) and weak prevention (absolute)
59	00111010	joint prevention (absolute)
60	00111011	weak causation (absolute) and weak prevention (absolute)
61	00111100	weak causation (absolute) and weak prevention (absolute)
62	00111101	weak causation (relative to Q) and weak prevention (absolute)
63	00111110	weak causation (absolute) and weak prevention (relative to Q)
64	00111111	weak causation (absolute) and weak prevention (absolute)
65	01000000	non-connection due to incontingency of P and/or R, and of Q
66	01000001	non-connection due to incontingency of P and/or R
67	01000010	strong prevention
68	01000011	joint prevention (absolute)
69	01000100	non-connection due to incontingency of P and/or R, and of Q
70	01000101	non-connection due to incontingency of P and/or R
71	01000110	joint prevention (absolute)
72	01000111	joint prevention (absolute)
73	01001000	strong prevention, with Q incontingent
74	01001001	joint prevention (relative to notQ)

Table 18.6 (detail) – Interpretations of the individual moduses (continued)

ID	summary	interpretations of the individual moduses
75	01001010	strong prevention
76	01001011	joint prevention (relative to notQ)
77	01001100	joint prevention (absolute), with Q incontingent
78	01001101	joint prevention (absolute)
79	01001110	joint prevention (absolute)
80	01001111	joint prevention (absolute)
81	01010000	non-connection due to incontingency of P and/or R
82	01010001	non-connection due to incontingency of P and/or R
83	01010010	strong prevention
84	01010011	joint prevention (absolute)
85	01010100	non-connection due to incontingency of P and/or R
86	01010101	non-connection due to incontingency of P and/or R
87	01010110	joint prevention (relative to Q)
88	01010111	joint prevention (absolute)
89	01011000	strong prevention
90	01011001	joint prevention (relative to notQ)
91	01011010	strong prevention
92	01011011	joint prevention (relative to notQ)
93	01011100	joint prevention (absolute)
94	01011101	joint prevention (absolute)
95	01011110	joint prevention (relative to Q)
96	01011111	joint prevention (absolute)
97	01100000	non-connection due to incontingency of P and/or R
98	01100001	joint causation (relative to notQ)
99	01100010	joint prevention (absolute)
100	01100011	weak causation (relative to notQ) and weak prevention (absolute)
101	01100100	joint causation (absolute)
102	01100101	joint causation (relative to notQ)
103	01100110	weak causation (absolute) and weak prevention (absolute)
104	01100111	weak causation (relative to notQ) and weak prevention (absolute)
105	01101000	joint prevention (relative to Q)
106	01101001	weak causation (rel. to Q and notQ) and weak prevention (rel. to Q and notQ)
107	01101010	joint prevention (relative to Q)
108	01101011	weak causation (relative to notQ) and weak prevention (relative to Q and notQ)
109	01101100	weak causation (absolute) and weak prevention (relative to Q)
110	01101101	weak causation (relative to Q and notQ) and weak prevention (relative to Q)
111	01101110	weak causation (absolute) and weak prevention (relative to Q)
112	01101111	weak causation (relative to notQ) and weak prevention (relative to Q)
113	01110000	non-connection due to incontingency of P and/or R
114	01110001	joint causation (absolute)
115	01110010	joint prevention (absolute)
116	01110011	weak causation (absolute) and weak prevention (absolute)
117	01110100	joint causation (absolute)
118	01110101	joint causation (absolute)
119	01110110	weak causation (absolute) and weak prevention (relative to Q)
120	01110111	weak causation (absolute) and weak prevention (absolute)

Table 18.6 (detail) – Interpretations of the individual moduses (continued)

ID	summary	interpretations of the individual moduses
121	01111000	joint prevention (relative to Q)
122	01111001	weak causation (relative to Q) and weak prevention (relative to Q and notQ)
123	01111010	joint prevention (relative to Q)
124	01111011	weak causation (absolute) and weak prevention (relative to Q and notQ)
125	01111100	weak causation (absolute) and weak prevention (relative to Q)
126	01111101	weak causation (relative to Q) and weak prevention (relative to Q)
127	01111110	weak causation (absolute) and weak prevention (relative to Q)
128	01111111	weak causation (absolute) and weak prevention (relative to Q)
129	10000000	non-connection due to incontingency of P and/or R, and of Q
130	10000001	strong causation
131	10000010	non-connection due to incontingency of P and/or R
132	10000011	joint causation (absolute)
133	10000100	strong causation, with Q incontingent
134	10000101	strong causation
135	10000110	joint causation (relative to notQ)
136	10000111	joint causation (relative to notQ)
137	10001000	non-connection due to incontingency of P and/or R, and of Q
138	10001001	joint causation (absolute)
139	10001010	non-connection due to incontingency of P and/or R
140	10001011	joint causation (absolute)
141	10001100	joint causation (absolute), with Q incontingent
142	10001101	joint causation (absolute)
143	10001110	joint causation (absolute)
144	10001111	joint causation (absolute)
145	10010000	non-connection due to incontingency of P and/or R
146	10010001	joint causation (absolute)
147	10010010	joint prevention (relative to notQ)
148	10010011	weak causation (absolute) and weak prevention (relative to notQ)
149	10010100	joint causation (relative to Q)
150	10010101	joint causation (relative to Q)
151	10010110	weak causation (rel. to Q and notQ) and weak prevention (rel. to Q and notQ)
152	10010111	weak causation (relative to Q and notQ) and weak prevention (relative to notQ)
153	10011000	joint prevention (absolute)
154	10011001	weak causation (absolute) and weak prevention (absolute)
155	10011010	joint prevention (relative to notQ)
156	10011011	weak causation (absolute) and weak prevention (relative to notQ)
157	10011100	weak causation (relative to Q) and weak prevention (absolute)
158	10011101	weak causation (relative to Q) and weak prevention (absolute)
159	10011110	weak causation (relative to Q) and weak prevention (relative to Q and notQ)
160	10011111	weak causation (relative to Q) and weak prevention (relative to notQ)
161	10100000	non-connection due to incontingency of P and/or R
162	10100001	strong causation
163	10100010	non-connection due to incontingency of P and/or R
164	10100011	joint causation (absolute)
165	10100100	strong causation
166	10100101	strong causation

Table 18.6 (detail) – Interpretations of the individual moduses (continued)

ID	summary	interpretations of the individual moduses
167	10100110	joint causation (relative to notQ)
168	10100111	joint causation (relative to notQ)
169	10101000	non-connection due to incontingency of P and/or R
170	10101001	joint causation (relative to Q)
171	10101010	non-connection due to incontingency of P and/or R
172	10101011	joint causation (absolute)
173	10101100	joint causation (absolute)
174	10101101	joint causation (relative to Q)
175	10101110	joint causation (absolute)
176	10101111	joint causation (absolute)
177	10110000	non-connection due to incontingency of P and/or R
178	10110001	joint causation (absolute)
179	10110010	joint prevention (absolute)
180	10110011	weak causation (absolute) and weak prevention (absolute)
181	10110100	joint causation (relative to Q)
182	10110101	joint causation (relative to Q)
183	10110110	weak causation (relative to Q and notQ) and weak prevention (relative to Q)
184	10110111	weak causation (relative to Q and notQ) and weak prevention (absolute)
185	10111000	joint prevention (absolute)
186	10111001	weak causation (relative to Q) and weak prevention (absolute)
187	10111010	joint prevention (absolute)
188	10111011	weak causation (absolute) and weak prevention (absolute)
189	10111100	weak causation (relative to Q) and weak prevention (absolute)
190	10111101	weak causation (relative to Q) and weak prevention (absolute)
191	10111110	weak causation (relative to Q) and weak prevention (relative to Q)
192	10111111	weak causation (relative to Q) and weak prevention (absolute)
193	11000000	non-connection due to incontingency of P and/or R, and of Q
194	11000001	joint causation (absolute)
195	11000010	joint prevention (absolute)
196	11000011	weak causation (absolute) and weak prevention (absolute)
197	11000100	joint causation (absolute), with Q incontingent
198	11000101	joint causation (absolute)
199	11000110	weak causation (relative to notQ) and weak prevention (absolute)
200	11000111	weak causation (relative to notQ) and weak prevention (absolute)
201	11001000	joint prevention (absolute), with Q incontingent
202	11001001	weak causation (absolute) and weak prevention (relative to notQ)
203	11001010	joint prevention (absolute)
204	11001011	weak causation (absolute) and weak prevention (relative to notQ)
205	11001100	weak causation (absolute) and weak prevention (absolute), Q incontingent
206	11001101	weak causation (absolute) and weak prevention (absolute)
207	11001110	weak causation (absolute) and weak prevention (absolute)
208	11001111	weak causation (absolute) and weak prevention (absolute)
209	11010000	non-connection due to incontingency of P and/or R
210	11010001	joint causation (absolute)
211	11010010	joint prevention (relative to notQ)
212	11010011	weak causation (absolute) and weak prevention (relative to notQ)

Table 18.6 (detail) – Interpretations of the individual moduses (continued)

ID	summary	interpretations of the individual moduses
213	11010100	joint causation (absolute)
214	11010101	joint causation (absolute)
215	11010110	weak causation (relative to notQ) and weak prevention (relative to Q and notQ)
216	11010111	weak causation (relative to notQ) and weak prevention (relative to notQ)
217	11011000	joint prevention (absolute)
218	11011001	weak causation (absolute) and weak prevention (relative to notQ)
219	11011010	joint prevention (relative to notQ)
220	11011011	weak causation (absolute) and weak prevention (relative to notQ)
221	11011100	weak causation (absolute) and weak prevention (absolute)
222	11011101	weak causation (absolute) and weak prevention (absolute)
223	11011110	weak causation (absolute) and weak prevention (relative to Q and notQ)
224	11011111	weak causation (absolute) and weak prevention (relative to notQ)
225	11100000	non-connection due to incontingency of P and/or R
226	11100001	joint causation (relative to notQ)
227	11100010	joint prevention (absolute)
228	11100011	weak causation (relative to notQ) and weak prevention (absolute)
229	11100100	joint causation (absolute)
230	11100101	joint causation (relative to notQ)
231	11100110	weak causation (relative to notQ) and weak prevention (absolute)
232	11100111	weak causation (relative to notQ) and weak prevention (absolute)
233	11101000	joint prevention (absolute)
234	11101001	weak causation (relative to Q and notQ) and weak prevention (relative to notQ)
235	11101010	joint prevention (absolute)
236	11101011	weak causation (relative to notQ) and weak prevention (relative to notQ)
237	11101100	weak causation (absolute) and weak prevention (absolute)
238	11101101	weak causation (relative to Q and notQ) and weak prevention (absolute)
239	11101110	weak causation (absolute) and weak prevention (absolute)
240	11101111	weak causation (relative to notQ) and weak prevention (absolute)
241	11110000	non-connection due to incontingency of P and/or R
242	11110001	joint causation (absolute)
243	11110010	joint prevention (absolute)
244	11110011	weak causation (absolute) and weak prevention (absolute)
245	11110100	joint causation (absolute)
246	11110101	joint causation (absolute)
247	11110110	weak causation (relative to notQ) and weak prevention (relative to Q)
248	11110111	weak causation (relative to notQ) and weak prevention (absolute)
249	11111000	joint prevention (absolute)
250	11111001	weak causation (relative to Q) and weak prevention (relative to notQ)
251	11111010	joint prevention (absolute)
252	11111011	weak causation (absolute) and weak prevention (relative to notQ)
253	11111100	weak causation (absolute) and weak prevention (absolute)
254	11111101	weak causation (relative to Q) and weak prevention (absolute)
255	11111110	weak causation (absolute) and weak prevention (relative to Q)
256	11111111	weak causation (absolute) and weak prevention (absolute)
count	256	

This is new material that has not been previously researched. I will begin by listing some statistics drawn from it: First, apart from the formally impossible modus (#1), there are 48 moduses signifying non-connection due to the incontingency of P and/or R. In some of these cases Q is also incontingent; but if both P and R are contingent, the incontingency of Q does not impede a connection, however tenuous, between P and R. This leaves us with 207 moduses signifying a connection of some sort, whether causation only, prevention only or a mix of both, as the following extract shows:

Detail from Table 18.6 – Interpretation of the 3-Item Moduses.

causation and/or prevention	stats
causation only	63
prevention only	63
both causation and prevention	81
neither causation nor prevention	49
total	256

Next, we see that, for each of causation or prevention (their behavior must be similar, since they are mirror images of each other), there are ten logically possible causative formulas – **mn**, **mq** abs only, etc., and each of these has a certain frequency of occurrence in the moduses, as shown on the next table detail. Note that the joint strong-weak relations may be relative to Q or to notQ, and also that they are rather rare, compared to the absolutes. The most frequent relation is **pq** abs only (79 moduses). The total number of such general relations is 144, note. This is all true, to repeat, for causation and again for prevention.

Detail from Table 18.6 – Interpretation of the 3-Item Moduses.

for each of causation or prevention	stats
mn	9
mq abs only	19
np abs only	19
pq abs only	79
mq rel to Q	4
np rel to Q	4
pq rel to Q	1
mq rel to notQ	4
np rel to notQ	4
pq rel to notQ	1
total	144

The next detail table shows us the variety and frequency of *lone* determinations in causation and prevention respectively (again the patterns are as can be expected repetitive).

Detail from Table 18.6 – Interpretation of the 3-Item Moduses.

lones in each of causation or prevention	stats
m-alone rel to Q and notQ	19
n-alone rel to Q and notQ	19
m-alone rel to Q only	4
n-alone rel to Q only	4
m-alone rel to notQ only	4
n-alone rel to notQ only	4
p-alone rel to Q and **q-alone** rel notQ	4
q-alone rel to Q and **p-alone** rel notQ	4
p-alone rel to Q	7
q-alone rel to Q	7
p-alone rel to notQ	7
q-alone rel to notQ	7
total	90

Note well that there are no moduses signifying absolute lone determination, as already established in Chapter 12.2. All the lones that do arise (90 in all) are relative. They may arise relative to Q (30 cases) or to notQ (30 cases) or even to both (38 cases). The latter should not surprise – it is logically consistent, and indeed most common. Here as well, all this is true for causation and again for prevention.

The next statistical table shows the conjunctions possible between the joint determinations and the lone determinations, for each of causation and prevention. This teaches us that – except for 54 cases, viz. **mn** (9 cases), some cases (43) of **pq** absolute and all cases (2) of **pq** relative, all joint determinations occur in tandem with some lone determination(s), and conversely no lone determination occurs without a joint determination underlying it.

Detail from Table 18.6 – Interpretation of the 3-Item Moduses.

formulae found in each of causation or prevention		stats
mn		9
mq abs only	**m-alone** rel to Q and notQ	19
mq rel to Q	**m-alone** rel to notQ only	4
mq rel to notQ	**m-alone** rel to Q only	4
np abs only	**n-alone** rel to Q and notQ	19
np rel to Q	**n-alone** rel to notQ only	4
np rel to notQ	**n-alone** rel to Q only	4
pq abs only	**p-alone** rel to Q and **q-alone** rel to notQ	4
pq abs only	**q-alone** rel to Q and **p-alone** rel to notQ	4
pq abs only	**p-alone** rel to Q	7
pq abs only	**p-alone** rel to notQ	7
pq abs only	**q-alone** rel to Q	7
pq abs only	**q-alone** rel to notQ	7
pq abs only	no lone	43
pq rel to Q		1
pq rel to notQ		1
total		144

Moreover, as the table above reveals, these conjunctions follow certain patterns. For instances, **mq** abs only (i.e. without **q** rel to Q or to notQ) is always paired off with **m-alone** relative to Q and notQ; if **mq** rel to Q occurs, it is always accompanied by **m-alone** rel to notQ only; if **mq** rel to notQ occurs, it is always accompanied by **m-alone** rel to Q only; and similarly for **np** abs, and in other cases.

Thus, of the 144 cases of causation, 54 are without adjacent lones. Note also that there are a total of 36 cases of **pq** abs only (i.e. not implied by **pq** rel) conjoined with lones, as against 43 not so conjoined. All the same can be said for prevention, of course. Now consider the following table:

Detail from Table 18.6 – Interpretation of the 3-Item Moduses.

all lones, with **sw** rel, **sw** abs or **pq** abs	stats
all lones, in causation and/or prevention	162
lones without **sw** rel, **sw** abs or **pq** abs	0
lones with **sw** rel to Q or notQ	32
sw rel to Q or notQ, without lone	0
lones with **sw** abs	76
sw abs, without lone	0
lones with **pq** abs	54
-of which, in both causation and prevention	18
lones with **pq** rel to Q or notQ	0
pq abs or **pq** rel, without lone	27

This table, counting both causation and prevention, shows us again that no lone occurs without an associated joint strong-weak (**sw**) connection, whether relative or absolute, or without at least a **pq** absolute connection (162 = 32 + 76 + 54). Note that the joint determinations **sw** rel and **sw** abs never occur without conjunction of one or more lones. On the other hand, lones are never conjoined with **pq** relative to Q or notQ. Note also that the 54 cases of lones with **pq** abs in this table coincide with the 36 cases for causation and 36 more for prevention in the preceding table; this just tells us that there is overlap between 18 such cases of causation and prevention.

The last row of the above table tells us of **pq** abs (23 cases) or **pq** rel (4 cases) that are without lone. These 27 cases reappear in the next table, together with 54 cases of absolute causation and/or prevention associated with one to four lone determinations (which are always associated with **pq** abs):

Detail from Table 18.6 – Interpretation of the 3-Item Moduses.

analysis in the 3fw of modus 16 of the 2fw	stats
lone causation and/or prevention with 1-4 lones (with **pq** abs)	54
pq relative causation or prevention (the other = **pq** abs)	4
only one of causation or prevention with **pq** abs	0
absolute causation and prevention without lone	23
total	81

Now, it is interesting to note that the 81 moduses of the 3-item framework corresponding to item #16 of the 2-item framework (identified in Table 17.4) coincide with the 81 moduses where causation and prevention overlap in Table 18.6. Examining the latter table, we see that these 81 moduses include the 4 moduses #s 190, 232, 127, 220, in which one side has "**pq** rel to Q" or "**pq** rel to notQ" and the other side has "**pq** abs"; and the 77 moduses with "**pq** abs" on both sides, of which 54 also involve lone determinations, on one side and/or the other, while 23 moduses involve no lone determination.

While the other 54 + 4 cases signify a relation to the third item Q and/or its negation notQ, the 23 cases make no mention of Q or notQ. Note, too, that the 3-item modus #256 is among those 23. The said 23 moduses are of especial interest, because they will help us solve the earlier described problem of apparently having no modus with which to account for non-connection between contingent items.

In any case, it is now evident (looking at table 18.6) that this problem is not limited to the 2-item framework, but recurs in the 3-item framework. Here too, we see that none of the 256 moduses refer to non-connection between contingent items. They all refer to either incontingencies or to causative and/or preventive connections. We shall have to deal with this issue in more detail in the next chapter.

Chapter 19. DEFINING CAUSATION.

1. Back to the Beginning.

In the present chapter, I propose to deal with some of the difficulties that have become apparent in the previous two chapters. Before doing so, however, it is perhaps wise to review our basic definitions of the four generic determinations of causation – complete, necessary, relative partial and relative contingent causation – and their two derivative concepts, viz. absolute partial and absolute contingent causation.

1.	**complete causation** by P of R	symbol: **m**	
a)	If P, then R	(P + notR) is impossible	
b)	if notP, not-then R	(notP + notR) is possible	<=> 2(c)
c)	P is possible	(P + R) is possible	<=> 2(b)
2.	**necessary causation** by P of R	symbol: **n**	
a)	If notP, then notR	(notP + R) is impossible	
b)	if P, not-then notR	(P + R) is possible	<=> 1(c)
c)	notP is possible	(notP + notR) is possible	<=> 1(b)
3.	**partial causation** by P (with Q) of R	symbol: **p** rel	=> not-1
a)	If (P + Q), then R	(P + Q + notR) is impossible	<= 1(a)
b)	if (notP + Q), not-then R	(notP + Q + notR) is possible	=> 1(b)
c)	if (P + notQ), not-then R	(P + notQ + notR) is possible	=> not-1(a)
d)	(P + Q) is possible	(P + Q + R) is possible	=> 1(c)
4.	**contingent causation** by P (with Q) of R	symbol: **q** rel	=> not-2
a)	If (notP + notQ), then notR	(notP + notQ + R) is impossible	<= 2(a)
b)	if (P + notQ), not-then notR	(P + notQ + R) is possible	=> 2(b)
c)	if (notP + Q), not-then notR	(notP + Q + R) is possible	=> not-2(a)
d)	(notP + notQ) is possible	(notP + notQ + notR) is possible	=> 2(c)
5.	partial causation by P of R (abs)	symbol: **p** abs	<= 3, => not-1
a)	If P, not-then R	(P + notR) is possible	<=> not-1(a)
b)	if notP, not-then R	(notP + notR) is possible	<=> 1(b), 6(c)
c)	if P, not-then notR	(P + R) is possible	<=> 1(c), 6(b)

6.	contingent causation by P of R (abs)	symbol: **q** abs	<= 4, => not-2
a)	If notP, not-then notR	(notP + R) is possible	<=> not-2(a)
b)	if P, not-then notR	(P + R) is possible	<=> 2(b), 5(c)
c)	if notP, not-then R	(notP + notR) is possible	<=> 2(c), 5(b)

Let us now explain and justify these definitions. To claim complete causation (**m**) implies we know (or think or believe) that one thing P is invariably accompanied by another thing R, i.e. that P without R is impossible (in the mode of modality concerned – be it logical, extensional natural, or whatever). However, that P implies R cannot by itself signify causation. We need to also know that notP does not imply R, i.e. that notP without R is *not* impossible, for if both P and notP implied R, then R would be independent of them. Thirdly, we need to know that P is possible, so as to ground the first implication in actuality; and given that P is possible and that P implies R, it follows that R is also possible, i.e. that the conjunction of P and R is possible.

I go into this in detail to make clear to readers that these definitions were not pulled out of the blue or arbitrarily imposed, but are the product of reasoning. Necessary causation (**n**) is very similar to complete causation, except that the polarities of all the items involved are inversed. It is a statement that without P, R cannot occur; i.e. that the conjunction of notP and R is impossible; in such cases P is called a *sine qua non* (without which not) of R. Here again we must on logical grounds add two more propositions to the definition to make it applicable correctly.

Note that complete and necessary causation share the last two of their defining clauses, but differ in their first clause. However, since these first clauses do not, according to the laws of thought, exclude each other, it follows that the generic determinations of complete and necessary causation can be combined into one specific determination **mn**. However, they do not formally have to be so combined; i.e. **m** may be true without **n** being true, and vice versa. This brings us to the concepts of partial and contingent causation.

The relationship of partial causation (**p**) is designed to resemble that of complete causation, except that the cause is not one thing P, but a conjunction of two things P and Q, the latter being called the complement of the former. The first clause in our definition is a claim that P and Q together bring about R. But for this to be true, we must also ascertain that Q without P and P without Q are not also always followed by R; otherwise one or both of them might be accidental to the occurrence of R (i.e. P or Q might alone cause R, or R might be independent of their conjunction). The second and third clauses in the definition guarantee the dependence of R on P and Q together. The fourth clause serves to ground the hypothetical relationship implied by the first; and together they tell us that the conjunction of the three items P, Q and R is possible.

Contingent causation (**q**) is similarly constructed, but by analogy to necessary causation. The partial and contingent forms of causation are called weak determinations, in comparison to the complete and necessary forms (called strong determinations), because in the former (unlike in the latter) the cause P (or for that matter its complement Q) is not by itself strong enough to bring about the effect R. It is clear from the definitions of **p** and **q** that these relations are true *relative to* a specific complement Q. If we put notQ in place of Q, P and R remain cause and effect in a similar sense, but their exact relationship is considerably modified, note well. The complement Q (or alternatively notQ) signifies the conditions under which the (weak) causative relation between P and R comes into play.

Note too that **p** and **q** (relative to Q) do *not* share any defining clauses, unlike **m** and **n**. Since they refer to the possibility and impossibility of different sets of conjunction, there is no conflict between them, and they (as generic determinations) can logically be combined as (the specific determination) **pq** without infringing any law of thought.

Now compare the above listed definitions and implications of partial and complete causation. It is of course noteworthy that **m** and **n** involve only two items P and R, whereas **p** and **q** involve three items P, Q and R; but this does not prevent logical comparisons. We see that clause 1(a) formally implies clause 3(a), and clauses 3(b) and 3(d) respectively imply 1(b) and 1(d), but clause 3(c) negates clause 1(a). This means that **p** and **m** are on the whole contrary to each other, though they do share some elements of information. Similarly, **q** is incompatible with **n**, though they have some common aspects.

Now compare **m** and **q**. We see that 4(d) implies 1(b), and 4(b) implies 1(c), but no clause in **q** conflicts with 1(a), and none in **m** conflicts with 4(a) or for that matter 4(c). Similarly in the comparison between **n** and **p**, we find no notable opposition between them. This means that, formally speaking, nothing prevents the specific combinations of strong and weak forms **mq** and **np** from occurring (separately, of course).

Let us now turn our attention to the last two forms[118] – *absolute* partial causation (**p** abs) and *absolute* contingent causation (**q** abs), not to be confused with the preceding two forms of **p** and **q** relative to Q (or eventually to notQ), henceforth symbolized by **p** rel and **q** rel. The idea of absolute weak causation forms was generated by two related considerations. First, we wanted to express the weak determinations in terms of two items rather than three, for purposes of matricial analysis and direct comparisons to the strong determinations; and second, we wanted to express the weak determinations without regard to whether the complement is Q or notQ, or anything else for that matter.

Thus, the qualification of weak causations as 'absolute' here is only intended to mean that they are *not relative*, note well. It does not signify some stronger relationship, but on the contrary (as is soon apparent) *a weaker relationship*! Comparing the above definitions of **p** abs to **p** rel, we see that 5(a) is implied by 3(c), 5(b) is implied by 3(b), and 5(c) is implied by 3(d); but these implications are not mutual. Thus, **p** abs is a derivative of, i.e. a restatement of *some but not all* of the information in it. Notice especially the absence in **p** abs of any of the information contained in clause 3(a) of **p** rel, though this clause is *the crucial part* of it, the part most indicative of causation! All the same can be said of **q** abs and **q** rel, *mutatis mutandis*.

It is also noteworthy that if we change Q to notQ and vice versa in the clauses of the definition of **p** rel, the implied **p** abs is exactly the same. That is, **p** relative to Q and **p** relative to notQ yield the same subaltern **p** abs. This is of course to be expected, since neither Q nor notQ are mentioned in it. But additionally, **p** abs does not mention any other eventual third item – and so is identical for all eventual third items, X, Y, Z or whatever. Whence the characterization of it as 'absolute'. Now, this should cause us alarm; how can we know something so general from so little information, we might well ask. But the truth is that in fact **p** abs tells us exactly nothing about Q or notQ or any other third item! All the same can be said of **q** abs and **q** rel, *mutatis mutandis*.

---

118 The definitions of complete and necessary causation are first given in chapter 2.1. Those of relative partial and relative contingent causation are introduced in chapter 2.3. The definitions here put forward of absolute partial and absolute contingent causation are not found till chapter 13.4, although the concepts are developed much earlier, as of chapter 11.3.

Now compare **p** abs with **m**. We see that 5(a) contradicts 1(a), though 5(b) and 5(c) are identical with 1(b) and 1(c) respectively; this tells us that, albeit their having some common ground, **p** abs and **m** are contrary to each other. Also compare **p** abs with **n**. We see that 5(b) is the same as 2(c) and 5(c) is the same as 2(b), while 5(a) and 2(a) do not affect each other; this means that **p** abs and **n** are compatible and can be conjoined. Similar results are obtained comparing **q** abs with **n**, and **q** abs and **m**. Thus, the compounds **mq** abs and **np** abs are logically conceivable.

As for the oppositions between **p** abs and **q** abs, 5(b) is identical with 6(c) and 5(c) is identical with 6(b), whereas 5(a) and 6(a) do not impinge on each other; thus the two forms are compatible, i.e. can be conjoined in a compound form **pq** abs. What does this compound form tell us? Simply, that each of the four conceivable combinations of P and R, viz. P+R, P+notR, notP+R, notP+notR, is possible.

The above six definitions for (i) causation by P or R can be modified to define (ii) prevention by P of R (by replacing R by notR, and notR by R, throughout them), (iii) inverse causation by P of R (by replacing P by notP, and notP by P, and R by notR, and notR by R, throughout them), and (iv) inverse prevention by P of R (by replacing P by notP, and notP by P, throughout them). Note in passing that **pq** abs has the same value in causation (and inverse causation) and in prevention (and inverse prevention), since it always just means that the four conjunctions of P, R, notP and notR, are all possible.

All this has been said before but is here repeated briefly to enable us to once and for all resolve a certain difficulty mentioned earlier. We shall see that the difficulty in question is upon closer inspection more apparent than real.

2. The Puzzle of No Non-connection.

Looking at the interpretation table for the moduses in a 2-item framework (Table 18.1, page 6), we see that only seven of the moduses refer to connection – and apparently not even one refers to '**non-connection albeit contingency**'! Incontingency counts as non-connection, of course; but what interests us here is to logically conceive *non*-connection between two *contingent* items. Apparently, judging by the tabulated results, there is no such possibility! Note in passing that alternative words for connection and non-connection are *dependence* and *independence*. Are we to think that all contingent items are mutually dependent in some way or other? Surely not! What does this mean, then? This result is indeed so surprising that I shall call it 'the puzzle of no non-connection'.

Considering that the logic of causation as here presented, i.e. through microanalytic tabulation, is *entirely a formal product of the laws of thought*, this is indeed mysterious. This result to fix in an *a-priori* manner a detail about reality, by mere logical analysis, without need for empirical observation. Although some philosophers, indeed many of them across history, have adopted this position, it does not make sense. It would mean we cannot even imagine or theoretically conceive of non-connection between contingent items, which certainly goes against our commonsense impression that we at least comprehend such non-connection. All our concepts need contradictories to be intelligible. If we cannot even hypothetically formally define non-connection between contingent items, the concept of connection itself becomes doubtful.

My discovery of this mystery is not new to phase III; I had already encountered it and made an effort to explain it in phase II (see Chapters 13.2 and 16.2). Here, I will succeed in going deeper into the question and remove all lingering doubt once and for all.

No doubt, seeing this puzzling result, believers in **extreme determinism** (which include many materialists and behaviorists still today) will rush to judgment and say: "See, we told you, since we cannot logically define indeterminism, it is not even open to debate – everything in the universe is determined, and there is no place in it for natural spontaneity or human freewill or any other indeterminism." But if we consider the matter more closely (again, look at Table 18.1 page 6), we see that the seven cases with both items contingent refer to *varying degrees of causation*: 2 cases are **mn** (maximal determination), 4 cases are **mq** abs or **np** abs (medium determination), and 1 case is **pq** abs (minimal determination). Thus, only two relations are fully determining, whereas five others are partly undetermined, and we cannot draw an extreme determinist conclusion.

Another group likely to welcome this puzzling result are believers in the Buddhist viewpoint that everything is causatively related to everything else in an inextricable web of '**interdependence**' (or 'dependent origination' or, in Sanskrit, *Pratityasamutpada*). They will say: "See! Since there is no such thing as non-connection between some pairs of contingent items, *any* two contingent items taken at random may be considered, without any recourse to experience, as causatively related, at the very least through partial contingent causation (and similarly prevention), i.e. **pq** abs." But such jubilation is premature and unjustified, as we shall now go on to show.

The simplest answer is that what we have called 'partial contingent causation' is not really causation! To see the truth of this, let us return to our initial definitions of **p** abs and **q** abs, in the previous section. What distinguishes these forms (numbered 5 and 6) from those preceding them (1-4) is that they lack an if-then clause. They each specify the possibility of three combinations of P, R and their negations, but they distinctively do *not* specify the *im*possibility of any such combinations. Yet such if-then or impossibility of conjunction constitutes *the main clause* of the definitions of strong causation and relative weak causation.

Thus, the absolute weak determinations are not forms of causation in the usual sense. This does not mean we ought to, or even can, just discard these two concepts. For it is clear that we formed them out of a real need. They do in fact play a role in causative relations – but their role is a supporting one. In combination with **m** or **n**, i.e. in **mq** abs or **np** abs, they are indicative of actual causation; but taken apart from the strong determinations, i.e. in the combination **pq** abs (i.e. $\mathbf{p}_{abs}\mathbf{q}_{abs}$), all they tell us is that the four basic conjunctions, viz. P+R, P+notR, notP+R, and notP+notR, are all possible, which is not a statement of actual causation but still leaves open the logical possibility of causation at a deeper level (as evident in Table 18.6).

Remember, **p** abs is contrary to **m**, and **q** abs is contrary to **n**. When **m** is combined with **q** abs, we have the important information that, though there is causation, it is not of the powerful **mn** sort. Similarly with regard to **np** abs – the **p** abs part serves to deny the conjunction of **m** to the **n** part. It is significant to remember, too, that there are no absolute lone determinations, that is: absolute **m-alone**, i.e. **m** conjoined to neither **n** nor **q** abs, is logically impossible; similarly, absolute **n-alone** is impossible, and so are the absolute weaks alone. Thus, **p** abs and **q** abs are formally needed for causative discourse in a 2-item framework.

However, though these absolute weak determinations are implied by the corresponding relative determinations, they do not in turn imply them. They are mere subalterns, not equivalents. At best, **p** abs tells us that **p** rel *might* occur, and likewise **q** abs tells us that **q** rel *might* occur; the former certainly do not imply that the latter are bound to occur. And the issue here is not merely that we do not know whether Q or notQ is the applicable complement. As the definitions in the previous section make clear, **p** abs and **q** abs remain the same, even if we change the polarity of Q to notQ and of notQ to Q in **p** rel and **q** rel. But, moreover, as Table 18.6 makes clear, **p** abs and **q** abs can be true without implying either **p** rel or **q** rel in relation to Q or notQ !

The latter finding should by itself cause alarm: how could we, using a PQR matrix only, know about a weak causative relation between P and R through an intermediary other than Q or notQ ? Such a thing is unthinkable in deductive logic – there are no magical leaps, no windfall profits – we can only conclude things already given in the premises. But if we look more closely at instances of **pq** abs only, we see that they do not tell us anything about causation involving some unstipulated fourth item other than Q or notQ, because they do not imply that some causation between P and R (and/or their negations) is indeed operative. They merely specify the various possibilities of conjunction between these two items; this is valuable information, but it is not causation.

Thus, although **p** abs and **q** abs are relevant to causation in the compound propositions **mq** abs and **np** abs, they are not definitely indicative of causation as **pq** abs, in the 2-item framework as modus #16 (see Table 18.1), or in the 3-item framework as the 23 moduses #s 52, 56, 60-61, 64, 103, 116, 120, 154, 180, 188, 196, 205-208, 221-222, 237, 239, 244, 253, 256 (see Table 18.6).

Note that the 2-item modus #16 unfolds as 81 distinct moduses in the 3-item framework. Among those 81, only the just mentioned 23 moduses (which include the last modus 256, note) are in turn empty of causative information. The remaining 58 moduses all involve some definite causation, whether through relative lone determinations (54 cases) or relative partial contingent causation or prevention (4 cases). For this reason, we can rightly say that the 2-item modus #16 is ambiguous as to whether there is or not some causation or prevention deeper down in a 3-item framework.

Similarly, each of the 23 said 3-item moduses may or may not at a deeper level become a connection of some sort, *ad infinitum*. Thus, to call **pq** abs 'causation' (or 'prevention', as the case may be) is a misnomer – it is excessive, inaccurate, misleading to do so, because though this compound is sometimes expressive of causation – it is sometimes not so. Thus, the solution to our problem is that to regard **pq** abs as a form of connection is to misuse the term. We should therefore, strictly speaking, refer to the 2-item modus #16 as ***possible connection and possible non-connection*** (as I suggested in phase II); and likewise for each of the 23 above listed 3-item moduses (as now proven in phase III).

We have thus clearly located where non-connection between contingent items can be placed. Let me further explain this as follows, so it is fully understood. ***The essence of connection (causation or prevention) lies in the limitations of possibility*** to be found in nature or logic. When we say that an item, say P, 'causes' (or 'prevents') another item, say R, in some way, to some degree, we mean that in the presence or absence of P, the presence or absence of R is somewhat *restricted*. It is not the occurrence of the latter item or its negation that signify causation, but the fact that *some other avenue* of occurrence has been naturally (in some cases, volitionally) or logically blocked.

Thus, the 'force' of causality lies not so much in positive events as in the restrictions in the degrees of freedom offered to an item by the interference of another; i.e. in the *negative* boundaries the one sets on the other. In more formal terms, we can say: it is not so much the '1s' (the bases) that matter as the '0s' (the connections). Roughly stated, the more zeros, the stronger the causal relation; the less zeros, the weaker the causal relation. If no zeros are to be found at any depth, there is no causal relation. In cases involving strong causation, the restrictions are very evident, whereas in cases involving only weak causation, the restrictions are not always evident – and by extrapolation, we may at least conceive of cases without restriction.

We can also put it as follows, to show that it makes perfect sense. For two items to be connected in some way, there has to be some incompatibility between them and/or their negations, some conflict that forces one or the other of them to behave in an special manner. If the items and their negations are every which way compatible, then they do not impinge upon each other but coexist harmoniously. Thus, the **pq** abs compound, which signifies such thorough compatibility, is essentially indicative of non-connection, though some connection at a deeper level is not excluded by it offhand.

Once this crucial new insight is grasped, it is easy to see why some modus(es) in any framework (such as the last modus in the 2-item framework or the stated 23 in the 3-item one) are the reasonable place where non-connection (in whatever sense) between contingent items may be found. Partial contingent causation or prevention are indeed possibly housed in such modus(es); but we must admit that diverse forms of non-connection are possibly housed there too. *Their correct interpretation is thus ambiguous*, and it is an error to interpret them only one way – as only connection, or for that matter as only non-connection.

Furthermore, we should point out that the 2-item modus #16 and the analogous 23 moduses in the 3-item framework signify *both* **pq** abs of causation and **pq** abs of prevention, and not merely one or the other. This fact should not be swept aside as insignificant, although of course it does not go against the laws of thought. It is, however, unthinkable that something might be both a partial contingent causative and a partial contingent preventative of something else, relative to the same complement or even contradictory complements. This we know by looking at Tables 18.5 and 18.5 (pages 7-8), which teach us that the four forms **pq** rel to Q, **pq** rel to notQ, for causation, and **pq** rel to Q, **pq** rel to notQ, for prevention, have each only one modus, namely respectively moduses 190, 232, 127 and 220, and no modus in common. Causation and prevention are thus essentially antithetical, not only in their stronger forms but even in their weakest form.

This shows us that, even if **pq** abs of causation and **pq** abs of prevention are superficially compatible (indeed, they are *identical*, having in common the 2-item modus #16 and all their 81 moduses in the 3-item framework), such compatibility must not be interpreted as meaning that they can ever be realized together relative to any specific complement(s) Q and/or notQ. Such realization (i.e. going from absolute to relative) is logically impossible, so that the apparent compatibility between causation and prevention is purely illusory. Thus, the conceptual joining of **p** abs and **q** abs is, from the causative point of view, an abstraction without concrete referents. The generic forms are valid abstractions, because they can be validly joined to **n** and **m**, respectively, in the specific causative forms **np** abs and **mq** abs; but they do not produce a common causative form **pq** abs. The latter is meaningful (as a statement of possibility of conjunction every which way), but not as causation or prevention, and least of all as both causation and prevention.

It should also be stressed that when we here refer to the possible non-connection between two specified items P and R, we are in no way making a general claim about the non-connection of each of these items to some other unspecified items. The contingent item P may be unconnected to the contingent item R, but still be connected to one or many *other* contingent items X, Y, Z. Non-connection does not imply universal non-connection: it is here clearly intended as a characterization of the relation between a *specified* pair of contingent items.

Thus, this finding about the logically possible existence of non-connection must not be taken as an a-priori statement that 'some contingent things are not connected to any others', or more extremely that 'nothing is connected to anything else'. These would be generalizations beyond what we have sought to establish here – which is only that, taking any two contingent items at random, there is no logical necessity that they be connected in a real sense (i.e. one stronger than the misnamed **pq** abs). The said moderate and extreme generalizations do however remain open to debate.

The extreme proposition 'nothing is connected to anything else' has been put forward in philosophy by Nagarjuna, David Hume, and others. I firmly reject it on the formal ground that they do not explain how all the other logical possibilities – i.e. those of connection between contingents – have been excluded from consideration by them; such skepticism is manifestly arbitrary.

The moderate position 'some contingent things are not connected to any others' is certainly not deductively proven here, either, but it remains quite conceivable, since we have identified the moduses within which such disconnection might occur and we do not claim an exclusive universal application. It formally opens the door to claims of occasional natural spontaneity (as in Niels Bohr's interpretation of the uncertainty principle), and to claims of circumscribed human freewill and similar powers of volitions (which most people adhere to).

The antithesis to this would be the claim that 'every contingent thing is connected to some other(s)'. Many philosophers throughout history have advocated this determinist thesis, calling it 'the law of causation' – but it is important to realize that, from a formal point of view, it is just a hypothesis. Moreover, what does 'connected' mean here – i.e. what degree of connection is intended? The extreme version of this thesis would affirm that 'for any given contingent item R, there must be some item P that is a complete and necessary cause of it'.

A more moderate version might be postulated, however, that affirms such strong connection in most cases, but allows for exceptions, whereat natural spontaneity and/or volition may come into play next to determinism. I personally believe such combination of theses is the most credible alternative, being closest to commonsense belief. Our causative logic is thus, in any event, quite capable of assimilating all philosophical discourse concerning causation, note well.

3. The Definition of Causation.

In the preceding section, we saw that moduses that mean no more than "**pq** abs" (i.e. $\mathbf{p}_{abs}\mathbf{q}_{abs}$) cannot rightly be counted as signifying a causative connection, but at best only a possible connection, which is also a possible non-connection. We saw the truth of this with reference to the 2-item modus #16, which was found to give rise to 81 moduses in the 3-item framework, of which 58 moduses were indicative of some causation or prevention (as well as **pq** abs), whereas 23 moduses signified no more than **pq** abs.

However, here we must admit that such ambiguity cannot be tolerated. If we want to produce a clear definition of causation, which is one of the goals of our study, we must make up our minds and declare moduses that mean "**pq** abs only" to signify either a connection or a non-connection. So far in our tables, we have opted for the designation of the 2-item modus #16 and its equivalents 3-item moduses to signify connection. But in view of our analysis in the

preceding section, we must now reverse this policy if we wish to produce an accurate definition. This is reasonable, since two items related only by way of **pq** abs cannot be guaranteed to be causatively related, and so may be counted as not so related (unless or until more specific conditions are specified that imply them to be causatively related).
On this basis, the tables concerning the broad concepts of causation, prevention and connection, and their respective negations, must be rewritten with all cases of **pq** abs only moved over from the positive to the negative side, whether manually or by modifying the calculation formulae as appropriate. Thus, for instance, the 2-item modus #16 must be moved from the columns of causation, prevention and connection to those of non-causation, non-prevention and non-connection. Similarly for 23 moduses in the 3-item framework. We shall tag these new columns as concerning '**strict**' causation, prevention and connection and their negations – so that the corresponding old columns can be left unchanged, except that we understand that they concern causation etc. in a 'loose' sense.
The outcome of this revision are the following two tables, derived from earlier ones as just explained, which are posted at the website as usual:

- Table 19.1 – 2-item PR Moduses of Forms – Strict Moduses. (1 page in pdf file).
- Table 19.2 – 3-item PQR Moduses of Forms – Strict Moduses. (5 pages in pdf file).

Having done this, we can now proceed with constructing definitions of the concepts of causation, prevention and connection in their strict sense (i.e. with '**pq** abs only' not counted as causation, etc.). The following extract from Table 19.1 suffices for this purpose:

Details from Table 19.1 – Causation, prevention and connection.

relation	summary moduses			notable features
strict causation	1001	1011	1101	outers both 1, inners one or both zero
strict non-causation	all other moduses, except #1			
strict prevention	0110	0111	0110	inners both 1, outers one or both zero
strict non-prevention	all other moduses, except #1			
strict connection	strict causation or strict prevention			features of both
strict non-connection	all other moduses, except #1			

We see here that, strictly speaking, causation is applicable to three moduses (Nos. 10, 12, 14, to be specific), whose common features are that their summary moduses start with a 1 (for P+R) and end with a 1 (for notP+notR), and have one or two 0s in the middle (for P+notR or notP+R). Similarly, strict prevention concerns three moduses (Nos. 7, 8, 15), featuring two 1s on the inside and one or two 0s on the outside. Connection accordingly covers these six moduses, and is thus definable by the sum of their features. The negations of these relations refer to all remaining moduses, except #1 (consisting of four 0s, which is universally impossible). Modus #16 (consisting of four 1s) always falls in the negative relation (strictly speaking) – its lack of any 0 puts in doubt any causative relation in it.

We may express these results concerning strict causation in words as follows: ***causation is the relation between two items, if and only if they are found to have the following set of features: (a) the first cannot occur without the second and/or the second cannot occur without the first, and in any case (b) the first and second can occur together and their negations can occur together.*** If these conditions are satisfied, this first item is called cause and the second is called effect. The relation of prevention refers to causation of negation; and the relation of connection refers to either causation or prevention. The negations of all these relations can accordingly be defined. Note well that if the two items and their negations are *compatible together every which way*, they cannot strictly be said to be causatively related in any way; for such relation to be recognized, *some incompatibility* between the items and/or their negations must be established.

Of course, the here stated definition of causation (and thence those of prevention and connection) could be argued to be rather rough, being based on Table 19.1 only, that is to say on the configuration of 'absolute' causation between two items, comprising strong causation (**mn**) and its combinations with absolute weak causation (**mq** abs and **np** abs). It ignores causation relative to a third item, which is more complex and difficult to define. The simplest way to do it would be to say: 'relative' causation requires a more complicated and subtle definition, and rather than try and formulate one I refer you to Table 19.2. Alternatively, we could try and construct a verbal definition with reference to the original forms listed in section 1 of the present chapter.

But I do not see the value of such a wordy and intricate definition in practice. Definitions should effectively lead us to the intended object, and not mystify us by their complexity. I think the rough definition proposed here suffices for most purposes; and when we do need to get very precise, we can just point to the original forms or the said table, without attempting a formal summary. One more thing needs doing, however – we need to explain the application of the proposed definition of causation (and its derivatives) in terms of generic 'possibility' and 'impossibility' to the different *modes of causation*, and say more about *the way knowledge of causation is acquired* in them.

With regard to the logical or 'de dicta' mode of causation, the modal specifications of 'possibility' and 'impossibility' refer of course to some or no 'contexts of knowledge'. In this domain, our inductive practice is to assume modus #16 to be true, until and unless we manage to demonstrate another relation to be true. The truth of this principle can be seen in the theory of 'opposition', where we assume two propositional forms to be fully compatible (i.e. neutral to each other) if we do not manage to specifically prove them (if only by some logical insight) to be contradictory or contrary or subcontrary or implicant or subalternative.

Turning our attention now to the 'de re' modes, we can say: in extensional causation, 'possibly' means in some cases and 'impossibly' means in no cases; in natural, temporal and spatial causation, these modalities refer respectively to some or no circumstances, some or no times, and some or no places. In these modes, our inductive practice is the exact opposite of that for the logical mode. That is to say, here we assume the items concerned to be incompatible if we do not succeed in directly or indirectly finding empirical grounds to consider them as compatible. For example, we do not affirm that 'some X are Y' if we have not directly observed any such cases, or at least (more indirectly) empirically confirmed a theory that implies this proposition.

Thus, modus #16 is not taken for granted as easily for the de re modes as it is for the logical mode. In the logical mode, it is used as the *default* option when no other option is established. Whereas, in the de re modes we are not allowed to make such assumptions offhand, but rather remain in a state of ignorance until some good reason to accept modus #16, or any other modus, whether of causation or of non-causation, is found. In this sense, the logical mode is more 'a priori' and the de re modes are more 'a posteriori'. But as regards their formalities they differ little.

I think we need not belabor this topic further, except to point out, once again, how much more accurate our definitions are from those implied by David Hume and from other past attempts.

4. Oppositions and Other Inferences.

Once we have analyzed each and every possible form of causation and its sources and derivatives in matricial analysis, it is very easy to compare forms and determine their oppositions, eductions, syllogisms and any other sorts of inference.

We can formulate general rules of **opposition**, from which the oppositions between any pair of forms can be determined, as follows[119].

- *Implicance*: two forms *all* of whose alternative moduses are identical may be said to imply each other; i.e. they are implicants. For example, **m** in causation and **n** in inverse causation are equivalent, having the exact same moduses (2-item moduses #s 10, 12), no more and no less. It follows necessarily, note, that their negations are also implicants. For example, **not-m** in causation and **not-n** in inverse causation are equivalent (2-item moduses #s 2-9, 11, 13-16).
- *Subalternation*: if one form has more moduses than another, and its list of moduses includes *all* the moduses of that other and *none* of the moduses of its negation, the second form may be said to imply but not be implied by the first; i.e. they are subalternatives: respectively, subalternant and subaltern. Note well that it is the (narrower ranging, more precise) form with *less* moduses that implies the (broader ranging, vaguer) form with *more* moduses, and not vice versa. For example, "P is a complete cause of R" (2-item moduses #s 10, 12) subalternates "if P, not-then notR" (moduses 9-16). It follows necessarily, note, that their negations are also subalternatives, though in the opposite direction. For example, "if P, then notR" (moduses 2-8) subalternates "P is a not complete cause of R" (moduses 2-9, 11, 13-16).
- *Contradiction*: if two forms do not share any modus and if their moduses together make up the total number of moduses in the framework concerned (minus the universally impossible first modus), they may each be said to imply the other's negation (i.e. to be incompatible) and their negations each to imply the other's affirmation (i.e. to be exhaustive); that is, they are contradictories. For example, **m** has 2-item moduses #s 10, 12 and **not-m** has moduses 2-9, 11, 13-16; therefore, **m** and **not-m** are contradictory.
- *Contrariety*: if two forms *do not* have any modus in common, and if their moduses together *do not* add up to the total number of moduses in the framework concerned (minus the universally impossible first modus), their affirmations may each be said to imply the other's negation, though their negations do not each imply the other's affirmation; that is, they are incompatible but not exhaustive, i.e. contraries. For example, **m** (2-item moduses 10, 12) and **p** abs (moduses 14, 16) are contrary forms. Note that if two forms are contrary, their negations are necessarily subcontrary.

119 See chapter 13.3 for applications of this technique in phase II.

- *Subcontrariety*: if two forms *do* have some modus(es) in common, and if their moduses together *do* add up to the total number of moduses in the framework concerned (minus the universally impossible first modus), their negations may each be said to imply the other's affirmation, though their affirmations do not each imply the other's negation; that is, they are exhaustive but not incompatible, i.e. subcontraries. For example, **not-m** (2-item moduses 2-9, 11, 13-16) and **not-p** abs (moduses 2-13, 15) are subcontrary forms. Note that if two forms are subcontrary, their negations are necessarily contrary.
- *Unconnectedness*: if two forms have some modus(es) in common, *and* their negations have some modus(es) in common, *and* the affirmation of each of them has some modus(es) in common with the negation of the other, these forms may be said to be unconnected with each other, for this simply means that the four stated combinations are possible, i.e. that each form and its negation is compatible with the other form and its negation. For example, "if P, then R" (2-item moduses #s 2-4, 9-12) and "if P, not-then R" (moduses 5-8, 13-16) are both unconnected to both "if notP, then notR" (moduses 2, 5-6, 9-10, 13-14) and "if notP, not-then notR" (moduses 3-4, 7-8, 11-12, 15-16).
  Remember, this last category of opposition, viz. unconnectedness, also called 'neutrality', means that the forms concerned do not imply each other, and their negations do not imply each other, and their affirmations do not imply their negations, and their negations do not imply their affirmations; i.e. the two forms are compatible in every which way and exhibit no incompatibility in any way – that is why they are said to be unconnected or neutral. This covers all leftover cases, i.e. it applies when neither implicance, nor subalternation either way, nor contradiction, nor contrariety, nor subcontrariety relate the two forms under scrutiny.

Let me remark here: the word 'opposition' was initially intended (in everyday parlance) to mean 'conflict' – i.e. it referred to contradiction or contrariety. The sense was then slightly enlarged by logicians so as to include subcontrariety (which refers to contrariety of negations). Then, it was further enlarged to enable the inclusion of implicance and subalternation; this changed the meaning of 'opposition' to 'face-off'. Finally, the theory of opposition naturally called for a further concept, one denying all the preceding forms of opposition – i.e. a concept of 'unconnectedness' or neutrality (see my *Future Logic*, chapter 6.1). This relation too, though negative, can and must be regarded a form of 'opposition' *in an enlarged sense* (i.e. face-off).[120]

Note that the above definition of unconnectedness in terms of moduses justifies my thesis earlier in the present chapter that there has to be room in causation theory for non-connection, since it demonstrates that there is one more relation of 'opposition' than the six traditionally listed. For opposition theory (and more broadly, inference theory) is nothing other than causation theory in the realm of logical modality[121]; it concerns causes in the special sense of

---

[120] I must in passing deride the couple of people who have written scholarly-looking articles where they seem to deny my concept of 'unconnectedness' to be a logically possible relation between propositional forms and a needed category of 'opposition'! This is not an issue open to choice, but (to repeat) a natural demand to exhaust the logical alternatives. Such people allow themselves to be misled by mere words, thinking that opposition must needs signify conflict since that is the popular sense of the term. Or they are pettily annoyed that this additional category does not fit into their pretty 'squares of opposition'. This is the kind of silliness that focus on trivia produces.

[121] Implicance and subalternation each way are logical causation; and contradiction, contrariety and subcontrariety are logical prevention. In each case, the determinations are respectively **mn**, **mq** and **np**.

'reasons'. What is true for this de dicta mode of modality is equally true for the de re modes, since there is no formal difference between them in the present context.

Eduction is immediate inference from one (or more) forms with identical terms. When one form implies another, the latter can be educed from the former. When one form is incompatible with another, the negation of either can be educed from the affirmation of the other. When two forms are exhaustive, the affirmation of either can be educed from the negation of the other. From these principles we can likewise, with reference to moduses, determine all possible eductions.

We can similarly work out all **syllogism** (i.e. mediate inference, through a middle term) with reference to moduses, as already explained in chapter 14.1 and demonstrated thereafter. If the premises have no moduses in common, or if the premises do have some moduses in common but these moduses imply contradictory conclusions (i.e. some imply one conclusion and others the negation of it), they are incompatible and therefore cannot make up a syllogism. But otherwise, the conclusion is generally the common ground of the premises, i.e. the moduses they have in common.

Thus, matricial analysis – more precisely, microanalysis – provides us with a practical way to correctly interpret all conceivable situations in causative logic.

Chapter 20. CONCERNING COMPLEMENTS.

1. Reducing Numerous Complements to Just Two.

To fully understand partial and contingent causation, we need to return to the issue of complementary causes. In chapter 2.3, where these concepts were first introduced, we showed how any number of complements can be reduced to just two. I wish to here review this important doctrine and further develop it with reference to matricial analysis.

Why this doctrine is important is worth reiterating here. Remember, we originally defined partial and/or contingent causation with reference to only two complementary causes (say, P and Q) for a certain effect (say, R). Focusing on one of the complements (say, P) as the first item and 'main' cause (or the cause mainly of interest to us in a given context), the other complement (here, Q) could be regarded as embodying the 'surrounding conditions' for the partial and/or contingent causation of the third item (R). That is, even though the second item Q is in our definition presented as a single item, it is intended to signify any number of 'surrounding conditions' Q1, Q2, Q3, etc.

The question arises: what is the exact relation between the underlying numerous surrounding conditions, and their single representative or stand-in term Q? Obviously we mean Q to be a putative 'collective effect' of Q1, Q2, Q3... – we conceive of a causative relation between the numerous underlying causes and their single representative. Indeed, *very often we invent a new term* Q to stand in for a number of terms Q1, Q2, Q3…, viewing Q as an 'abstraction' implied by its constituent items Q1, Q2, Q3, etc. We *define* Q by saying: "let that be the collective effect of Q1, Q2, Q3, etc."

This is indeed *one of the usual avenues of concept formation*. In other words, Q need not be something concretely observed in isolation, but may be an abstraction produced *ad hoc* to facilitate a causative statement. It may, however, thenceforth acquire a life of its own in our discourse. For example, the definition of force as mass times acceleration tells us that the abstraction called 'force' can be calculated by measuring the mass of the physical body concerned and the acceleration the force is assumed to effect in it, and multiplying the two quantities together; we do not observe and measure 'force' as such, but derive it from more concrete items. But henceforth, force becomes an oft-used term in other equations, as if it was directly experienced.

Thus, a complementary item such as Q may be viewed as itself the effect of a partial and/or contingent causation whose causes are, say, Q1 and Q2. If there are more than these two underlying items, then one of them (say Q2) may in turn be viewed as the product of two deeper items, say Q3 and Q4. And so on, successively, till all the relevant surrounding conditions are exhausted. Conversely, any number of 'underlying items' or 'surrounding conditions' can, by successive mergers of two into one, be represented by just one overall representative item Q.

This is at least true theoretically, because this is not really how we proceed in practice. In practice, we rather think more globally, as already described in chapter 2.3 (see there for more details). This is the thought process and inductive method of 'changing one thing at a time, while keeping all other things equal, and observing the effect of that single change'. This can be described in more formal terms as follows for partial causation:

If (Q1 + Q2 + Q3 + …), then Q;
If (notQ1 + Q2 + Q3 + …), not-then Q;
If (Q1 + notQ2 + Q3 + …), not-then Q;
If (Q1 + Q2 + notQ3 + …), not-then Q;
Etc. (as per number of factors involved);
And (Q1 + Q2 + Q3 + …) is indeed possible.

For contingent causation, similar clauses can be used, with the polarities of all the terms involved must be reversed. Thus, the first clause would be: 'If (notQ1 + notQ2 + notQ3 + …), then notQ', and so on. That is to say: where Q is a partial cause, with P of R, then Q1, Q2, Q3, etc. are in turn partial causes of Q; and where Q is a contingent cause, with P of R, then Q1, Q2, Q3, etc. are in turn contingent causes of Q. In cases where both partial and contingent causation are involved, Q is the collective effect of the underlying items Q1, Q2, Q3, etc., on both the positive and negative sides.

We can also describe such multiple weak causations by means of nesting; this is of course the meaning of the successive reductions of many conditions to just one mentioned above. Nesting of 'if (Q1 + Q2 + Q3 + …), then Q' would have the form 'if (Q1 + (Q2 + (Q3 + …))), then Q'; and similarly for the other clauses. But to repeat, the nesting approach complicates matters perhaps unduly, and is here mentioned just to promote theoretical understanding rather than as a practical means.

Of course, nothing forces us to limit ourselves to just two complementary causes, other than the limitations in our computer (hardware and software) resources. A large organization, such as WHO (the World Health Organization), for which I worked for some years long ago, does I think have the computer, personnel and financial resources to investigate any number of complementary causes of any health factor or disease, social problem or solution, or whatever.

Note that nothing is said or intended here with regard to the relative part played by the various complements in causing the effect concerned. The *quantitative* role of these factors is not being examined here, only *the fact that* they are factors in the causation. They may be very widely different in their degrees of involvement; one complement may be the major determinant, while the other(s) is/are minor factors, or they may all be more or less on an equal footing[122]. I leave aside the issue of proportion, here – without, however, intending to deny its great importance. It is, rightly, regarded as crucial in modern science, but our study here is only concerned with the 'whether', not the 'how much'.

2. Dependence Between Complements.

While on the topic of complementary causes, a question worth asking is: what of interconnected complements – i.e. can the two complements be causatively related at all, or

122 Eventual variations in proportions in time and/or space should, I think, be considered as due to more phenomenal underlying factors. For example, the 'age' of an organism may be a causal factor; but the significance of aging at the cellular level or deeper would have to be investigated to understand why 'time' seems to play a role.

are they always independent of each other? That is, given that P, Q are partial and/or contingent causes of R, does it follow that P, Q are unconnected (i.e. compatible every which way, as explained in the previous chapter) – or may they in some cases be connected?
This question can be answered by looking at the 3-item moduses found in relative partial and contingent causation and prevention (90 moduses, all told), and seeing whether any of them signify some implication between the complements P, Q and/or their negations. It is found that in 32 cases, one or the other of the four possible implications are involved (i.e. 8 cases for each, symmetrically), the remaining 58 cases signifying only that the various conjunctions of the complements and their negations are just contingent. The list of cases concerned may be seen in Table 20.1, posted at the website:

- Table 20.1 – 3-item PQR Moduses of Forms – Dependent Complements. (2 pages in pdf file).

Let us take one of these cases for further examination, say modus 23 (ditto for 31, 55, 63). This concerns relative contingent prevention, i.e. "P with complement Q is a contingent cause of notR", whose clauses are "if notP and notQ then R; if P and notQ, not-then R; if notP and Q, not-then R; and notP and notQ is possible". Now, according to our table, modus 23 also corresponds to "P and Q is impossible (i.e. if P, then notQ)". As can be seen, these two propositions are not in conflict; meaning that the relation of dependence between P and Q does not impinge on their stated causative relation to notR (which nowhere mentions "P and Q"). Obviously, this is not the sort of dependence we are looking for; we seek implications between the complements that affect the causative relation to the third item somewhat.
Similarly, modus 42 (ditto for 46, 58, 62), which refers to relative contingent causation, i.e. "P with complement Q is a contingent cause of R" and to "P implies notQ" has no notable impact. And likewise, *mutatis mutandis*, in six other sets of four moduses (I won't bother listing them; their first ones are highlighted on the said table). So in fact, by this simple method, we find no significant dependence between the complements, i.e. one having a differential impact on the relative causation or prevention concerned.
Obviously, when such generic relative causations or preventions are conjoined to strong determinations, nothing is changed, since the latter involve only two items (e.g. P and R). However, what happens when they are conjoined to each other (when compatible, of course)? We know that this option (i.e. **pq** rel to Q or notQ, for R and notR) concerns only four 3-item moduses altogether, viz. 127, 190, 220, 232; and these, as our table shows, are not among the 32 relevant moduses; therefore, no significant impact arises here, either.
Thus, to conclude, although conjunctions of the complements (P, Q, notP, notQ) are not always possible, the cases where they are impossible do not affect the causative relation concerned. Note that this only concerns implication and does not exclude the possibility that the complements might have a weaker causative relation mediated by some additional item. This issue might be further investigated using the 4-item matrix, but I will not attempt it here, having shown how the job can be done. There is no real need, for this investigation is moved merely by curiosity – since all the valid moduses of forms generated by matricial analysis are thereby known to involve no internal inconsistency.

3. Exclusive Weak Causation.

Another, more valuable, investigation I wish to launch here is into the features of exclusive causation. With regard to the strong determinations, this would take the forms: "If and only if P, then R" or "If and only if notP, then notR". If we think about them, we realize these both mean "If P, then R; and if notP, then notR" – i.e. **mn**. Nothing new here, since we have already studied the properties of **mn** in considerable detail.

I would just take this opportunity to remind readers of the danger of ambiguity when we say: "Only if P, R" or "Only if notP, notR" (notice my removal of the words "if and" and "then"). Though statements of this sort often signify exclusive strong causation in the sense just defined (i.e. "if and only if –, then –), often what is intended is much weaker, namely: "If P, possibly R"; and if notP, necessarily notR" and "If notP, possibly notR; and if P, necessarily R", respectively. In such cases, we are only informing that the consequent R (or notR, as the case may be) is only possible with the antecedent P (or notP) – but we are not claiming that P brings about R (or notP brings about notR). For this reason, it is wise to use the more precise wording (which modern logicians abbreviate to "iff –, then –"[123]).

Let us now turn our attention to the weak determinations, and ask what is meant by "If and only if P and Q, then R" or "If and only if notP and notQ, then notR", which forms we will respectively label as **p** ex and **q** ex (or $\mathbf{p}_{ex}$ and $\mathbf{q}_{ex}$) – the suffix 'ex' standing for exclusive, of course. Note my use here of the harder "iff" sort of exclusion just explained; also, to avoid all ambiguity, note that our intent here is to apply this operator to the conjunction (P and Q) or (notP and notQ), and not merely to the first mentioned complement (i.e. to P or notP). Thus, what I have in mind is, roughly put, the following propositions:

7.	<u>**Exclusive partial causation** by P and Q of R</u>		symbol: **p** ex
	If (P + Q), then R		((P + Q) + notR) is impossible
	if not(P + Q), then notR	(etc.)	(not(P + Q) + R) is impossible
8.	<u>**Exclusive contingent causation** by P and Q of R</u>		symbol: **q** ex
	If (notP + notQ), then notR		((notP + notQ) + R) is impossible
	if not(notP + notQ), then R	(etc.)	(not(notP + notQ) + notR) is possible

I have (for our purposes here) numbered these forms 7 and 8, to indicate continuation of the list given in Chapter 19.1. They are necessarily 'relative' (i.e. have at least 3 items); they do not have 'absolute' (2-item) versions. Needless to say, the number of complements involved in them need not only be two; any number might be considered, but we shall here focus our investigation on just two complements as usual, so that we can refer to 3-item matricial analysis to answer questions that arise.

Clearly, the second clauses of forms 7 and 8 can each be expanded into three clauses, as can be proved by means of syllogisms using the clause not(P + Q) or not(notP + notQ) as our middle thesis as the case may be. Furthermore, though we do not mention this above, each

[123] This valuable word, "iff", has unfortunately not passed over into general usage. The reason for that is, I think, obvious: it is a word that is distinguishable in written language, but not in spoken language.

implication in causation has a base (i.e. the possibility that the three terms it mentions be conjoined). Thus, each of the above two forms could have been defined more precisely and usefully with reference to eight clauses, as follows:

7.	**Exclusive partial causation** by P and Q of R	symbol: **p** ex	=> not-8
a)	If (P + Q), then R	(P + Q + notR) is impossible	<=> 8(d)
b)	if (notP + Q), then notR	(notP + Q + R) is impossible	=> not-8(i)
c)	if (P + notQ), then notR	(P + notQ + R) is impossible	=> not-8(h)
d)	if (notP + notQ), then notR	(notP + notQ + R) is impossible	<=> 8(a)
g)	(P + Q) is possible	(P + Q + R) is possible	<=> 8(j)
h)	(notP + Q) is possible	(notP + Q + notR) is possible	=> not-8(c)
i)	(P + notQ) is possible	(P + notQ + notR) is possible	=> not-8(b)
j)	(notP + notQ) is possible	(notP + notQ + notR) is possible	<=> 8(g)

8.	**Exclusive contingent causation** by P and Q of R	symbol: **q** ex	=> not-7
a)	If (notP + notQ), then notR	(notP + notQ + R) is impossible	<=> 7(d)
b)	if (P + not Q), then R	(P + notQ + notR) is impossible	=> not-7(i)
c)	if (notP + Q), then R	(notP + Q + notR) is impossible	=> not-7(h)
d)	if (P + Q), then R	(P + Q + notR) is impossible	<=> 7(a)
g)	(notP + notQ) is possible	(notP + notQ + notR) is possible	<=> 7(j)
h)	(P + notQ) is possible	(P + notQ + R) is possible	=> not-7(c)
i)	(notP + Q) is possible	(notP + Q + R) is possible	=> not-7(b)
j)	(P + Q) is possible	(P + Q + R) is possible	<=> 7(g)

Now, these definitions show us that of the eight possible combinations of P, Q, R and their negations, four combinations are impossible and four others are possible, in each form. We see that, although some clauses are identical in both the forms **p** ex and **q** ex, there are serious conflicts between them; namely, clauses b and c of each are incompatible with clauses i, h respectively of the other. Thus, these two forms are contrary and can never be conjoined as **pq** ex for the same items PQR. This is a reasonable result, the essence of the forms **p** ex and **q** ex being that they mimic complete-necessary causation (**mn**), with reference to more than two items; they are thus intermediate degrees of causation, behaving somewhat like strongs and somewhat like weaks.

Let us now compare these forms to their predecessors, listed in Chapter 19.1. We see that, as we would expect, **p** ex is incompatible with **m** (see 1a and 7i) and **q** rel (see 4b and 7c, 4c and 7b), but compatible with **n** and **p** rel (they have no conflicting clauses). Indeed, **p** ex implies **n** since (7b + 7d) = 2a[124], 7g => 2b, and 7h or 7j => 2c; and **p** ex also implies **p** rel since 7a = 3a, 7h = 3b, 7i = 3c, and 7g = 3d. Whence it follows that **p** ex implies the joint form **np** rel. Similarly, **q** ex is incompatible with **n** and **p** rel, but compatible with **m** and **q** rel. Indeed, **q** ex implies **m** and **q** rel, i.e. **q** ex implies **mq** rel.

Can we now prove the converse, i.e. that **np** rel implies **p** ex and that **mq** rel implies **q** ex? The answer is no! The clauses 7c and 7j cannot be drawn from **np** rel; and similarly, the clauses 8c

124 This is easily established by dilemmatic argument : given "if (notP + Q), then notR" and "if (notP + notQ), then notR", the conclusion is "if notP, then notR".

and 8j cannot be drawn from **mq** rel. Therefore, the forms **p** ex and **q** ex are in fact stronger determinations than the specific forms **np** rel and **mq** rel (and not just stronger than the generic forms **p** rel and **q** rel).
The next obvious question is: what are the oppositions between these various forms when the items concerned are given different polarities? This is best investigated more mechanically by means of matricial analysis. The results are given in the following table, which can be viewed at the website:

- Table 20.2 – 3-item PQR Moduses of Forms – Exclusive Weak Causations. (4 pages in pdf file).

The results of this table are interesting, since they show that each of these exclusive forms yields only one modus. Thus, the above mentioned two initial forms **p** ex and **q** ex have respectively modus #s 150 and 170. This is comparable in degree of specificity to the single modus of **pq** rel. The full list of forms and their corresponding moduses can be summarized as follows:

Summary of Table 20.2 – Moduses of exclusive weak forms.

Main exclusive forms	modus
exclusive partial causation (**p** ex) PQR	150
exclusive contingent causation (**q** ex) PQR	170
exclusive partial causation (**p** ex) PnotQR	102
exclusive contingent causation (**q** ex) PnotQR	167
exclusive partial causation (**p** ex) PQnotR (prevention PQR)	107
exclusive contingent causation (**q** ex) PQnotR (prevention PQR)	87
exclusive partial causation (**p** ex) PnotQnotR (prevention PnotQR)	155
exclusive contingent causation (**q** ex) PnotQnotR (prevention PnotQR)	90
Inverse exclusive forms	**modus**
exclusive partial causation (**p** ex) notPnotQnotR	170
exclusive contingent causation (**q** ex) notPnotQnotR	150
exclusive partial causation (**p** ex) notPQnotR	167
exclusive contingent causation (**q** ex) notPQnotR	102
exclusive partial causation (**p** ex) notPnotQR (prevention notPnotQnotR)	87
exclusive contingent causation (**q** ex) notPnotQR (prevention notPnotQnotR)	107
exclusive partial causation (**p** ex) notPQR (prevention notPQnotR)	90
exclusive contingent causation (**q** ex) notPQR (prevention notPQnotR)	155

Notice that the inverses have the same items with opposite polarities; and that the modus of their form **p** ex becomes that of **q** ex, and vice versa. Now, the fact that each of these exclusive forms is expressive of only one modus should be useful for working out their oppositions and interpretations. For a start, we note that all 8 main forms are contrary to each other, since they have no modus in common; the inverses are of course their respective equivalents, with the already stated changes.

For the rest, our macroanalysis above seems after all to suffice; microanalysis adds nothing much more. For instance, regarding modus 150 we see, in Table 18.6, pages 3-4, that it is one of four moduses (the others being 149, 181, 182) which mean "**np** rel to Q (and **n-alone** rel to notQ only) in causation". This does not mean that modus 150 is identical to the other three, but only that it has this common implication, i.e. **np** rel (etc.) which we knew already (save for the implied lone). What this does tell us, however, is that our interpretations of the moduses thus far were somewhat lacking, since they reveal no difference between the more restrictive exclusive forms (like #150)) and their more ordinary cousins (viz. #s 149, 181, 182, in this example). This shows that our introduction of these additional specifications was useful and important.

Upon reflection, we should have expected the exclusive forms to be represented by only one modus, since they are defined by eight clauses! Indeed, any modus could be represented in words by eight clauses concerning the possibility or impossibility of each combination of the three items and their negations. The peculiarity of the exclusive forms is that they do this succinctly and are popularly used.

4. The Need for an Additional Item (or Two).

The important thing to note in the first section of the present chapter is that our 3-item format of partial and/or contingent causative propositions was from the start *intended to cover all eventual numbers of complements*. We have not used it as merely the simplest, most accessible, format – but as *an all-inclusive format*, to which all other weak causations can in principle be reduced when necessary. Thus, our investigation into the logic of causation with reference to only one complement Q (to P in the causation of R) was not intended to be supplemented later by consideration of more and more complements. Three items were supposed to do the trick.

Why then do we need to consider a fourth (and even possibly a fifth) item, now, in phase III ? For the simple reason that, when we consider causative syllogism we must look into cases with the major and/or the minor premise involving a complement. Since the minor, middle and major terms of our syllogism already take up three items (P, Q, R), we need an additional item S (and maybe even two of them, S and T) to investigate syllogisms with one (or both) premises about relative weak causation.[125]

Note well that eventual 4-item (or even 5-item) *syllogisms* are all composed of 3-item *propositions* (at least, as regards their premises, though some conclusions may conceivably involve four items). A syllogism requires at least three terms (the major, the middle and then minor) deployed in two premises (the major and minor premises, which share the middle term) and a conclusion (which relates the major and minor terms). This allows for only two terms per proposition. If one (or both) of the premises has a third term (i.e. a complement of weak causation), then the syllogism will have four (or respectively, five) terms. The conclusion will then be expected to have a third (and even fourth) term.

---

[125] The subsidiary term (S) is mentioned in phases I and II in the following places: chapters 5.3 and 9.4, where the various possible subfigures of the syllogism are tabulated; chapters 14.3 and 15.1, where it is stressed that the syllogisms here developed are not 4-item ones – i.e. that their full elucidation requires 4-item research; chapter 16.1, where the problem and the way to the solution of 4-item syllogism are presented.

Based on past experience with syllogistic reasoning, we certainly need at least one additional item, the fourth (or subsidiary) term S; for we can well expect a weak premise combined with a strong one to yield a weak conclusion. Regarding a possible fifth item T, it is probable that we do not need one, because it is unlikely that two weak premises can yield any conclusion at all; but this must of course be formally established in some way (and I doubt any way other than microanalysis can do the trick).

The introduction of a fourth item (S) means dealing with a grand matrix of 65,536 moduses each of which is defined by 16 digits; this is in the realm of the possible given my current computer resources (hardware and software capabilities) – just about. But these material resources are quite insufficient to deal with a fifth item (T), which would require a grand matrix of 2^{32} = 4,294,967,296 moduses of 2^5 = 32 digits each; therefore I can only speculate about the probable results of a study of the latter.

More will be said concerning the fourth item S in the next chapter, when we consider 4-item syllogisms.

Chapter 21. CAUSATIVE SYLLOGISMS.

Most of the limitations mentioned in the present chapter are overcome in the next chapter. Likewise, most the results obtained here are improved upon there. Nevertheless, I have not tried to rewrite the present chapter, considering that showing the process through which the study progressed is a good thing. This chapter has to be read to fully understand the next, because it breaks much new ground, uncovering issues and how to deal with them, and setting the stage for the finale in the next.

1. Methodology.

As we saw in phase II, as of chapter 14, validating or invalidating syllogistic arguments using the moduses identified through microanalysis is simple enough. In principle, any two or more premises might be put together, and their potential for 'conclusion' is the list of moduses they have in common. In practice, things are a bit more demanding.

For a start, the premises must have some item(s) in common; or else they will have no moduses in common. In the event that the argument is a sorites involving more than two premises, each premise must have at least one item in common with at least one other premise; and indeed the series must form a continuous chain reducible to two-premise syllogisms. Secondly, given that the two premises do have some item(s) in common, we need to check that they do indeed have moduses in common; if they have none, it means that the premises are incompatible and so cannot be put together to form an argument. Thirdly, if the premises do have some moduses in common (i.e. logically intersect), two things can happen.

It may be that these moduses are too 'scattered' – i.e. that some of them suggest a certain verbal conclusion and others of them suggest a contradictory verbal conclusion, the outcome being we have effectively no formal conclusion. This is not an inconsistency in the premises, but a sort of indecision in their joint implications. Alternatively, the moduses obtained from the conjunction of the premises all point to the same formal conclusion; in that case, we have a valid syllogism. Note well, to yield a valid conclusion, the common moduses of the premises must *all* be included in the list of moduses of the putative formal conclusion.

What I mean by a 'formal conclusion' is any propositional *form*. As we have seen, every propositional form 'has' a number (one or more) of alternative moduses. This means that *each of* 'its' moduses is enough to imply the proposition. It follows that if the premises jointly yield a certain set of moduses (one or more), and all of these moduses are included in the list of moduses for a given propositional form, that form is their conclusion. This is true, because each of their moduses is capable by itself of implying the form and they are all agreed in this implication. On the other hand, if *any* (one or more) of the moduses shared by the premises is *not* included in the list of moduses that imply the putative formal conclusion, it is not a valid conclusion.

The premises and conclusion of a syllogism may in principle be any relational statement; that is, each of them may be a conjunction of items (e.g. 'P and R is possible'), or a conditional statement (e.g. 'if P then R'), or a causative proposition (e.g. 'P is a complete cause of R').

Likewise, the premises and conclusion may have either polarity, and they may have any combination of positive and negative items. In the present chapter, I am limiting our attentions to *positive causative* premises and (positive or negative) *causative* conclusions, to avoid an excess of information. I just want to demonstrate how syllogisms are validated or invalidated using a spreadsheet. But in principle, this limitation is artificial and we must study all conceivable combinations of premises and conclusions.

The main difficulty in researching syllogism through matricial analysis is in finding a conclusion that includes all the moduses the premises have in common. These moduses are easily 'calculated' using simple formulae; but to find the appropriate formal conclusion we must look at all available forms and see which one(s) include all the common moduses. This seems hard to do mechanically with a mere spreadsheet program; more sophisticated software or ad hoc programming seems required[126]. For this reason, I do not at this stage try to search for conclusions mechanically, but instead am content for now to verify the conclusions established manually in previous phases of the present work.

Given a putative conclusion, it is easy to test it in a spreadsheet program. We just write a formula for the conjunction of the two premises and the putative conclusion. If the result is a number of moduses (one or more), we know the putative conclusion is applicable. But we cannot yet be sure that it is a valid conclusion. We must still try to conjoin the same two premises *with the negation of* the putative conclusion. If the latter trial conjunction yields one or more moduses in common, our putative conclusion is invalid, for reasons already explained. If, however, the said trial conjunction yields *zero* moduses in common, then our putative conclusion is finally proven valid.

The advantage of doing this work with a spreadsheet is the speed of calculation and the increased certainty in the results obtained. Assuming no error is made in formula writing, once we have a formula in the first cell of a column, we just copy it all the way down the column and the work is done. Moreover, we can copy a given formula from column to column and make changes to it as appropriate. What manually takes days and weeks of painstaking work can now be done in a few minutes or hours, and the results are more credible. Of course, errors in formula writing are possible, but they can usually be readily spotted by comparing the number of moduses obtained in similar columns and checking whether they are symmetrical.

I should add that the results obtained by me mechanically in the present phase were all compared to results obtained manually in previous phases of the research, and I can report that they are consistent. This shows both that the earlier manual calculations were all accurate and that the present formula based calculations were all accurate. The three phases have, thus, I am happy to say, verified and confirmed each other's results.

2. 3-Item Syllogisms.

Having already in phases I and II analyzed 3-item causative syllogism in considerable detail, it was easy to reproduce them in phase III and check the results. Regarding such syllogism, which is the main object of our research, there are, in each of the three figures (ignoring the fourth figure, as usual), 64 conceivable moods with positive causative premises involving

[126] Actually, I do finally manage to do the work of scanning for conclusions by means of (many and bulky) spreadsheets – in the next chapter.

positive items. All the pairs of premises listed are compatible, and so the remaining question concerning them is only whether they yield a formal conclusion or do not.

Note that in 3-item syllogism the premises and conclusions concerning weak causation (**mq**, **np**, **pq**, **p**, **q**) are all about *absolute* causation; relative causation can only be dealt with as of 4-item syllogism.

Having already found the applicable causative conclusions (mostly positive, though some negative) in previous phases of the research, our task here is just to verify them. This is done, firstly, by checking that the conjunction of the two given premises and the proposed conclusion yields *one or more* common moduses, and that these moduses are indeed all included under the putative conclusion. Secondly, the same is attempted with the negation of the proposed conclusion, and this should yield *no* common modus. If both these conditions are satisfied, the proposed conclusion is validated; otherwise, it is not.

The following four tables – which are all as usual on display in the The Logician website – show the results obtained (mechanically) in phase III and their full consistency with results previously listed in Table 14.4 (obtained by manual method).

- Table 21.1 – 3-Item Syllogisms – First Figure Moods. (4 pages in pdf file).
- Table 21.2 – 3-Item Syllogisms – Second Figure Moods. (4 pages in pdf file).
- Table 21.3 – 3-Item Syllogisms – Third Figure Moods. (4 pages in pdf file).
- Table 21.4 – 3-Item Syllogisms – Formulas Used. (4 pages in pdf file).

The first three of these tables list the valid moods for the three figures of the syllogism, and the fourth table shows the formulae used to produce the first three tables. The verification of these results was indeed done by me, by modifying the formulae in the three tables, so that the contradictory of the proposed conclusion (be it positive or negative) is tried instead; and I can report that in all cases, the result was zero common moduses. I have not bothered to produce additional pdf files showing these zero results, so as not to needlessly clutter the presentation of evidence; the reader can take my word for it or try doing the job independently.

It must be stressed that I have not here verified "nil" conclusions with equal meticulousness. Such non-conclusions from certain combinations of premises are, as already explained, due to the moduses found to be shared by the premises having scattered implications – some of them implying one formal conclusion and others implying a contradictory conclusion, so that no *uniform* conclusion from them is possible. All I have done here in such cases is list and count the moduses in common to the premises concerned. But I have not gone on to check that these moduses are indeed, as previously ascertained, too scattered for any finite conclusion[127]. I trust my previous manual check and see no point in repeating them.

The object of the present phase is to mechanize solutions to problems, remember. In the case of verification of "nil" conclusions, there is no doubt that such mechanization is technically feasible. This would proceed as follows. If the intersection of two premises result in a number of moduses (one or more), the program would have to check whether *all* these moduses fall within the modus list of any causative (or more broadly, propositional) form(s). If they do, we have a formal conclusion; if no such form is found, then we have no formal conclusion. This is not an easy task to perform with a spreadsheet program, since the program would have to automatically repeat the search and compare tasks for the full range of defined forms, before it could declare that the conclusion it found to be complete or that there was no conclusion.

---

127 To repeat, this work is done in the next chapter.

For this reason, I limit the present stage of research to verification of previously manually validated conclusions. Moods previously found invalid are here accepted as such. And indeed no effort is made to expand the research and look for eventual conclusions of any form other than causative (positive or negative). It may be, for all I know, that where some or no causative conclusion is possible some other form of conclusion (whether pre-causative, or causative with some or all items of negative polarity) might be found valid; i.e. there might be a not purely causative form that includes all the moduses shared by the premises. This is certainly an important question, which ought to eventually be investigated in detail. But I have not attempted to do it here, because (to repeat) it does not seem mechanically feasible with my present resources[128].

As for premises about inverse causation, prevention and inverse prevention (separately or mixed together), they are also ignored here, so as to avoid a surfeit of information; the same can be said for pre-causative premises – i.e. possible conjunctions of items or conditional propositions. This does not mean that such moods are ultimately less interesting or important than purely causative moods; but only that there is no real need here to present all possible combinations of premises. There should be no difficulty for anyone to investigate such syllogisms, using the method of inference through moduses that we have here demonstrated with reference to purely causative premises. Indeed, it is possible to do the job merely by successive changes in the polarities of the terms in already established causative syllogisms.

To sum up, then, all I have done in the above mentioned tables is to verify previously identified formal conclusions from the main 3-item premises (positive causatives only). I have not tried to enlarge the research, but merely wished for now to demonstrate how syllogism can be validated using spreadsheet software. Of course, this was made easy thanks to the work already done (in the preceding chapters) in mechanically identifying the moduses corresponding to each and every form of proposition in the preceding chapters.

A statistical note: in each figure of 3-item causative syllogism, we have found 23 positive conclusions, 16 negative ones (**not-m** or **not-n**) and 25 nil conclusions. This being out of a total of 64 moods in each figure, the percentages were respectively: 36%, 25% and 39%. Although the total numbers of valid and invalid moods are the same in the three figures, the specific conclusions from superficially similar premises are of course not always the same.

3. 4-Item Syllogisms.

I have adopted the same minimalist approach for 4-item positive causative syllogism as I did for the 3-item arguments (in the preceding section). I tabulated the already known 4-item syllogisms, obtained in phase I through matricial analysis (i.e. through macroanalysis in this case – see detailed listings in chapter 6; or Tables 7.3, 7.4 and 7.5), and was content to here test mechanically whether the conclusions previously identified were reliable; they indeed all were. Actually, I went a bit further than this, and in certain cases sought out an additional conclusion, as will presently be explained. But some negative conclusions and all nil conclusions were simply *passed over* from 3-item syllogism to 4-item syllogism without attempt at improvement, as will be presently explained.

The work done is shown in the following five tables, which are all as usual on display in the The Logician website:

---

128 Again note: I belie this assumption in the next chapter.

- Table 21.5 – 4-Item Syllogisms – 3 Phases Compared Results. (4 pages in pdf file).
- Table 21.6 – 4-Item Syllogisms – First Figure Moods. (714 pages in pdf file).
- Table 21.7 – 4-Item Syllogisms – Second Figure Moods. (718 pages in pdf file).
- Table 21.8 – 4-Item Syllogisms – Third Figure Moods. (718 pages in pdf file).
- Table 21.9 – 4-Item Syllogisms – Formulae Used. (6 pages in pdf file).

The first of these tables, Table 21.5, summarizes the premises and conclusions for the three figures. The following is an extract from it for you:

Table 21.5 (detail) –Phases II (3-items) and III (4-items) Results Compared

First Figure

Fig 1. mood #	111	112	113	114	115	116	117	118
major QR or QSR	**mn**	**mn**	**mn**	**mn**	**mn**	**mn**	**mn**	**mn**
minor PSQ or PQ	**mn**	**mq**	**np**	**pq**	**m**	**n**	**p**	**q**
3-item concl. PR	**mn**	**mq abs**	**np abs**	**pq abs**	**m**	**n**	**p abs**	**q abs**
4-item concl. PR or PSR	**mn**	**mq rel**	**np rel**	**pq rel**	**m**	**n**	**p rel**	**q rel**

Fig 1. mood #	121	122	123	124	125	126	127	128
major QR or QSR	**mq**	**mq**	**mq**	**mq**	**mq**	**mq**	**mq**	**mq**
minor PSQ or PQ	**mn**	**mq**	**np**	**pq**	**m**	**n**	**p**	**q**
3-item concl. PR	**mq abs**	**mq abs**	**nil**	**not-n**	**mq abs**	**nil**	**nil**	**not-n**
4-item concl. PR or PSR	**mq rel**	**m**	**"nil"**	**not-n**	**m**	**"nil"**	**"nil"**	**not-n**

Fig 1. mood #	131	132	133	134	135	136	137	138
major QR or QSR	**np**	**np**	**np**	**np**	**np**	**np**	**np**	**np**
minor PSQ or PQ	**mn**	**mq**	**np**	**pq**	**m**	**n**	**p**	**q**
3-item concl. PR	**np abs**	**nil**	**np abs**	**not-m**	**nil**	**np abs**	**not-m**	**nil**
4-item concl. PR or PSR	**np rel**	**"nil"**	**n**	**not-m**	**"nil"**	**n**	**not-m**	**"nil"**

Fig 1. mood #	141	142	143	144	145	146	147	148
major QR or QSR	**pq**	**pq**	**pq**	**pq**	**pq**	**pq**	**pq**	**pq**
minor PSQ or PQ	**mn**	**mq**	**np**	**pq**	**m**	**n**	**p**	**q**
3-item concl. PR	**pq abs**	**not-n**	**not-m**	**nil**	**not-n**	**not-m**	**nil**	**nil**
4-item concl. PR or PSR	**pq rel**	**not-n**	**not-m**	**"nil"**	**not-n**	**not-m**	**"nil"**	**"nil"**

Table 21.5 (detail) –Phases II (3-items) and III (4-items) Results Compared

Fig 1. mood #	151	152	153	154	155	156	157	158
major QR or QSR	**m**	**m**	**m**	**m**	**m**	**m**	**m**	**m**
minor PSQ or PQ	**mn**	**mq**	**np**	**pq**	**m**	**n**	**p**	**q**
3-item concl. PR	**m**	**mq abs**	**nil**	**not-n**	**m**	**nil**	**nil**	**not-n**
4-item concl. PR or PSR	**m**	**m**	**"nil"**	**not-n**	**m**	**"nil"**	**"nil"**	**not-n**

Fig 1. mood #	161	162	163	164	165	166	167	168
major QR or QSR	**n**	**n**	**n**	**n**	**n**	**n**	**n**	**n**
minor PSQ or PQ	**mn**	**mq**	**np**	**pq**	**m**	**n**	**p**	**q**
3-item concl. PR	**n**	**nil**	**np abs**	**not-m**	**nil**	**n**	**not-m**	**nil**
4-item concl. PR or PSR	**n**	**"nil"**	**n**	**not-m**	**"nil"**	**n**	**not-m**	**"nil"**

Fig 1. mood #	171	172	173	174	175	176	177	178
major QR or QSR	**p**	**p**	**p**	**p**	**p**	**p**	**p**	**p**
minor PSQ or PQ	**mn**	**mq**	**np**	**pq**	**m**	**n**	**p**	**q**
3-item concl. PR	**p abs**	**nil**	**not-m**	**nil**	**nil**	**not-m**	**nil**	**nil**
4-item concl. PR or PSR	**p rel**	**"nil"**	**not-m**	**"nil"**	**"nil"**	**not-m**	**"nil"**	**"nil"**

Fig 1. mood #	181	182	183	184	185	186	187	188
major QR or QSR	**q**	**q**	**q**	**q**	**q**	**q**	**q**	**q**
minor PSQ or PQ	**mn**	**mq**	**np**	**pq**	**m**	**n**	**p**	**q**
3-item concl. PR	**q abs**	**not-n**	**nil**	**nil**	**not-n**	**nil**	**nil**	**nil**
4-item concl. PR or PSR	**q rel**	**not-n**	**"nil"**	**"nil"**	**not-n**	**"nil"**	**"nil"**	**"nil"**

Second Figure

Fig 2. mood #	211	212	213	214	215	216	217	218
major RQ or RSQ	**mn**	**mn**	**mn**	**mn**	**mn**	**mn**	**mn**	**mn**
minor PSQ or PQ	**mn**	**mq**	**np**	**pq**	**m**	**n**	**p**	**q**
3-item concl. PR	**mn**	**mq abs**	**np abs**	**pq abs**	**m**	**n**	**p abs**	**q abs**
4-item concl. PR or PSR	**mn**	**mq rel**	**np rel**	**pq rel**	**m**	**n**	**p rel**	**q rel**

Table 21.5 (detail) –Phases II (3-items) and III (4-items) Results Compared

Fig 2. mood #	221	222	223	224	225	226	227	228
major RQ or RSQ	**mq**	**mq**	**mq**	**mq**	**mq**	**mq**	**mq**	**mq**
minor PSQ or PQ	**mn**	**mq**	**np**	**pq**	**m**	**n**	**p**	**q**
3-item concl. PR	**np abs**	**nil**	**np abs**	**not-m**	**nil**	**np abs**	**not-m**	**nil**
4-item concl. PR or PSR	**n + not-p rel (not np)**	**"nil"**	**n**	**not-m**	**"nil"**	**n + not-p rel (not np)**	**not-m**	**"nil"**

Fig 2. mood #	231	232	233	234	235	236	237	238
major RQ or RSQ	**np**	**np**	**np**	**np**	**np**	**np**	**np**	**np**
minor PSQ or PQ	**mn**	**mq**	**np**	**pq**	**m**	**n**	**p**	**q**
3-item concl. PR	**mq abs**	**mq abs**	**nil**	**not-n**	**mq abs**	**nil**	**nil**	**not-n**
4-item concl. PR or PSR	**m + not-q rel (not mq)**	**m**	**"nil"**	**not-n**	**m + not-q rel (not mq)**	**"nil"**	**"nil"**	**not-n**

Fig 2. mood #	241	242	243	244	245	246	247	248
major RQ or RSQ	**pq**	**pq**	**pq**	**pq**	**pq**	**pq**	**pq**	**pq**
minor PSQ or PQ	**mn**	**mq**	**np**	**pq**	**m**	**n**	**p**	**q**
3-item concl. PR	**pq abs**	**not-n**	**not-m**	**nil**	**not-n**	**not-m**	**nil**	**nil**
4-item concl. PR or PSR	**not-p rel + not-q rel**	**not-n**	**not-m**	**"nil"**	**not-n**	**not-m**	**"nil"**	**"nil"**

Fig 2. mood #	251	252	253	254	255	256	257	258
major RQ or RSQ	**m**	**m**	**m**	**m**	**m**	**m**	**m**	**m**
minor PSQ or PQ	**mn**	**mq**	**np**	**pq**	**m**	**n**	**p**	**q**
3-item concl. PR	**n**	**nil**	**np abs**	**not-m**	**nil**	**n**	**not-m**	**nil**
4-item concl. PR or PSR	**n**	**"nil"**	**n**	**not-m**	**"nil"**	**n**	**not-m**	**"nil"**

Fig 2. mood #	261	262	263	264	265	266	267	268
major RQ or RSQ	**n**	**n**	**n**	**n**	**n**	**n**	**n**	**n**
minor PSQ or PQ	**mn**	**mq**	**np**	**pq**	**m**	**n**	**p**	**q**
3-item concl. PR	**m**	**mq abs**	**nil**	**not-n**	**m**	**nil**	**nil**	**not-n**
4-item concl. PR or PSR	**m**	**m**	**"nil"**	**not-n**	**m**	**"nil"**	**"nil"**	**not-n**

Table 21.5 (detail) –Phases II (3-items) and III (4-items) Results Compared

Fig 2. mood #	271	272	273	274	275	276	277	278
major RQ or RSQ	**p**	**p**	**p**	**p**	**p**	**p**	**p**	**p**
minor PSQ or PQ	**mn**	**mq**	**np**	**pq**	**m**	**n**	**p**	**q**
3-item concl. PR	**q abs**	**not-n**	**nil**	**nil**	**not-n**	**nil**	**nil**	**nil**
4-item concl. PR or PSR	**not-q rel**	**not-n**	**"nil"**	**"nil"**	**not-n**	**"nil"**	**"nil"**	**"nil"**

Fig 2. mood #	281	282	283	284	285	286	287	288
major RQ or RSQ	**q**	**q**	**q**	**q**	**q**	**q**	**q**	**q**
minor PSQ or PQ	**mn**	**mq**	**np**	**pq**	**m**	**n**	**p**	**q**
3-item concl. PR	**p abs**	**nil**	**not-m**	**nil**	**nil**	**not-m**	**nil**	**nil**
4-item concl. PR or PSR	**not-p rel**	**"nil"**	**not-m**	**"nil"**	**"nil"**	**not-m**	**"nil"**	**"nil"**

Third Figure

Fig. 3 mood #	311	312	313	314	315	316	317	318
major QR or QSR	**mn**	**mn**	**mn**	**mn**	**mn**	**mn**	**mn**	**mn**
minor QSP or QP	**mn**	**mq**	**np**	**pq**	**m**	**n**	**p**	**q**
3-item concl. PR	**mn**	**np abs**	**mq abs**	**pq abs**	**n**	**m**	**q abs**	**p abs**
4-item concl. PR or PSR	**mn**	**n + not-p rel (not np)**	**m + not-q rel (not mq)**	**not-p rel + not-q rel**	**n**	**m**	**not-q rel**	**not-p rel**

Fig. 3 mood #	321	322	323	324	325	326	327	328
major QR or QSR	**mq**	**mq**	**mq**	**mq**	**mq**	**mq**	**mq**	**mq**
minor QSP or QP	**mn**	**mq**	**np**	**pq**	**m**	**n**	**p**	**q**
3-item concl. PR	**mq abs**	**nil**	**mq abs**	**not-n**	**nil**	**mq abs**	**not-n**	**nil**
4-item concl. PR or PSR	**mq rel**	**"nil"**	**m**	**not-n**	**"nil"**	**m**	**not-n**	**"nil"**

Table 21.5 (detail) –Phases II (3-items) and III (4-items) Results Compared

Fig. 3 mood #	331	332	333	334	335	336	337	338
major QR or QSR	**np**	**np**	**np**	**np**	**np**	**np**	**np**	**np**
minor QSP or QP	**mn**	**mq**	**np**	**pq**	**m**	**n**	**p**	**q**
3-item concl. PR	**np abs**	**np abs**	**nil**	**not-m**	**np abs**	**nil**	**nil**	**not-m**
4-item concl. PR or PSR	**np rel**	**n**	**"nil"**	**not-m**	**n**	**"nil"**	**"nil"**	**not-m**

Fig. 3 mood #	341	342	343	344	345	346	347	348
major QR or QSR	**pq**	**pq**	**pq**	**pq**	**pq**	**pq**	**pq**	**pq**
minor QSP or QP	**mn**	**mq**	**np**	**pq**	**m**	**n**	**p**	**q**
3-item concl. PR	**pq abs**	**not-m**	**not-n**	**nil**	**not-m**	**not-n**	**nil**	**nil**
4-item concl. PR or PSR	**pq rel**	**not-m**	**not-n**	**"nil"**	**not-m**	**not-n**	**"nil"**	**"nil"**

Fig. 3 mood #	351	352	353	354	355	356	357	358
major QR or QSR	**m**	**m**	**m**	**m**	**m**	**m**	**m**	**m**
minor QSP or QP	**mn**	**mq**	**np**	**pq**	**m**	**n**	**p**	**q**
3-item concl. PR	**m**	**nil**	**mq abs**	**not-n**	**nil**	**m**	**not-n**	**nil**
4-item concl. PR or PSR	**m**	**"nil"**	**m + not-q rel (not mq)**	**not-n**	**"nil"**	**m**	**not-n**	**"nil"**

Fig. 3 mood #	361	362	363	364	365	366	367	368
major QR or QSR	**n**	**n**	**n**	**n**	**n**	**n**	**n**	**n**
minor QSP or QP	**mn**	**mq**	**np**	**pq**	**m**	**n**	**p**	**q**
3-item concl. PR	**n**	**np abs**	**nil**	**not-m**	**n**	**nil**	**nil**	**not-m**
4-item concl. PR or PSR	**n**	**n + not-p rel (not np)**	**"nil"**	**not-m**	**n**	**"nil"**	**"nil"**	**not-m**

Fig. 3 mood #	371	372	373	374	375	376	377	378
major QR or QSR	**p**	**p**	**p**	**p**	**p**	**p**	**p**	**p**
minor QSP or QP	**mn**	**mq**	**np**	**pq**	**m**	**n**	**p**	**q**
3-item concl. PR	**p abs**	**not-m**	**nil**	**nil**	**not-m**	**nil**	**nil**	**nil**
4-item concl. PR or PSR	**p rel**	**not-m**	**"nil"**	**"nil"**	**not-m**	**"nil"**	**"nil"**	**"nil"**

Table 21.5 (detail) –Phases II (3-items) and III (4-items) Results Compared

Fig. 3 mood #	381	382	383	384	385	386	387	388
major QR or QSR	q	q	q	q	q	q	q	q
minor QSP or QP	mn	mq	np	pq	m	n	p	q
3-item concl. PR	**q abs**	**nil**	**not-n**	**nil**	**nil**	**not-n**	**nil**	**nil**
4-item concl. PR or PSR	**q rel**	**"nil"**	**not-n**	**"nil"**	**"nil"**	**not-n**	**"nil"**	**"nil"**

The next three tables (notice their lengths: remember, each has 65,536 rows!) show the lists of moduses obtained for each 4-item syllogism investigated, and the last table shows the formulae used to produce these three tables. Examining the results obtained here mechanically, we see that they perfectly match earlier results obtained manually (with a few exceptions explained below). The contradictory conclusion test was carried out throughout and further guaranteed the known conclusions (though here again, I ask you to take my word for it, because I do not want to publish too many tables, and especially not empty tables!).

Notice that both 3-item and 4-item syllogisms are listed in these tables, in order to compare results. Of course, the 3-item syllogisms involving absolute weak determinations in either or both premises yield *absolute* weak conclusions (if any), whereas the 4-item syllogisms here considered concern *relative* weak premises and conclusions (if any) – relative, that is, to a fourth item S – so they are not quite comparable. Nevertheless, when the 3-item (absolute) conclusions differ in form from the analogous 4-item (relative) ones, we would naturally want to double check the latter to be sure. This I did, and found the results obtained in the past essentially correct. Or more precisely put: none were incorrect, but some were incomplete.

In many cases, the absolute (i.e. irrespective of any complement) conclusion was found to have a relative (to complement S) analogue. For example, mood number 112, 1/**mn**/**mq**, yields the conclusion $\mathbf{mq}_{abs}$ if its minor premise is absolute (3-item syllogism) and $\mathbf{mq}_{rel}$ if its minor premise is relative (4-item syllogism). In some cases, this continuity does not hold. For example, mood number 125, 1/**mq**/**m**, yields the conclusion $\mathbf{mq}_{abs}$ if its major premise is absolute (3-item syllogism) but only **m** if its major premise is relative (4-item syllogism). What do I mean by "only **m**"? I mean that, even though the 3-item conclusion $\mathbf{mq}_{abs}$ remains valid, we cannot predict the conjunction of **m** with either $\mathbf{q}_{rel}$ or **not-**$\mathbf{q}_{rel}$.

This concerns 4-item syllogism, remember. In some cases, we can go further than this and predict the conjunction of **m** with **not-**$\mathbf{q}_{rel}$ – meaning that the possibility of $\mathbf{mq}_{rel}$ is formally excluded even though the 3-item conclusion $\mathbf{mq}_{abs}$ is still valid. An example of this is mood number 231, 2/**np**/**mn**. In some other cases, as we shall later see (in Table 21.10), the results are split up. Here, I am referring to moods that mix absolute and relative premises. In some of these cases, the results are the same when the absolute premise is the major and the relative premise is the minor, and vice versa. But in certain cases, the results differ – one way yielding an indefinite "only **m**" type conclusion, and the other way yielding a definite "**m** + **not-**$\mathbf{q}_{rel}$" type of conclusion; I have labeled such conclusions "at least **m**". An example of this is mood number 232, 2/**np**/**mq**.

It should be noted that such details were not brought up in our phase I analysis of 4-item syllogism, for the simple reason that we were unable at that time to deal with negative causative propositions. For the same reason, many moods that seemed inconclusive in phase I (for 4-item syllogism) were found to yield a negative conclusion like **not-m** or **not-n** in phase

II (for 3-item syllogism). Now, in phase III, we are able to mechanically generate such negative conclusions (for 4-item syllogism), as well as conclusions that negate relative weaks, i.e. which involve **not-p_{rel}** or **not-q_{rel}**.

In case it is not clear to you, let me underline the following: the form **q_{abs}** is compatible with both the forms **q_{rel}** and **not-q_{rel}** – note well, though **q_{rel}** implies **q_{abs}** and **not-q_{rel}** does not imply **q_{abs}** – the latter two forms are quite compatible. Similarly, **p_{rel}** and **not-p_{rel}** are compatible propositions. With reference to matricial analysis, remember, compatible propositions have one or more moduses in common. For this reason, a 3-item syllogism with a valid conclusion consisting of or involving an absolute weak causation (**p_{abs}** and/or **q_{abs}**) neither implies nor excludes the validity of a 4-item syllogism with a conclusion consisting of or involving a relative weak causation (**p_{rel}** and/or **q_{rel}**).

Obviously, *any* absolute conclusion (whether positive or negative, strong or weak) found valid in 3-item syllogism remains valid in 4-item syllogism – since the relative premises of 4-item syllogism formally imply the absolute premises of 3-item syllogism. This is direct reduction. But (to repeat) it does not follow that the corresponding *relative* conclusions are valid (since an absolute proposition does not imply a relative one). It is also obvious that, although the 4-item syllogism can yield a *more precise* conclusion (i.e. one signified by fewer moduses) than the analogous 3-item one, it cannot in any case yield a contradictory or contrary conclusion.

Diagram 21.1. Nil Conclusions are Not Reducible.

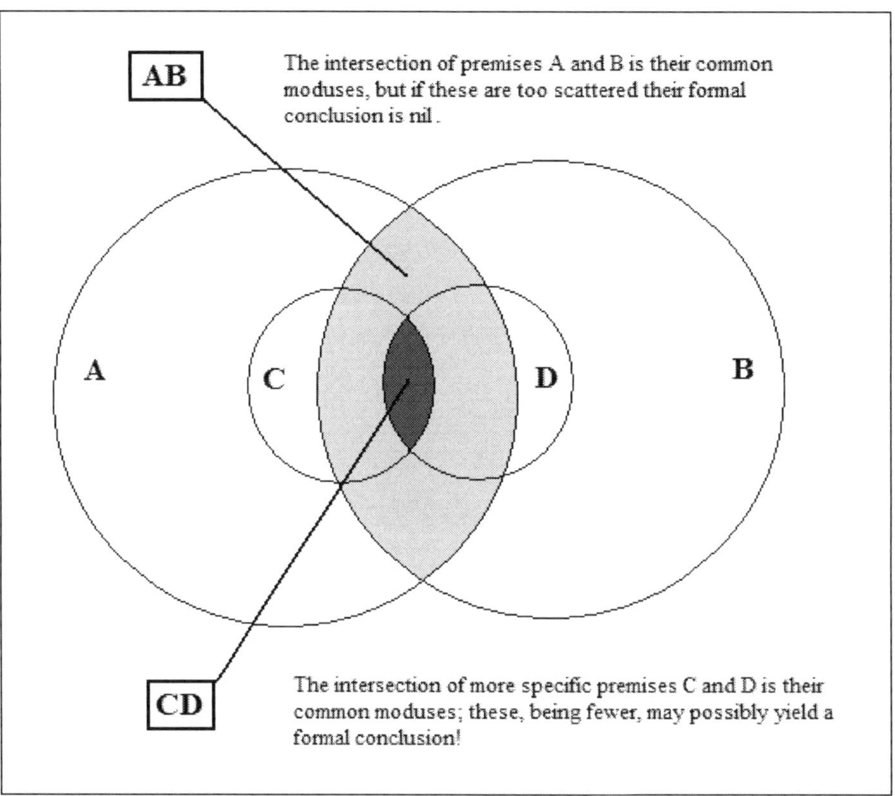

It does *not* follow from the above, however, that any conclusion which is found to be invalid in 3-item syllogism is bound to be invalid in 4-item syllogism. We cannot use indirect reduction (i.e. *reduction ad absurdum*) in such cases, for if the latter's conclusion is denied it does not follow that the former's conclusion is also denied.

This is made clear by above diagram. Consider two premises A and B, and suppose their intersection (the area AB they share) signifies a number of moduses which are too scattered to form a conclusion (i.e. no form includes them all); call these moduses *v, w, x, y, z*. It is still conceivable that there are two more specific premises, say C and D, whose intersection (the area CD they share) yields a formal conclusion; for CD may signify only the moduses *v, w* (excluding *x, y, z*) and these may happen to be included in the same form, say E. Note well that this *may* happen, but does not necessarily happen; the conjunction CD may also be inconclusive. All that may equally be understood with reference to implications. If A, B are or involve absolute weaks, and C, D are or involve analogous relative weaks, then C implies A, and D implies B. Given that C + D have a certain relative (or even absolute) conclusion E, it does not follow that A + B have that same conclusion E or any subaltern conclusion to E. Thus, AB may well be invalid while CD is valid.

Notwithstanding all that, I have here assumed that when a mood of 3-item syllogism yields no conclusion, then the 4-item syllogism need not be investigated further. That is, I have arbitrarily (or rather, speculatively) declared, for the time being, at least, that the latter's conclusion will likewise be nil. Similarly, by the way, when a 3-item syllogism yielded a **not-m** or **not-n** conclusion, I generally accepted the identical conclusion for the corresponding 4-item syllogism to be all that can be inferred, even while aware that in some cases some additional formal conclusion(s) might conceivably be found. Such assumptions may be taken as inductive probabilities, which have yet to be proven right or wrong deductively when appropriate mechanical means are devised[129].

Just as with 3-item syllogism, no attempt was made to extend the research and consider other premises and other conclusions than those already investigated, so here with 4-item syllogism further explorations are kept to a minimum. This restraint is largely due to the fact that my computer (hardware and software) capabilities have been stretched to their limit when dealing with four items, in addition to the more general need (to repeat) to find a mechanical way to systematically look for conclusions. In view of these limitations, I must relinquish at this time the ambition to be exhaustive and be satisfied with mere demonstration, i.e. with showing the way putative conclusions can be validated[130].

A notable change in the treatment of 4-item syllogism in the present phase III compared to the earlier phase I is the issue of moods with five items. In the earlier phase, when both premises involved weak causation, I assigned the fourth item (S) to one of them, and *took one of the other three items as the complement for the other two*. For example, see mood #124 in chapter 6.2, which has item P as both complement (of Q) in the major premise and minor term (complemented by S) in the minor premise. My intent there was obviously to construct a 5-item syllogism at all costs, even if it was rather illusory or at any rate a very rare or special occurrence. But now I think that this is a misleading approach, in that it diverts our attention from the *more general* problem, viz. that of dealing systematically with five distinct items (P, Q, R, S and T).

Remember that we can always in such cases draw the corresponding 3-item conclusion (if any) anyway, i.e. an absolute weak conclusion; the issue for us is whether we can do better than that. Lacking the means to investigate 5-item syllogism when *both* premises are partly or wholly weak, I have not here as in the past (as just explained) adopted the special case where the fifth item T is identical with one of the other four items. What I have done instead is to

129 All this is indeed fully dealt with in the next chapter.

130 Here again, I must stress that all these issues are successfully resolved in the next chapter.

consider cases where either one of the premises is taken as relative (to the fourth item S) and the other is taken as absolute (i.e. without a fifth item needing to be specified). The latter artifice does yield some further conclusions.

I have again chosen to narrow the problem somewhat, by considering only some of the moods concerned. The moods susceptible to having mixtures of absolute and relative premises are those with some weak causation in both premises. Looking at Table 21.5, we see that there are 25 such moods in each figure. Looking at the results of 3-item syllogism, we see that, of these 25 moods per figure, 15 yield no conclusion at all, 8 yield only a negative conclusion and only 2 yield a positive absolute conclusion.

Let us suppose, as we did before, that the nil conclusions and negative conclusions found in 3-item syllogism are also applicable to 4-item syllogism. This assumption is admittedly presumptuous, but as already discussed it is inductively reasonable in the context of our current technical limitations. This leaves us with only (3 times 2 =) 6 moods to study more closely, whose numbers are 122, 133, 223, 232, 323, 332. These 6 become 12 moods, since each may have the major premise absolute and the minor one relative to S (this is called subfigure b), or the major premise relative to S and the minor one absolute (this is called subfigure c).

Our task is to see whether any of these 12 moods give us a relative conclusion, i.e. a more specific conclusion than the absolute conclusion we already know they give when both the premises are absolute (i.e. through 3-item syllogisms). This job is done in the following table, which is as usual on display in the The Logician website:

- Table 21.10 – 4-Item Syllogisms – Mixing Absolute and Relative Premises. (800 pages in pdf file).

As this table shows, 8 moods do not give a new conclusion (i.e. one relative to S, instead of absolute) – but, surprisingly, 4 moods do give us a new conclusion, though not the conclusions $\mathbf{q}_{rel}$ (with **m**) or $\mathbf{p}_{rel}$ (with **n**), but *their negations*! The 4 moods are numbers 223c, 232c (in the second figure) and 323b, 332b (in the third figure). When we try a positive weak conclusion for them, we obtain zero moduses; whereas when we try for the contradictory conclusion, we obtain 16 moduses in each case. With regard to the 8 moods that do not yield conclusions (of which 4 are in the first figure), both the positive trial and the negative trial result in a number of moduses (namely, 4 and 12 respectively), so we do not know which way to lean, and so we must remain content with just the absolute given by 3-item syllogism.

To conclude, the following little table (not numbered) summarizes the validity rate found in the 4-item causative syllogisms we have studied in this section. Please refer to Table 21.5 again for an overview of the results.

Statistics for 4-item syllogism	Fig. 1	Fig. 2	Fig. 3	sum	percent
new valid positive conclusion	10	5	5	20	10
new valid negative conclusion	0	9	9	18	9
same valid positive conclusion	13	9	9	31	16
same valid negative conclusion	16	16	16	48	25
no formal conclusion (assumed)	25	25	25	75	39
total moods considered	64	64	64	192	100

This summary includes all moods in all three figures of 4-item syllogism. Out of the total of 192 moods, 31 (16%) yielded the same positive conclusion in 4-item syllogism as in 3-item syllogism, meaning a strong causation (with or without an absolute weak, as the case may be); and 48 (25%) yielded the same negative conclusion in 4-item syllogism as in 3-item syllogism, meaning a denied strong causation. More interesting were the 38 (17%) moods giving *new* conclusions, i.e. inferences peculiar to 4-item syllogism; these included 20 positive conclusions (10%) and 18 negative ones (9%)[131]. Finally, 75 moods (39%) gave no conclusions.

The important moods in the present context, I would say, are the 20 'new' positive moods, and to a lesser degree the 18 'new' negative moods, because they concern weak causation relative to S. These 38 moods show the need for 4-item microanalysis – i.e. the value of this work for causative logic. The remaining 79 'same' valid moods and 75 invalid moods are not specific to 4-item syllogism, but rather validated or invalidated by 3-item syllogism.

The 20 new positive valid moods of 4-item syllogism are the following: in figure one: the 10 moods numbered 112, 113, 114, 117, 118, 121, 131, 141, 171, 181; in figure two: the 5 moods numbered 212, 213, 214, 217, 218; and in figure three: the 5 moods numbered 321, 331, 341, 371, 381. The 18 new negative valid moods are: in figure one: none; in figure two: the 9 moods numbered 221, 223c, 226, 231, 232c, 235, 241, 271, 281; and in figure three: the 9 moods numbered 312, 313, 314, 317, 318, 323b, 332b, 353, 362. These most significant valid moods have been highlighted in Table 21.5. Take a good look at them!

4. About 5-item Syllogism.

I would like to say a few words now concerning 5-item syllogism, i.e. moods whose premises are relative to two different complements, say S and T. To put things in a wider perspective, I would today modify Table 5.2 (see in Chapter 5.3) to look like this:

Table 21.1 (modified). Subfigures of each figure.

Subfigures	**a**	**b**	**c**	**d**
Figure 1	QR PQ PR	QR P(S)Q P(S)R	Q(S)R PQ P(S)R	Q(T)R P(S)Q P(ST)R
Figure 2	RQ PQ PR	RQ P(S)Q P(S)R	R(S)Q PQ P(S)R	R(T)Q P(S)Q P(ST)R
Figure 3	QR QP PR	QR Q(S)P P(S)R	Q(S)R QP P(S)R	Q(T)R Q(S)P P(ST)R

131 Concerning the latter conclusions, note that for 4 of them (2 in each of figures 2 and 3) I only here intend as new their negative element (negation of relative weak causation). In fact, they include positive elements (strong causation) already found in 3-item syllogism, which are here glossed over; if we wanted to count them, we would have to add them to the earlier category of 'same valid positives' and there would be some overlap. See Table 21.10.

The main change here is the insertion of a fifth complement (T) instead of (P) in the major premise of subfigures (d); notice too the consequent change in the conclusions' complements from (S) to (ST). Subfigures (d), thus, concern premises and conclusions all involving some relative weak determination (whether or not conjoined to a strong determination). Notice that the conclusion is in such cases presumably one with two complements (S and T) – though we can speculate that a conclusion with only one complement might in some cases be feasible.

Subfigures (a) are dealt with in 3-item syllogism. Note that they concern not only "both premises strong" as stated in the original definition, but are equally applicable to moods with some absolute weak premise(s) and/or conclusion[132]. Similarly, we should consider the "strong only" premise in subfigures (b) and (c) as possibly an absolute weak premise. In short, a premise or conclusion with only two terms need not be strong only, but may be a mixture of strong and absolute weak or entirely absolute weak. At the time I wrote phase I, the concept of absolute partial or contingent causation was perhaps not yet fully formed, if at all, in my mind.

We can state the following generalities concerning the relation between syllogisms with more or less items. If a mood with fewer (e.g. 3 or 4) items is valid, the corresponding mood with more (e.g. respectively 4 or 5) items may or may not be valid. Why? Referring again to Diagram 21.1, suppose the vaguer premises A and B constitute a valid mood, i.e. suppose all the moduses they have in common (the area of intersection AB in our diagram) are included in some causative form. There is no guarantee that the more precise premises C and D (which respectively imply, i.e. fall within, A and B) will likewise constitute a valid mood. Granting they have a common item, these premises will intersect (forming the area CD) and do so within the larger area AB (as on our diagram).

However, it does not follow from the fact that the more numerous moduses of AB do yield a vague formal conclusion that the fewer moduses of CD must yield an analogous, more precise formal conclusion. The moduses of CD may be too scattered to be all included in the list of moduses of a specific conclusion. In such case, the mood CD will still of course yield the same vague conclusion as AB – but it will not be able to produce *a more precise one*. For example, as already seen, mood number 125, 1/**mq**/**m**, yields the conclusion $\mathbf{mq}_{abs}$ if its major premise is absolute (3-item syllogism), yet it does not yield the conclusion $\mathbf{mq}_{rel}$ if its major premise is relative (4-item syllogism).

So direct reduction of the latter to the former is impossible. In other words, we cannot readily *derive* syllogisms with more items from syllogisms with less items. We must develop ad hoc means for each increase in the number of items. And these means increase exponentially our need for computer space and power, in hardware and software. Nevertheless, 5-item causative syllogism research is technically within the realm of the possible. The necessary computing resources are sure to be found in many existing research centers.[133]

The problem of 5-item syllogism can only be adequately solved when we are able to develop 5-item matrices. To do so, we would need software or a program, and hardware capabilities, able to deal with tables with over 4 billion rows and dozens of columns. My computing resources are barely able to handle 4-item syllogism (which have 65,536 rows). To generate the latter, I had to delete all data (i.e. columns) not directly needed to produce the tabulated conclusions, and I had to use a separate spreadsheet (i.e. a separate file) for each figure;

---

[132] Note that when I do not mention the subfigure, it is probably subfigure (a).

[133] Note that I do not go further than this in the next chapter. 5-item syllogism remains an open issue in the present volume.

otherwise, the software got stuck. It follows that matricial analysis with 5-items is not feasible for me at this time.

This limitation does not worry me greatly, because I intuitively doubt 5-item syllogism will yield any (or many) new conclusions; I may be wrong, of course. By 'new conclusions' I mean conclusions that are specific to 5-item syllogism. As can be seen in the tables of the preceding section, most (though not all) of the conclusions obtained in 4-item syllogisms were in fact conclusions of the corresponding 3-item syllogisms. We can expect 5-item syllogism to yield similarly rarified results (which still leaves us with some hope of new conclusions).

If we examine Table 21.5 again, we see that in each figure there are 25 moods capable of giving rise to 5-item syllogism (being in subfigure d). We know regarding the 3-item syllogisms corresponding to these 25 moods in each figure that 2 (8%) yield a positive absolute conclusion, 8 (32%) yield a negative conclusion, and the remaining 15 (60%) yield no conclusion at all. Further investigation of the 6 moods (in the 3 figures) yielding positive absolute conclusions, through 4-item syllogisms with mixed absolute-relative or relative-absolute premises (in which event they become 12 moods), shows that their conclusions can be differentiated, with 8 of them yielding a bare **m** or **n** conclusion, while 4 of them yield a more precise $\mathbf{m}+\mathbf{q}_{rel}$ or $\mathbf{n}+\mathbf{p}_{rel}$ conclusion.

Admittedly, as already discussed at length, many of the moods without conclusion in 3-item syllogism may upon further scrutiny turn out to have a conclusion in 4-item or even in 5-item syllogism. Similarly, some negative or even positive conclusions might at a more specific level yield additional information. But the trend as we increase the number of items seems to be one of diminishing returns, so I believe that the chances of getting some new inference in 5-item syllogism, though not zero, are very slim indeed. This, I think, is as far as we can go for now trying to predict 5-item syllogism.

Chapter 22. SCANNING FOR CONCLUSIONS.

In the present chapter, thank G-d, I finally find the way to achieve the major task of this whole study of causative logic, namely discovering and validating ALL causative (and preventive) positive (and negative) conclusions of 3- and 4-item positive causative syllogism!

1. Methodology.

After completing the preceding chapter, it occurred to me that with a bit (i.e. a lot!) more patience and effort on my part, I could validate or invalidate all possible conclusions to causative syllogism. I did not need to wait to develop an ad hoc computer program for this job, but could do it using spreadsheet software, provided I cut up the task into sufficiently small segments. That is, instead of awaiting a fully 'mechanical' solution to the problem, I could make-do with a semi-mechanical one. The present chapter is the result of this work – this crucial work, for after all it was one of the main goals of all my research into causative logic.

I am happy to say that I succeeded entirely. *Eureka!* I found all valid positive and negative causative and preventive conclusions to positive causative syllogism, and proved all remaining such propositions to be invalid conclusions. This for all three figures, for 3-item and for 4-items syllogism. Other work done in this context is presented further on.

The method I used is essentially no different from the 'manual' one I previously used, except that here I apply it more systematically. To determine the valid and invalid conclusions from a pair or set of premises (a 'mood' of the syllogism, in the language of logicians), we must 'scan' over a 'range' of possible conclusions, and see which fit and which do not. How is this done? Simply:

a) Find the moduses that the given conjunction of premises, AB, yields, using matricial analysis as demonstrated already. Suppose their number is *x*.
b) Find the moduses, if any, of the same conjunction of premises with a proposed conclusion, C, again by matricial analysis. Suppose that the conjunction ABC yields *y* moduses.
c) Now, *if y=x*, then C is indeed a valid conclusion of AB (since the conjunctions AB and ABC have exactly the same moduses); that is to say, C is logically implied by AB. On the other hand, *if y=0* (zero), then C is an invalid conclusion for AB, i.e. not-C is a valid conclusion of AB (since the conjunction of AB with not-C is bound to have the same moduses, x in number, as AB alone); that is to say, C is logically denied by AB. *If y neither =x nor =0*, then AB neither implies nor denies C – i.e. neither C nor not-C is a valid or invalid conclusion of AB.

This, then, is what I mean by 'scanning for conclusions'. Knowing the moduses corresponding to a given set of premises (here labeled AB), we mechanically test each form (such as C) within a chosen range of forms, to see whether when conjoined to the premises the resulting moduses are identical, entirely different, or in between. If the moduses are identical, we have a valid conclusion; if they are entirely different, then the contradictory of the form tested must be a valid conclusion; and if neither of these results is obtained, it means the moduses are too scattered for any conclusion of that form or its negation to be drawn.

One small improvement in this method is that there is no need to test both a form C and its negation not-C. If we start by finding the moduses for the conjunction of premises (AB) alone, then we need only test positive forms (C) and the status of their negations (not-C) follows as just explained. This saves us half the work. Moreover, we need only find the moduses of the premises AB once, and then compare them to the moduses with various conclusions (C1, C2, C3, etc.) conjoined to them. This simplifies formula writing in the spreadsheets.

The 'range of conclusions' investigated is here symbolized by C1, C2, C3, etc. In the preceding chapter, I had limited the research to causative conclusions of the form PR or PSR (i.e. the six forms **m**, **n**, **p** rel to S, **q** rel to S, **p** abs, **q** abs); this was the range there chosen, minimal in its ambition. In the present chapter, the range has been considerably expanded, since I have added causative conclusions with a negative complement, i.e. PnotSR forms (**p** rel to notS, **q** rel to notS), and a similar range of preventive conclusion, i.e. eight forms with items PnotR, PSnotR, and PnotSnotR.

Note that I did not look into what I call 'pre-causative' conclusions to causative syllogism, i.e. possible conclusions of conjunctive or conditional form, but only for causative conclusions (in the broad sense, meaning causative and/or preventive). This may be justified by saying that the central goal of causative reasoning is, after all, to look for causative conclusions – to look for other forms of conclusion constitutes at best an additional and lesser interest. It is quite feasible with additional work (using the same techniques), but I do not bother to do it here.

All the above mentioned work concerns positive causative syllogism, i.e. syllogism with both premises having positive causative form, note well. As will be seen, none of the results obtained here contradict past results; i.e. no errors were found here in past conclusions. All past conclusions are, happily, here confirmed – but more conclusions are now discovered, thus justifying the whole elaborate procedure. Throughout the work, I kept reminding myself that this would be a one-time effort on my part for all humanity and for all time. Once these syllogistic problems are solved, they are solved for all time; so the effort invested is worthwhile.

As we shall presently see, most of the additional conclusions found are negative; only a very few are positive[134]. The positive conclusions previously obtained (using a narrower range) remain the main conclusions, though all negative conclusions are also significant. What is important, anyway, is that at the end of this chapter we can confidently say that we have a definitive, complete list of positive and negative valid conclusions from all positive causative syllogisms – and we know for sure which conclusions are invalid and which moods yield no conclusions at all. This was, to repeat, one of the principal goals of our research.

This is true for all 3-item and 4-item moods in the three figures. Needless to say, 5-item moods are still not treated here, and there is definitely no possibility of dealing with them using spreadsheets in a personal computer – the number of moduses involved is just too great for these tools. I leave this also important task to future researchers!

---

[134] These are not earth-shattering news, but they are interesting in that they explain (i.e. imply) previously known negative conclusions (obtained in the narrower range of consideration).

2. Forms Studied and their Oppositions.

The premises and conclusions studied here are given in the following tables, which can be viewed online at my website (www.thelogician.net) as usual:

- Table 22.1 – 3-Item Causative Syllogisms – Premises and Conclusions Scanned. (16 pages in pdf file.)
- Table 22.2 – 4-Item Causative Syllogisms – Premises and Conclusions Scanned. (3160 pages in pdf file.)

These two tables contain the basic data, which will be used in subsequent tables. They show, for each of the three figures, the moduses of the positive causative premises used and the moduses of the chosen range (causative and preventive) of putative conclusions from these premises. The data here displayed is simply selected, cut and pasted from past tables (namely, 18.4 and 18.9).

Moreover, we here examine the oppositions between the conclusions scanned. The reason for this is that, in order to reduce the conclusions obtained from a given set of premises to a minimum, we have to distinguish between primary conclusions and subaltern conclusions. The latter are formally implied by the former, so that their implication by the conjunction of premises is incidental. Thus, to better understand and rationalize the results of scanning work, we need to know the oppositions between all possible forms of conclusion – i.e. which of them imply which, and which of them are not-implied by the others.

This knowledge can be developed by a technique akin to scanning. Say we take two forms K and L, to know how they relate to each other, we must look into all their conjunctions, i.e. K and L (11), K and not-L (10), not-K and L (01), and not-K and not-L (00). If the forms concerned have no moduses in common, it means they are incompatible; inversely, they are compatible if they do have some moduses (no matter how many) in common. The results obtained by this technique are listed in the preceding two tables (22.1 and 22.2). The following two tables, which can be viewed online at my website as usual, show the formulas actually used in them, for the record:

- Table 22.3 – 3-Item Syllogism – Formulas Used for Oppositions. (2 pages in pdf file.)
- Table 22.4 – 4-Item Syllogism – Formulas Used for Oppositions. (2 pages in pdf file.)

As it turns out, for the ranges of possible conclusions studied here, some KL conjunctions are impossible, but no remarkable cases were found where K implies L or vice-versa (other of course than obvious implications, like **p** rel implying **p** abs), and no pair of forms were found exhaustive (i.e. such that the conjunction of both their negations was impossible). This means we can often infer a form's negation from another form's affirmation – but we can never draw useful inferences from a form's negation (other than obvious cases like **not-p** abs implying **not-p** rel).

This is evidenced in the following summary table, including possible positive causative and preventive conclusions from both 3-Item and 4-Item syllogisms:

- Table 22.5 – 3 and 4 Item Syllogisms – Oppositions between Conclusions. (1 page in pdf file.)

This table can be viewed online at my website as usual. The results in it of practical interest to us are the following:

- Finding the causative (PR) conclusion **m**, then the causatives **p** rel to S, **p** rel to notS, **p** abs, as well as all preventive (PnotR) forms, viz. **m**, **n**, **p** or **q** rel to S, **p** or **q** rel to notS, **p** or **q** abs are all denied. The remaining causative (PR) forms, **n**, **q** rel to S, **q** rel to notS, **q** abs, are neither implied nor denied by **m** (PR).

- Similarly for **n** (PR), mutatis mutandis.
- Finding the causative (PR) conclusion **p** rel to S, then the causative **p** abs is implied, and the causatives **m**, **p** rel to notS are denied, as are the preventive (PnotR) forms **m**, **p** rel to S, **q** rel to notS. The remaining causative (PR) forms, **n**, **q** rel to S, **q** rel to notS, **q** abs, and likewise the remaining preventive (PnotR) forms, **n**, **q** rel to S, **p** rel to notS, **p** abs, **q** abs, are neither implied nor denied by **p** rel to S (PR).
- Similarly for **q** rel to S (PR), mutatis mutandis.
- Also note: although separately there are no implications between the causative and preventive forms of **p** abs and **q** abs, the compounds **pq** abs causative and **pq** abs preventive are identical.
- On the other hand, nothing can be inferred from denying **m**, **n**, **p** or **q** – except that denying **p** or **q** abs implies denial of, respectively, **p** or **q** rel to S or notS forms with the same items (i.e. PR or PnotR).

This knowledge, to repeat, helps us distinguish primary from subaltern conclusions, and so makes possible reducing statements of conclusion to as few words as possible.

3. 3-Item Syllogisms.

Scanning for conclusions was first applied to 3-item syllogism with positive causative premises (64 moods for each of the three figures, the moods being grouped in sets of eight, viz. #s 11-18, 22-28, etc. till 81-88). The collected results are given in the following summary table, which is posted online in my website as usual (with an extract printed further down here). This is the table requiring most of your attention:

- **Table 22.6-0 – 3-Item Causative Syllogisms – Summary of Conclusions Scanned. (9 pages in pdf file.)**

The main conclusions implied by the premises concerned are shown in pink; those denied by them are shown in grey; conclusions subaltern to these main positive or negative conclusions are identified as implied or denied but not colored pink or grey. Cells which are not labeled 'implied' or 'denied' signify that the form concerned is not part of the overall conclusion of that mood (obviously, if all the cells opposite a given mood are empty, it means there's no causative or preventive conclusion). That said for the first segment (3 pages) of the summary. In the second segment (next 3 pages), the results are repeated, all subaltern conclusions, if any, being eliminated, and color coding being removed; and the symbol for the valid positive or negative conclusions is written in place of the labels 'implied' and 'denied', respectively. The third segment of the table (last 3 pages) again repeats the results in the briefest possible way, i.e. stating only the essentials: these are collectively 'the conclusion' for the given mood ('nil' being used to signal no-conclusion).

The said summary table (22.6-0) is derived from the three tables listed next, where the actual scanning work is done. These tables can be viewed online at my website as usual:

- Table 22.6-1 – 3-Item Causative Syllogisms – Figure One – Scanning for Conclusions. (48 pages in pdf file.)
- Table 22.6-2 – 3-Item Causative Syllogisms – Figure Two – Scanning for Conclusions. (48 pages in pdf file.)

- Table 22.6-3 – 3-Item Causative Syllogisms – Figure Three – Scanning for Conclusions. (48 pages in pdf file.)

These three tables were produced in succession, using the first as the template or model for the next two, changing only the premises as appropriate (see Table 22.1) and then recalculating the whole spreadsheet. As you can see, the moduses are found for each mood of the syllogism (e.g. the positive causative premises **mn/mn**), and then for those same premises combined with eight different test conclusions (the positive causative (i.e. PR) forms **m**, **n**, **p** abs, **q** abs, and the positive preventive (i.e. PnotR) forms **m**, **n**, **p** abs, **q** abs, as earlier explained). The calculations in the spreadsheets are done by means of formulas, as usual. A sample set of formulas (those used in Table 22.6-1, i.e. for Figure 1) is given in the next table (which can be viewed online at my website as usual):

- Table 22.6-4 – 3-Item Causative Syllogisms – Scanning Template (Sample of Formulas Used). (9 pages in pdf file.)

Looking at the results obtained (return to Table 22.6-0 or see the extract from it below), and comparing them to past results (which are listed in the last segment of Table 22.6-0 for this purpose), we see – as earlier mentioned – that the latter are essentially confirmed. That is, as regards causative conclusions; but now we have some additional negative preventive conclusions we were not aware of before.

Extract from Table 22.6-0 – Scanning Results for 3-Item Causative Moods.

Mood #	Causative Premises	Figure One		Figure Two		Figure Three	
		Caus. (PR)	Prev. (PnotR)	Caus. (PR)	Prev. (PnotR)	Caus. (PR)	Prev. (PnotR)
11	**mn/mn**	**mn**		**mn**		**mn**	
12	**mn/mq abs**	**mq abs**		**mq abs**		**np abs**	
13	**mn/np abs**	**np abs**		**np abs**		**mq abs**	
14	**mn/pq abs**	**pq abs**		**pq abs**		**pq abs**	
15	**mn/m**	**m**		**m**		**n**	
16	**mn/n**	**n**		**n**		**m**	
17	**mn/p abs**	**p abs**		**p abs**		**q abs**	
18	**mn/q abs**	**q abs**		**q abs**		**p abs**	
21	**mq/mn**	**mq abs**		**np abs**		**mq abs**	
22	**mq/mq abs**	**mq abs**			**not-n**		**not-m**
23	**mq/np abs**		**not-m**	**np abs**		**mq abs**	
24	**mq/pq abs**	**not-n**	**not-m**	**not-m**	**not-n**	**not-n**	**not-m**
25	**mq/m**	**mq abs**			**not-n**		**not-m**
26	**mq/n**		**not-m**	**np abs**		**mq abs**	
27	**mq/p abs**		**not-m**	**not-m**	**not-n**	**not-n**	**not-m**
28	**mq/q abs**	**not-n**	**not-m**		**not-n**		**not-m**

Extract from Table 22.6-0 (continued).

31	np/mn	np abs		mq abs		np abs	
32	np/mq abs		not-n	mq abs		np abs	
33	np/np abs	np abs			not-m		not-n
34	np/pq abs	not-m	not-n	not-n	not-m	not-m	not-n
35	np/m		not-n	mq abs		np abs	
36	np/n	np abs			not-m		not-n
37	np/p abs	not-m	not-n		not-m		not-n
38	np/q abs		not-n	not-n	not-m	not-m	not-n
41	pq/mn	pq abs		pq abs		pq abs	
42	pq/mq abs	not-n	not-n	not-n	not-n	not-m	not-m
43	pq/np abs	not-m	not-m	not-m	not-m	not-n	not-n
44	pq/pq abs	*nil*		*nil*		*nil*	
45	pq/m	not-n	not-n	not-n	not-n	not-m	not-m
46	pq/n	not-m	not-m	not-m	not-m	not-n	not-n
47	pq/p abs	*nil*		*nil*		*nil*	
48	pq/q abs	*nil*		*nil*		*nil*	
51	m/mn	m		n		m	
52	m/mq abs	mq abs			not-n		not-m
53	m/np abs		not-m	np abs		mq abs	
54	m/pq abs	not-n	not-m	not-m	not-n	not-n	not-m
55	m/m	m			not-n		not-m
56	m/n		not-m	n		m	
57	m/p abs		not-m	not-m	not-n	not-n	not-m
58	m/q abs	not-n	not-m		not-n		not-m
61	n/mn	n		m		n	
62	n/mq abs		not-n	mq abs		np abs	
63	n/np abs	np abs			not-m		not-n
64	n/pq abs	not-m	not-n	not-n	not-m	not-m	not-n
65	n/m		not-n	m		n	
66	n/n	n			not-m		not-n
67	n/p abs	not-m	not-n		not-m		not-n
68	n/q abs		not-n	not-n	not-m	not-m	not-n
71	p/mn	p abs		q abs		p abs	
72	p/mq abs		not-n	not-n	not-n	not-m	not-m
73	p/np abs	not-m	not-m		not-m		not-n
74	p/pq abs	*nil*		*nil*		*nil*	
75	p/m		not-n	not-n	not-n	not-m	not-m
76	p/n	not-m	not-m		not-m		not-n
77	p/p abs	*nil*		*nil*		*nil*	
78	p/q abs	*nil*		*nil*		*nil*	

Extract from Table 22.6-0 (continued).

81	**q/mn**	**q abs**		**p abs**		**q abs**	
82	**q/mq abs**	**not-n**	**not-n**		**not-n**		**not-m**
83	**q/np abs**		**not-m**	**not-m**	**not-m**	**not-n**	**not-n**
84	**q/pq abs**	*nil*		*nil*		*nil*	
85	**q/m**	**not-n**	**not-n**		**not-n**		**not-m**
86	**q/n**		**not-m**	**not-m**	**not-m**	**not-n**	**not-n**
87	**q/p abs**	*nil*		*nil*		*nil*	
88	**q/q abs**	*nil*		*nil*		*nil*	

The following **statistics** are now applicable to each of the three figures: 23 moods (36%) yield a positive conclusion (always causative, some elementary, some compound); 16 moods (25%) yield two negative (**not-m** and/or **not-n**) conclusions only, one causative and one preventive (previously, the latter were unknown); 16 moods (25%) yield only a negative (**not-m** or **not-n**) preventive conclusion (previously all supposed nil); and 9 moods (14%) are without any causative or preventive conclusion (previously, 25 were thought inconclusive). The net validity rate for 3-item syllogism is thus 86%.

Clearly, we now have a wider range of possible conclusions and we can say with full confidence not only which forms are affirmed, but also which are denied and which are neither affirmed nor denied. We thus obtain *a truly definitive list* of valid and invalid conclusions. It is as if we previously used a microscope to observe a certain phenomenon, and now we have found a more powerful microscope capable of giving us a more accurate image of the phenomenon.

4. 4-Item Syllogisms.

Scanning for conclusions was next applied to 4-item syllogism with positive causative premises (64 moods per figure, as before). However, here the job is exponentially bigger, since we are dealing with 65,536 moduses per column (instead of a mere 256). This makes all opening, saving and closing of spreadsheet files very slow; and it especially slows down calculations, forcing us to split files into more manageable portions and perform calculations in smaller segments (otherwise, if too much is asked of the program, execution might even be blocked).

Moreover, although the grouping of moods into sets is maintained, in some cases (specifically, 5 out of 8 per figure) sets of 8 moods do not suffice, but must be expanded to sets of 13 moods. This occurs when moods have a weak causation (whether an element or part of a compound) in both premises; in such cases, as we saw earlier (see Table 21.10), a single mood becomes two (labeled 'b' and 'c', combining an absolute premise with a relative one, and a relative with an absolute, respectively). This is due, remember, to our dealing here with only four items, and not five (which my computer hardware and software cannot handle). In sum, here we are dealing with 89 moods per figure (instead of 64 before).

Furthermore, the workload is increased due to enlargement of the range of conclusions tested for each mood. For 3-item syllogism, we had 8 forms to test every time; here, in 4-item syllogism, we have 16 conclusions to consider, namely: the 8 causative (PR) forms **m**, **n**, **p** rel to S, **q** rel to S, **p** rel to notS, **q** rel to notS, **p** abs, **q** abs, and the similar 8 preventive (PnotR) forms. This ensures that every possible inference of causative and/or preventive form is found by us, and our results are truly exhaustive and definitive.

The collected results of such scanning work are given in the following summary table, which you can view in full online at my website as usual. I give an extract of it here (below), even though it is very long, because of its importance to our whole study. This the table requiring most of your attention in the present context. Look at the premises and the conclusions carefully in each row, and absorb the meaning of it all.

- **Table 22.7-0 – 4-Item Causative Syllogisms – Summary of Conclusions Scanned. (18 pages in pdf file.)**

Extract from Table 22.7-0 – Scanning Results for 4-Item Causative Moods.

Mood #	Causative Premises	Causative (PR) conclusions	Preventive (PnotR) conclusions
Figure One			
111	mn/mn	mn	
112	mn/mq (rel to S)	mq rel to S	
113	mn/np (rel to S)	np rel to S	
114	mn/pq (rel to S)	pq rel to S	
115	mn/m	m	
116	mn/n	n	
117	mn/p (rel to S)	p rel to S	
118	mn/q (rel to S)	q rel to S	
121	mq/mn (rel to S)	mq rel to S	
122 (b)	mq/mq (abs / rel S)	mq abs, not-q rel to notS	
122 (c)	mq/mq (rel S / abs)	mq abs, not-q rel to notS	
123 (b)	mq/np (abs / rel S)	not-p rel to notS	not-m, not-p rel to S, not-q rel to notS
123 (c)	mq/np (rel S / abs)	not-q rel to notS	not-m, not-q rel to S, not-p rel to notS
124 (b)	mq/pq (abs / rel S)	not-n, not-p rel to notS, not-q rel to notS	not-m, not-p rel to S, not-p rel to notS, not-q rel to notS
124 (c)	mq/pq (rel S / abs)	not-n	not-m

Extract from Table 22.7-0 (continued).

Mood #	Causative Premises	Causative (PR) conclusions	Preventive (PnotR) conclusions
125	**mq/m (rel to S)**	**mq abs, not-q rel to notS**	
126	**mq/n (rel to S)**	**not-q rel to notS**	**not-m, not-q rel to S, not-p rel to notS**
127 (b)	**mq/p (abs / rel S)**	**not-p rel to notS**	**not-m, not-p rel to S, not-q rel to notS**
127 (c)	**mq/p (rel S / abs)**		**not-m**
128 (b)	**mq/q (abs / rel S)**	**not-n, not-q rel to notS**	**not-m, not-p rel to notS**
128 (c)	**mq/q (rel S / abs)**	**not-n**	**not-m**
131	**np/mn (rel to S)**	**np rel to S**	
132 (b)	**np/mq (abs / rel S)**	**not-q rel to notS**	**not-n, not-q rel to S, not-p rel to notS**
132 (c)	**np/mq (rel S / abs)**	**not-p rel to notS**	**not-n, not-p rel to S, not-q rel to notS**
133 (b)	**np/np (abs / rel S)**	**np abs, not-p rel to notS**	
133 (c)	**np/np (rel S / abs)**	**np abs, not-p rel to notS**	
134 (b)	**np/pq (abs / rel S)**	**not-m, not-p rel to notS, not-q rel to notS**	**not-n, not-q rel to S, not-p rel to notS, not-q rel to notS**
134 (c)	**np/pq (rel S / abs)**	**not-m**	**not-n**
135	**np/m (rel to S)**	**not-p rel to notS**	**not-n, not-p rel to S, not-q rel to notS**
136	**np/n (rel to S)**	**np abs, not-p rel to notS**	
137 (b)	**np/p (abs / rel S)**	**not-m, not-p rel to notS**	**not-n, not-q rel to notS**
137 (c)	**np/p (rel S / abs)**	**not-m**	**not-n**
138 (b)	**np/q (abs / rel S)**	**not-q rel to notS**	**not-n, not-q rel to S, not-p rel to notS**
138 (c)	**np/q (rel S / abs)**		**not-n**

Extract from Table 22.7-0 (continued).

Mood #	Causative Premises	Causative (PR) conclusions	Preventive (PnotR) conclusions
141	**pq/mn (rel to S)**	**pq rel to S**	
142 (b)	**pq/mq (abs / rel S)**	**not-n**	**not-n**
142 (c)	**pq/mq (rel S / abs)**	**not-n, not-p rel to notS, not-q rel to notS**	**not-n, not-p rel to S, not-q rel to S, not-q rel to notS**
143 (b)	**pq/np (abs / rel S)**	**not-m**	**not-m**
143 (c)	**pq/np (rel S / abs)**	**not-m, not-p rel to notS, not-q rel to notS**	**not-m, not-p rel to S, not-q rel to S, not-p rel to notS**
144 (b)	**pq/pq (abs / rel S)**		
144 (c)	**pq/pq (rel S / abs)**		
145	**pq/m (rel to S)**	**not-n, not-p rel to notS, not-q rel to notS**	**not-n, not-p rel to S, not-q rel to S, not-q rel to notS**
146	**pq/n (rel to S)**	**not-m, not-p rel to notS, not-q rel to notS**	**not-m, not-p rel to S, not-q rel to S, not-p rel to notS**
147 (b)	**pq/p (abs / rel S)**		
147 (c)	**pq/p (rel S / abs)**		
148 (b)	**pq/q (abs / rel S)**		
148 (c)	**pq/q (rel S / abs)**		
151	**m/mn**	**m**	
152	**m/mq (rel to S)**	**mq abs, not-q rel to notS**	
153	**m/np (rel to S)**	**not-p rel to notS**	**not-m, not-p rel to S, not-q rel to notS**
154	**m/pq (rel to S)**	**not-n, not-p rel to notS, not-q rel to notS**	**not-m, not-p rel to S, not-p rel to notS, not-q rel to notS**
155	**m/m**	**m**	
156	**m/n**		**not-m**
157	**m/p (rel to S)**	**not-p rel to notS**	**not-m, not-p rel to S, not-q rel to notS**
158	**m/q (rel to S)**	**not-n, not-q rel to notS**	**not-m, not-p rel to notS**

Extract from Table 22.7-0 (continued).

Mood #	Causative Premises	Causative (PR) conclusions	Preventive (PnotR) conclusions
161	n/mn	n	
162	n/mq (rel to S)	not-q rel to notS	not-n, not-q rel to S, not-p rel to notS
163	n/np (rel to S)	np abs, not-p rel to notS	
164	n/pq (rel to S)	not-m, not-p rel to notS, not-q rel to notS	not-n, not-q rel to S, not-p rel to notS, not-q rel to notS
165	n/m		not-n
166	n/n	n	
167	n/p (rel to S)	not-m, not-p rel to notS	not-n, not-q rel to notS
168	n/q (rel to S)	not-q rel to notS	not-n, not-q rel to S, not-p rel to notS
171	p/mn (rel to S)	p rel to S	
172 (b)	p/mq (abs / rel S)		not-n
172 (c)	p/mq (rel S / abs)	not-p rel to notS	not-n, not-p rel to S, not-q rel to notS
173 (b)	p/np (abs / rel S)	not-m	not-m
173 (c)	p/np (rel S / abs)	not-m, not-p rel to notS	not-m, not-p rel to S
174 (b)	p/pq (abs / rel S)		
174 (c)	p/pq (rel S / abs)		
175	p/m (rel to S)	not-p rel to notS	not-n, not-p rel to S, not-q rel to notS
176	p/n (rel to S)	not-m, not-p rel to notS	not-m, not-p rel to S
177 (b)	p/p (abs / rel S)		
177 (c)	p/p (rel S / abs)		
178 (b)	p/q (abs / rel S)		
178 (c)	p/q (rel S / abs)		

Extract from Table 22.7-0 (continued).

Mood #	Causative Premises	Causative (PR) conclusions	Preventive (PnotR) conclusions
181	q/mn (rel to S)	q rel to S	
182 (b)	q/mq (abs / rel S)	not-n	not-n
182 (c)	q/mq (rel S / abs)	not-n, not-q rel to notS	not-n, not-q rel to S
183 (b)	q/np (abs / rel S)		not-m
183 (c)	q/np (rel S / abs)	not-q rel to notS	not-m, not-q rel to S, not-p rel to notS
184 (b)	q/pq (abs / rel S)		
184 (c)	q/pq (rel S / abs)		
185	q/m (rel to S)	not-n, not-q rel to notS	not-n, not-q rel to S
186	q/n (rel to S)	not-q rel to notS	not-m, not-q rel to S, not-p rel to notS
187 (b)	q/p (abs / rel S)		
187 (c)	q/p (rel S / abs)		
188 (b)	q/q (abs / rel S)		
188 (c)	q/q (rel S / abs)		

Figure Two

211	mn/mn	mn	
212	mn/mq (rel to S)	mq rel to S	
213	mn/np (rel to S)	np rel to S	
214	mn/pq (rel to S)	pq rel to S	
215	mn/m	m	
216	mn/n	n	
217	mn/p (rel to S)	p rel to S	
218	mn/q (rel to S)	q rel to S	

Extract from Table 22.7-0 (continued).

Mood #	Causative Premises	Causative (PR) conclusions	Preventive (PnotR) conclusions
221	mq/mn (rel to S)	np rel to notS	
222 (b)	mq/mq (abs / rel S)	not-q rel to notS	not-n, not-q rel to S, not-p rel to notS
222 (c)	mq/mq (rel S / abs)	not-p rel to S	not-n, not-q rel to S, not-p rel to notS
223 (b)	mq/np (abs / rel S)	np abs, not-p rel to notS	
223 (c)	mq/np (rel S / abs)	np abs, not-p rel to S	
224 (b)	mq/pq (abs / rel S)	not-m, not-p rel to notS, not-q rel to notS	not-n, not-q rel to S, not-p rel to notS, not-q rel to notS
224 (c)	mq/pq (rel S / abs)	not-m	not-n
225	mq/m (rel to S)	not-p rel to S	not-n, not-q rel to S, not-p rel to notS
226	mq/n (rel to S)	np abs, not-p rel to S	
227 (b)	mq/p (abs / rel S)	not-m, not-p rel to notS	not-n, not-q rel to notS
227 (c)	mq/p (rel S / abs)	not-m	not-n
228 (b)	mq/q (abs / rel S)	not-q rel to notS	not-n, not-q rel to S, not-p rel to notS
228 (c)	mq/q (rel S / abs)		not-n
231	np/mn (rel to S)	mq rel to notS	
232 (b)	np/mq (abs / rel S)	mq abs, not-q rel to notS	
232 (c)	np/mq (rel S / abs)	mq abs, not-q rel to S	
233 (b)	np/np (abs / rel S)	not-p rel to notS	not-m, not-p rel to S, not-q rel to notS
233 (c)	np/np (rel S / abs)	not-q rel to S	not-m, not-p rel to S, not-q rel to notS
234 (b)	np/pq (abs / rel S)	not-n, not-p rel to notS, not-q rel to notS	not-m, not-p rel to S, not-p rel to notS, not-q rel to notS
234 (c)	np/pq (rel S / abs)	not-n	not-m
235	np/m (rel to S)	mq abs, not-q rel to S	

Extract from Table 22.7-0 (continued).

Mood #	Causative Premises	Causative (PR) conclusions	Preventive (PnotR) conclusions
236	**np/n (rel to S)**	**not-q rel to S**	**not-m, not-p rel to S, not-q rel to notS**
237 (b)	**np/p (abs / rel S)**	**not-p rel to notS**	**not-m, not-p rel to S, not-q rel to notS**
237 (c)	**np/p (rel S / abs)**		**not-m**
238 (b)	**np/q (abs / rel S)**	**not-n, not-q rel to notS**	**not-m, not-p rel to notS**
238 (c)	**np/q (rel S / abs)**	**not-n**	**not-m**
241	**pq/mn (rel to S)**	**pq rel to notS**	
242 (b)	**pq/mq (abs / rel S)**	**not-n**	**not-n**
242 (c)	**pq/mq (rel S / abs)**	**not-n, not-p rel to S, not-q rel to S**	**not-n, not-q rel to S, not-p rel to notS, not-q rel to notS**
243 (b)	**pq/np (abs / rel S)**	**not-m**	**not-m**
243 (c)	**pq/np (rel S / abs)**	**not-m, not-p rel to S, not-q rel to S**	**not-m, not-p rel to S, not-p rel to notS, not-q rel to notS**
244 (b)	**pq/pq (abs / rel S)**		
244 (c)	**pq/pq (rel S / abs)**		
245	**pq/m (rel to S)**	**not-n, not-p rel to S, not-q rel to S**	**not-n, not-q rel to S, not-p rel to notS, not-q rel to notS**
246	**pq/n (rel to S)**	**not-m, not-p rel to S, not-q rel to S**	**not-m, not-p rel to S, not-p rel to notS, not-q rel to notS**
247 (b)	**pq/p (abs / rel S)**		
247 (c)	**pq/p (rel S / abs)**		
248 (b)	**pq/q (abs / rel S)**		
248 (c)	**pq/q (rel S / abs)**		
251	**m/mn**	**n**	
252	**m/mq (rel to S)**	**not-q rel to notS**	**not-n, not-q rel to S, not-p rel to notS**
253	**m/np (rel to S)**	**np abs, not-p rel to notS**	

Extract from Table 22.7-0 (continued).

Mood #	Causative Premises	Causative (PR) conclusions	Preventive (PnotR) conclusions
254	m/pq (rel to S)	not-m, not-p rel to notS, not-q rel to notS	not-n, not-q rel to S, not-p rel to notS, not-q rel to notS
255	m/m		not-n
256	m/n	n	
257	m/p (rel to S)	not-m, not-p rel to notS	not-n, not-q rel to notS
258	m/q (rel to S)	not-q rel to notS	not-n, not-q rel to S, not-p rel to notS
261	n/mn	m	
262	n/mq (rel to S)	mq abs, not-q rel to notS	
263	n/np (rel to S)	not-p rel to notS	not-m, not-p rel to S, not-q rel to notS
264	n/pq (rel to S)	not-n, not-p rel to notS, not-q rel to notS	not-m, not-p rel to S, not-p rel to notS, not-q rel to notS
265	n/m	m	
266	n/n		not-m
267	n/p (rel to S)	not-p rel to notS	not-m, not-p rel to S, not-q rel to notS
268	n/q (rel to S)	not-n, not-q rel to notS	not-m, not-p rel to notS
271	p/mn (rel to S)	q rel to notS	
272 (b)	p/mq (abs / rel S)	not-n	not-n
272 (c)	p/mq (rel S / abs)	not-n, not-q rel to S	not-n, not-q rel to notS
273 (b)	p/np (abs / rel S)		not-m
273 (c)	p/np (rel S / abs)	not-q rel to S	not-m, not-p rel to S, not-q rel to notS
274 (b)	p/pq (abs / rel S)		
274 (c)	p/pq (rel S / abs)		
275	p/m (rel to S)	not-n, not-q rel to S	not-n, not-q rel to notS
276	p/n (rel to S)	not-q rel to S	not-m, not-p rel to S, not-q rel to notS
277 (b)	p/p (abs / rel S)		

Extract from Table 22.7-0 (continued).

Mood #	Causative Premises	Causative (PR) conclusions	Preventive (PnotR) conclusions
277 (c)	p/p (rel S / abs)		
278 (b)	p/q (abs / rel S)		
278 (c)	p/q (rel S / abs)		
281	q/mn (rel to S)	p rel to notS	
282 (b)	q/mq (abs / rel S)		not-n
282 (c)	q/mq (rel S / abs)	not-p rel to S	not-n, not-q rel to S, not-p rel to notS
283 (b)	q/np (abs / rel S)	not-m	not-m
283 (c)	q/np (rel S / abs)	not-m, not-p rel to S	not-m, not-p rel to notS
284 (b)	q/pq (abs / rel S)		
284 (c)	q/pq (rel S / abs)		
285	q/m (rel to S)	not-p rel to S	not-n, not-q rel to S, not-p rel to notS
286	q/n (rel to S)	not-m, not-p rel to S	not-m, not-p rel to notS
287 (b)	q/p (abs / rel S)		
287 (c)	q/p (rel S / abs)		
288 (b)	q/q (abs / rel S)		
288 (c)	q/q (rel S / abs)		

Figure Three

311	mn/mn	mn	
312	mn/mq (rel to S)	np rel to notS	
313	mn/np (rel to S)	mq rel to notS	

Extract from Table 22.7-0 (continued).

Mood #	Causative Premises	Causative (PR) conclusions	Preventive (PnotR) conclusions
314	mn/pq (rel to S)	pq rel to notS	
315	mn/m	n	
316	mn/n	m	
317	mn/p (rel to S)	q rel to notS	
318	mn/q (rel to S)	p rel to notS	
321	mq/mn (rel to S)	mq rel to S	
322 (b)	mq/mq (abs / rel S)	not-p rel to S	not-m, not-q rel to S, not-p rel to notS
322 (c)	mq/mq (rel S / abs)	not-q rel to notS	not-m, not-q rel to S, not-p rel to notS
323 (b)	mq/np (abs / rel S)	mq abs, not-q rel to S	
323 (c)	mq/np (rel S / abs)	mq abs, not-q rel to notS	
324 (b)	mq/pq (abs / rel S)	not-n, not-p rel to S, not-q rel to S	not-m, not-p rel to S, not-q rel to S, not-p rel to notS
324 (c)	mq/pq (rel S / abs)	not-n	not-m
325	mq/m (rel to S)	not-q rel to notS	not-m, not-q rel to S, not-p rel to notS
326	mq/n (rel to S)	mq abs, not-q rel to notS	
327 (b)	mq/p (abs / rel S)	not-n, not-q rel to S	not-m, not-p rel to S
327 (c)	mq/p (rel S / abs)	not-n	not-m
328 (b)	mq/q (abs / rel S)	not-p rel to S	not-m, not-q rel to S, not-p rel to notS
328 (c)	mq/q (rel S / abs)		not-m
331	np/mn (rel to S)	np rel to S	
332 (b)	np/mq (abs / rel S)	np abs, not-p rel to S	
332 (c)	np/mq (rel S / abs)	np abs, not-p rel to notS	
333 (b)	np/np (abs / rel S)	not-q rel to S	not-n, not-p rel to S, not-q rel to notS

Extract from Table 22.7-0 (continued).

Mood #	Causative Premises	Causative (PR) conclusions	Preventive (PnotR) conclusions
333 (c)	np/np (rel S / abs)	not-p rel to notS	not-n, not-p rel to S, not-q rel to notS
334 (b)	np/pq (abs / rel S)	not-m, not-p rel to S, not-q rel to S	not-n, not-p rel to S, not-q rel to S, not-q rel to notS
334 (c)	np/pq (rel S / abs)	not-m	not-n
335	np/m (rel to S)	np abs, not-p rel to notS	
336	np/n (rel to S)	not-p rel to notS	not-n, not-p rel to S, not-q rel to notS
337 (b)	np/p (abs / rel S)	not-q rel to S	not-n, not-p rel to S, not-q rel to notS
337 (c)	np/p (rel S / abs)		not-n
338 (b)	np/q (abs / rel S)	not-m, not-p rel to S	not-n, not-q rel to S
338 (c)	np/q (rel S / abs)	not-m	not-n
341	pq/mn (rel to S)	pq rel to S	
342 (b)	pq/mq (abs / rel S)	not-m	not-m
342 (c)	pq/mq (rel S / abs)	not-m, not-p rel to notS, not-q rel to notS	not-m, not-p rel to S, not-q rel to S, not-p rel to notS
343 (b)	pq/np (abs / rel S)	not-n	not-n
343 (c)	pq/np (rel S / abs)	not-n, not-p rel to notS, not-q rel to notS	not-n, not-p rel to S, not-q rel to S, not-q rel to notS
344 (b)	pq/pq (abs / rel S)		
344 (c)	pq/pq (rel S / abs)		
345	pq/m (rel to S)	not-m, not-p rel to notS, not-q rel to notS	not-m, not-p rel to S, not-q rel to S, not-p rel to notS
346	pq/n (rel to S)	not-n, not-p rel to notS, not-q rel to notS	not-n, not-p rel to S, not-q rel to S, not-q rel to notS
347 (b)	pq/p (abs / rel S)		
347 (c)	pq/p (rel S / abs)		
348 (b)	pq/q (abs / rel S)		
348 (c)	pq/q (rel S / abs)		

Extract from Table 22.7-0 (continued).

Mood #	Causative Premises	Causative (PR) conclusions	Preventive (PnotR) conclusions
351	m/mn	m	
352	m/mq (rel to S)	not-p rel to S	not-m, not-q rel to S, not-p rel to notS
353	m/np (rel to S)	mq abs, not-q rel to S	
354	m/pq (rel to S)	not-n, not-p rel to S, not-q rel to S	not-m, not-p rel to S, not-q rel to S, not-p rel to notS
355	m/m		not-m
356	m/n	m	
357	m/p (rel to S)	not-n, not-q rel to S	not-m, not-p rel to S
358	m/q (rel to S)	not-p rel to S	not-m, not-q rel to S, not-p rel to notS
361	n/mn	n	
362	n/mq (rel to S)	np abs, not-p rel to S	
363	n/np (rel to S)	not-q rel to S	not-n, not-p rel to S, not-q rel to notS
364	n/pq (rel to S)	not-m, not-p rel to S, not-q rel to S	not-n, not-p rel to S, not-q rel to S, not-q rel to notS
365	n/m	n	
366	n/n		not-n
367	n/p (rel to S)	not-q rel to S	not-n, not-p rel to S, not-q rel to notS
368	n/q (rel to S)	not-m, not-p rel to S	not-n, not-q rel to S
371	p/mn (rel to S)	p rel to S	
372 (b)	p/mq (abs / rel S)	not-m	not-m
372 (c)	p/mq (rel S / abs)	not-m, not-p rel to notS	not-m, not-p rel to S
373 (b)	p/np (abs / rel S)		not-n
373 (c)	p/np (rel S / abs)	not-p rel to notS	not-n, not-p rel to S, not-q rel to notS
374 (b)	p/pq (abs / rel S)		
374 (c)	p/pq (rel S / abs)		

Extract from Table 22.7-0 (continued).

Mood #	Causative Premises	Causative (PR) conclusions	Preventive (PnotR) conclusions
375	p/m (rel to S)	not-m, not-p rel to notS	not-m, not-p rel to S
376	p/n (rel to S)	not-p rel to notS	not-n, not-p rel to S, not-q rel to notS
377 (b)	p/p (abs / rel S)		
377 (c)	p/p (rel S / abs)		
378 (b)	p/q (abs / rel S)		
378 (c)	p/q (rel S / abs)		
381	q/mn (rel to S)	q rel to S	
382 (b)	q/mq (abs / rel S)		not-m
382 (c)	q/mq (rel S / abs)	not-q rel to notS	not-m, not-q rel to S, not-p rel to notS
383 (b)	q/np (abs / rel S)	not-n	not-n
383 (c)	q/np (rel S / abs)	not-n, not-q rel to notS	not-n, not-q rel to S
384 (b)	q/pq (abs / rel S)		
384 (c)	q/pq (rel S / abs)		
385	q/m (rel to S)	not-q rel to notS	not-m, not-q rel to S, not-p rel to notS
386	q/n (rel to S)	not-n, not-q rel to notS	not-n, not-q rel to S
387 (b)	q/p (abs / rel S)		
387 (c)	q/p (rel S / abs)		
388 (b)	q/q (abs / rel S)		
388 (c)	q/q (rel S / abs)		

We shall return to these results in a moment. For now note that this table summarizes the findings of the following 48 tables (16 per figure), which are posted online in my website as usual. Each table concerns a specific set of 8 or 13 moods. For each set, table A tests 8 causative conclusions and table B tests 8 preventive conclusions. Each pdf file produced is over 1000 pages long (making almost 24'000 pages per figure, compared to a mere 48 pages before); seeing this, you can understand why it was necessary to split the job up into smaller pieces, and why I long hesitated to do it. Nevertheless, though much time was spent doing it and much attention was required, the job was not so hard, because once the two templates (the first couple of tables) were produced, it was easy to reproduce them with appropriately modified data and recalculate them.

Figure 1 - 16 tables (8 for causative conclusions and 8 for preventive ones).

- Table 22.7-11A – 4-Item Causative Syllogisms – Scanning Causative Conclusions to Moods 111-118. (1032 pages in pdf file.)
- Table 22.7-11B – 4-Item Causative Syllogisms – Scanning Preventive Conclusions to Moods 111-118. (1041 pages in pdf file.)
- Table 22.7-12A – 4-Item Causative Syllogisms – Scanning Causative Conclusions to Moods 121-128. (1720 pages in pdf file.)
- Table 22.7-12B – 4-Item Causative Syllogisms – Scanning Preventive Conclusions to Moods 121-128. (1735 pages in pdf file.)
- Table 22.7-13A – 4-Item Causative Syllogisms – Scanning Causative Conclusions to Moods 131-138. (1720 pages in pdf file.)
- Table 22.7-13B – 4-Item Causative Syllogisms – Scanning Preventive Conclusions to Moods 131-138. (1735 pages in pdf file.)
- Table 22.7-14A – 4-Item Causative Syllogisms – Scanning Causative Conclusions to Moods 141-148. (1720 pages in pdf file.)
- Table 22.7-14B – 4-Item Causative Syllogisms – Scanning Preventive Conclusions to Moods 141-148. (1735 pages in pdf file.)
- Table 22.7-15A – 4-Item Causative Syllogisms – Scanning Causative Conclusions to Moods 151-158. (1032 pages in pdf file.)
- Table 22.7-15B – 4-Item Causative Syllogisms – Scanning Preventive Conclusions to Moods 151-158. (1041 pages in pdf file.)
- Table 22.7-16A – 4-Item Causative Syllogisms – Scanning Causative Conclusions to Moods 161-168. (1032 pages in pdf file.)
- Table 22.7-16B – 4-Item Causative Syllogisms – Scanning Preventive Conclusions to Moods 161-168. (1041 pages in pdf file.)
- Table 22.7-17A – 4-Item Causative Syllogisms – Scanning Causative Conclusions to Moods 171-178. (1720 pages in pdf file.)
- Table 22.7-17B – 4-Item Causative Syllogisms – Scanning Preventive Conclusions to Moods 171-178. (1735 pages in pdf file.)
- Table 22.7-18A – 4-Item Causative Syllogisms – Scanning Causative Conclusions to Moods 181-188. (1720 pages in pdf file.)
- Table 22.7-18B – 4-Item Causative Syllogisms – Scanning Preventive Conclusions to Moods 181-188. (1735 pages in pdf file.)

Figure 2 - 16 tables (8 for causative conclusions and 8 for preventive ones).

- Table 22.7-21A – 4-Item Causative Syllogisms – Scanning Causative Conclusions to Moods 211-218. (1032 pages in pdf file.)
- Table 22.7-21B – 4-Item Causative Syllogisms – Scanning Preventive Conclusions to Moods 211-218. (1041 pages in pdf file.)
- Table 22.7-22A – 4-Item Causative Syllogisms – Scanning Causative Conclusions to Moods 221-228. (1720 pages in pdf file.)
- Table 22.7-22B – 4-Item Causative Syllogisms – Scanning Preventive Conclusions to Moods 221-228. (1735 pages in pdf file.)
- Table 22.7-23A – 4-Item Causative Syllogisms – Scanning Causative Conclusions to Moods 231-238. (1720 pages in pdf file.)
- Table 22.7-23B – 4-Item Causative Syllogisms – Scanning Preventive Conclusions to Moods 231-238. (1735 pages in pdf file.)
- Table 22.7-24A – 4-Item Causative Syllogisms – Scanning Causative Conclusions to Moods 241-248. (1720 pages in pdf file.)
- Table 22.7-24B – 4-Item Causative Syllogisms – Scanning Preventive Conclusions to Moods 241-248. (1735 pages in pdf file.)
- Table 22.7-25A – 4-Item Causative Syllogisms – Scanning Causative Conclusions to Moods 251-258. (1032 pages in pdf file.)
- Table 22.7-25B – 4-Item Causative Syllogisms – Scanning Preventive Conclusions to Moods 251-258. (1041 pages in pdf file.)
- Table 22.7-26A – 4-Item Causative Syllogisms – Scanning Causative Conclusions to Moods 261-268. (1032 pages in pdf file.)
- Table 22.7-26B – 4-Item Causative Syllogisms – Scanning Preventive Conclusions to Moods 261-268. (1041 pages in pdf file.)
- Table 22.7-27A – 4-Item Causative Syllogisms – Scanning Causative Conclusions to Moods 271-278. (1720 pages in pdf file.)
- Table 22.7-27B – 4-Item Causative Syllogisms – Scanning Preventive Conclusions to Moods 271-278. (1735 pages in pdf file.)
- Table 22.7-28A – 4-Item Causative Syllogisms – Scanning Causative Conclusions to Moods 281-288. (1720 pages in pdf file.)
- Table 22.7-28B – 4-Item Causative Syllogisms – Scanning Preventive Conclusions to Moods 281-288. (1735 pages in pdf file.)

Figure 3 - 16 tables (8 for causative conclusions and 8 for preventive ones).

- Table 22.7-31A – 4-Item Causative Syllogisms – Scanning Causative Conclusions to Moods 311-318. (1032 pages in pdf file.)
- Table 22.7-31B – 4-Item Causative Syllogisms – Scanning Preventive Conclusions to Moods 311-318. (1041 pages in pdf file.)
- Table 22.7-32A – 4-Item Causative Syllogisms – Scanning Causative Conclusions to Moods 321-328. (1720 pages in pdf file.)
- Table 22.7-32B – 4-Item Causative Syllogisms – Scanning Preventive Conclusions to Moods 321-328. (1735 pages in pdf file.)
- Table 22.7-33A – 4-Item Causative Syllogisms – Scanning Causative Conclusions to Moods 331-338. (1720 pages in pdf file.)

- Table 22.7-33B – 4-Item Causative Syllogisms – Scanning Preventive Conclusions to Moods 331-338. (1735 pages in pdf file.)
- Table 22.7-34A – 4-Item Causative Syllogisms – Scanning Causative Conclusions to Moods 341-348. (1720 pages in pdf file.)
- Table 22.7-34B – 4-Item Causative Syllogisms – Scanning Preventive Conclusions to Moods 341-348. (1735 pages in pdf file.)
- Table 22.7-35A – 4-Item Causative Syllogisms – Scanning Causative Conclusions to Moods 351-358. (1032 pages in pdf file.)
- Table 22.7-35B – 4-Item Causative Syllogisms – Scanning Preventive Conclusions to Moods 351-358. (1041 pages in pdf file.)
- Table 22.7-36A – 4-Item Causative Syllogisms – Scanning Causative Conclusions to Moods 361-368. (1032 pages in pdf file.)
- Table 22.7-36B – 4-Item Causative Syllogisms – Scanning Preventive Conclusions to Moods 361-368. (1041 pages in pdf file.)
- Table 22.7-37A – 4-Item Causative Syllogisms – Scanning Causative Conclusions to Moods 371-378. (1720 pages in pdf file.)
- Table 22.7-37B – 4-Item Causative Syllogisms – Scanning Preventive Conclusions to Moods 371-378. (1735 pages in pdf file.)
- Table 22.7-38A – 4-Item Causative Syllogisms – Scanning Causative Conclusions to Moods 381-388. (1720 pages in pdf file.)
- Table 22.7-38B – 4-Item Causative Syllogisms – Scanning Preventive Conclusions to Moods 381-388. (1735 pages in pdf file.)

The sorts of formulas used in above tables can be seen in the following two tables, which are posted online at my website as usual. The first of these (drawn from Table 22.7-11A) concerns tables with sets of 8 moods (viz. 11-18, 51-58, 61-68), and the second (from Table 22.7-12A) sets of 13 moods (viz. 21-28, 31-38, 41-48, 71-78, 81-88). With minimal appropriate changes, the formulas in these samples are easily adapted to different contexts.

- Table 22.7-4 – 4-Item Syllogisms – Scanning Template I (1st Sample of Formulas Used). (3 pages in pdf file.)
- Table 22.7-5 – 4-Item Syllogisms – Scanning Template II (2nd Sample of Formulas Used). (5 pages in pdf file.)

So much, thus far, for the technicalities. Let us now **compare and analyze** the results obtained. First let us compare the results shown for 4-item syllogism in these tables with those shown in the previous chapter. You might argue that such comparison is idle, or at best 'academic', since the rougher, older results are supplanted by present results, which are complete and definitive. But I think it is still worthwhile comparing them, insofar as it ensures I made no errors of inattention when writing the formulas that generated the latter results, and also because it helps justify the extra work I invested in producing them. Let us therefore look at Table 22.7-0; especially the last six pages, which show the new and old results opposite each other. We see that the conclusions newly obtained include all the conclusions previously obtained, so the results of both inquiries are consistent.

The large majority of additional conclusions obtained here are negative propositions, of forms not previously investigated: either the causatives **p**, **q** relative to notS, or any of the eight preventives. However, we do not find any new negations of **m** or **n** (causatives). But we *do* find a great many previously undiscovered negations of **p** and/or **q** relative to S (causatives). For instances, 222(c), 225, 233(c), 236, and so on. This is not surprising, in that no attempt was made to search for these before (due to the difficulties involved). We did however in four cases find such conclusions, viz. in moods 223(c), 232(c), 323(b), 332(b), which you may recall we looked into exploratively in Table 21.10.

There are no new positive causative conclusions of the forms **m**, **n**, **p** rel to S, **q** rel to S, **p** abs, **q** abs, and no positive preventive conclusions of any form. But there *are* some new positive causative conclusions of the form **p** and/or **q** rel to notS in the second and third figures; namely, for the 10 moods 221, 231, 241 271, 281, and 312, 313, 314, 317, 318. Concerning these conclusions, it should be noted that they all displace previously listed negative causative conclusions of the forms **p** and/or **q** relative to S. This is understandable, since **p**, **q** rel to notS deny **p**, **q** rel to S, respectively; i.e. the disappeared negations remain implicit as subaltern conclusions.

Another comparison worth making is between our final lists of conclusions for 3-item and 4-item syllogisms; these are posted together for comparison in the last six pages of Table 22.7-0. All conclusions obtained are unsurprisingly consistent. And, needless to say, conclusions are much richer in detail in 4-item syllogism than in 3-item syllogism. But more specifically, one question asked in the previous chapter can now be answered: are there any 'nil' conclusions in 3-item syllogism that become positive or negative conclusions in the corresponding moods of 4-item syllogism? This question concerns 9 moods in each figure, viz. 44, 47, 48, 74, 77, 78, 84, 87, 88. The answer is no – all these moods remain without valid positive or negative conclusion in 4-item syllogism, too. It is still of course conceivable that some or all of these moods yield some conclusion in 5-item syllogism, but at least we are now sure they do not yield any in 3- or 4-item syllogism.

Another question we can ask here is: are there any 4-item moods whose conclusions are not more informative than the corresponding 3-item moods? There are indeed such cases. Mostly, these occur because the 4-item mood is in fact a 3-item mood, so the conclusion has predictably the strong forms **m**, **n**, or their negations, alone or in combinations (note the preventive as well as the negative sides of each conclusion). In some cases, when a mood is split in two (abs/rel and rel/abs), we may find one of the 4-item moods has identical conclusion(s) to the 3-item mood, but the other is has additional negative conclusions. But on the whole we can say that 4-item syllogism generally adds some new conclusion, either raising a **p** or **q** conclusion from absolute to relative, or adding some negative conclusion in the causative and/or preventive column.

Let us now finish our analysis of Table 22.7-0 with some **statistics** for 4-item syllogism. We shall compute these with reference to 89 moods per figure (i.e. counting moods that are here split in two as abs/rel and rel/abs as two). There are 18 nil conclusions (20% of the 89 moods) in each figure – i.e. moods with no conclusions, whether positive or negative, causative or preventive[135]. So the number of moods with some sort of conclusion is 71, and the overall validity rate for 4-item syllogism is 80%.

Of these 71 valid moods, 65 involve some positive and/or negative causative proposition(s) and 46 involve some negative preventive proposition(s). Of the 71 valid moods, 25 are causative only (+ve and/or –ve), 6 are preventive only (all –ve), and 40 are mixtures of causative (+ve and/or –ve) and preventive (all –ve). Note that all positive conclusions are

[135] Though some or all of these moods might for all we know imply a mere conjunctive or conditional fragment of a conclusions. The matter is not here investigated, though it could be done with a certain amount of work, because our interest is centered in causative and preventive propositions (rather than their building blocks).

causative and none are preventive. Of the 65 causatives, 17 are positive only, 8 are positive and negative mixtures, and 40 are negative only.

As can be seen in the following little table (unnumbered), there are 25 moods (28%) per figure with some sort of positive conclusion. Note that in the first figure, relative weak conclusions are all 10 relative to S, whereas in the second and third figures 5 are relative to S and 5 are relative to notS. There are 8 absolute weak conclusions in each figure (4 in **mq** abs and 4 in **np** abs – none standing alone).

+ve conclusions	fig. 1	fig. 2	fig. 3
mn	1	1	1
mq rel to S	2	1	1
mq rel to notS	0	1	1
np rel to S	2	1	1
np rel to notS	0	1	1
pq rel to S	2	1	1
pq rel to notS	0	1	1
mq abs	4	4	4
np abs	4	4	4
pq abs	0	0	0
m	3	3	3
n	3	3	3
p rel to S	2	1	1
p rel to notS	0	1	1
q rel to S	2	1	1
q rel to notS	0	1	1
p abs	0	0	0
q abs	0	0	0
Total	25	25	25
percent out of 89 moods	28%	28%	28%

Note lastly that some moods have a conclusion composed of as many as 7 negative elements (maximum 3 on the causative side, and 4 on the preventive side). Some moods, of course, have only one negative element or even none. Although negative conclusions are not as interesting as positive ones, they are still logically significant, telling us not to expect the opposite positive proposition to arise. They are also useful for consistency checking, since, having reached a negative conclusion, if we discover that we elsewhere, earlier or later, uphold the opposite positive proposition, we know we have made a mistake in our reasoning or observation somewhere. So it is important to keep negative conclusions in mind. They can also, by the way, become premises in subsequent syllogisms, as we shall point out in the next chapter.

Chapter 23. EXPLORING FURTHER AFIELD.

1. Possible Forms of Premises.

It is clear that the scanning technique we have developed in the preceding chapter is *a thorough and foolproof method for finding the conclusion(s) from any set of premises with a common item*, i.e. any syllogism. We have there demonstrated the method with reference to positive causative premises and limited our search for conclusions to the range of all possible causative and preventive forms with the major and minor items concerned. But obviously, with appropriate changes in the data and/or the formulae used, and the required patience and effort, we can use the same method to solve any syllogistic problem whatever with full confidence.

We can use as premises any forms that can be entirely expressed in terms of moduses: viz. those I call 'pre-causative' forms, i.e. conjunctive propositions with two or more items, like 'P + Q is possible' or 'P + Q is impossible', or their derived conditional propositions, like 'if P, then Q' or 'if P, not-then Q', which constitute the building-blocks of causative propositions[136], or we can use any sort of causative proposition. Needless to add, all that is said here concerns not only propositions with two terms (like P, Q), but equally those involving a complement (i.e. a third term, be it positive or negative).

The term 'causative proposition' can be (and has here been) used in both a narrow and a broad sense, note well. In a narrow sense, 'causative propositions' signifies those of the form 'P is a (complete, necessary, partial and/or contingent) cause of Q' or the negation of such form, i.e. whose stated items are intended as positive. But in a broad sense, 'causative propositions' refers to all forms involving a causative relation 'is (or is not) a cause of', whatever the intended polarities of the items involved, thus including 'preventive propositions' (with items P and notQ), and inverse causative propositions (with notP, notQ) and inverse preventive propositions (with notP, Q).

The form of causation is not essentially affected by the polarities of the items involved. Every term, whether conceived as positive or negative, is just 'a term' to the causative relation involved. The issue of polarity of items only acquires significance when we compare two or more causative propositions having the same terms but with different polarities. Thus, the issue of polarity of items is merely comparative. But it is not without importance.

Preventive propositions, I remind you, can be stated (for instance) as 'P prevents Q', or as 'P causes notQ'; note the change of polarity of Q. As we have seen in matricial analysis, the

136 Please note that though I do not here develop the logic of pre-causative forms, I do not mean to imply that it is unimportant. The reason that I gloss over it is simply that this logic is not the topic of the current work. Moreover, the logic of conjunction and conditioning has been dealt with adequately in my *Future Logic*, and is in any case pretty well known and understood by logicians in general nowadays. Nevertheless, we could redo the whole of this pre-causative logic using the methods developed in the present volume for causative logic. The techniques of matricial analysis and scanning are very innovative, and would constitute a welcome renewal for pre-causative logic, which might well lead to previously unknown results. However, being human, I cannot do all this work, and I leave it to others.

moduses of 'causative' forms in the narrow sense and 'preventive' forms are not the same. This means that, although 'preventive' propositions are a species of 'causative' propositions in the broad sense, they stand in significant contrast to 'causative' propositions in the narrow sense. Therefore, it is often necessary to investigate those two sets of forms separately, if we want to be exhaustive in our treatment of logical processes.

On the other hand, matricial analysis has shown us that the moduses of 'inverse causation' are closely related to those of 'causation' (in the narrow sense), and likewise those of 'inverse prevention' are closely related to those of 'prevention'. This means that these two forms of causative proposition (in the broad sense) rarely require separate treatment, and their processes can be inferred from those of causative (in the narrow sense) and preventive propositions without need for bulky matricial analysis.

Another possible application of scanning is to syllogisms with one or both premises of 'vague' form, or of 'lone' form, or otherwise out of the ordinary. Unusual conjunctions of positive and negative forms can also be studied in this way. I have in the past (in phase II) looked into such moods briefly, but now we can do the job with more certainty and thoroughness. I will not however do it, but leave it to others.

As well, any mixture of forms of pre-causative or causative propositions can in principle be used as premises and/or conclusions in causative syllogism. However, we tend in our natural discourse not to mix such large categories of form too much. We would rarely if ever mix pre-causative and causative premises, and would rarely seek conclusions of pre-causative form to premises of causative form or vice versa. On the other hand, mixtures of different sorts of causative, preventive, inverse causative and inverse preventive forms, do occasionally occur in discourse, and should be considered.

2. Dealing with Negative Items.

Let us now, therefore, consider to what extent the syllogisms, 'causative' in the narrow sense, dealt with in the preceding chapter, cover the ground of 'causative syllogism' in the broadest sense. The moods dealt with so far have been with positive causative premises with positive items, in the three main figures. That is, they had the forms (sometimes relative to a subsidiary term S, not here shown) listed below:

Primary forms	Figure 1	Figure 2	Figure 3
Major premise::	Q-R	R-Q	Q-R
Minor premise:	P-Q	P-Q	Q-P
Conclusion(s):	P-R + P-notR	P-R + P-notR	P-R + P-notR

Note the large form of the conclusion, which might include not only causative propositions but also preventive ones (our chosen range having been expanded thus). We do not need to mention inverse conclusions like 'notP-notR' or 'notP-R' because these are entirely inferable from the eventual conclusions of form 'P-R' or 'P-notR'. For example, a conclusion **m** of form P-R is logically identical to a conclusion **n** of form notP-notR. We might thus view the inverse forms implicit in our eventual conclusions to represent subaltern moods of the same syllogism, and ignore them accordingly.

Let us now consider what happens when we change the polarity of the minor and/or major term in the premises, and so in the conclusion to those premises:

Negate major term	Figure 1	Figure 2	Figure 3
Major premise::	Q-notR	notR-Q	Q-notR
Minor premise:	P-Q	P-Q	Q-P
Conclusion(s):	P-notR + P-R	P-notR + P-R	P-notR + P-R

If we just change the polarity of the major item, we obtain, in the first and third figures, a preventive major premise, and in the second figure, an inverse preventive major premise, and in the conclusion, preventive and causative forms.

Negate minor term	Figure 1	Figure 2	Figure 3
Major premise::	Q-R	R-Q	Q-R
Minor premise:	notP-Q	notP-Q	Q-notP
Conclusion(s):	notP-R + notP-notR	notP-R + notP-notR	notP-R + notP-notR

If we just change the polarity of the minor item, we obtain, in the first and second figures, an inverse preventive minor premise, and in the third figure, a preventive minor premise, and in the conclusion, inverse preventive and inverse causative forms.

Negate extremes	Figure 1	Figure 2	Figure 3
Major premise::	Q-notR	notR-Q	Q-notR
Minor premise:	notP-Q	notP-Q	Q-notP
Conclusion(s):	notP-notR + notP-R	notP-notR + notP-R	notP-notR + notP-R

If we change the polarities of both the major and minor items, the premises are modified as above, become preventive or inverse preventive in form, and the conclusion becomes a conjunction of inverse causative and inverse preventive forms.

Negate middle term	Figure 1	Figure 2	Figure 3
Major premise::	notQ-R	R-notQ	notQ-R
Minor premise:	P-notQ	P-notQ	notQ-P
Conclusion(s):	P-R + P-notR	P-R + P-notR	P-R + P-notR

Similarly, if we negate the middle term in *both* the premises, no change occurs in the conclusion. And if we do this in addition to the above listed changes in major and/or minor premise, the conclusions found there remain the same. However, in all these cases, the categorization of the premises of course changes. Thus, where the polarity of the middle term is changed in both premises, but not that of the other terms, the premises all have preventive or inverse preventive forms (as shown immediately above). If now we change the polarity of the major and/or the minor term(s) as well, we will naturally again affect the forms of the premises in the various figures, some of them becoming causative or inverse causative. In this way, we gradually exhaust all possible forms of premises.

All the above concerns both 3-item and 4-item syllogism (and even 5-item syllogism). In 4-item (and likewise 5-item) syllogism, we must additionally take into consideration the complement(s) involved. When one of the premises is or involves a relative weak proposition with a complement S, we can similarly obtain an additional mood by replacing it with the complement notS. In such case, obviously, where the conclusion refers to S we will insert notS, and where it refers to notS we will insert S. Such substitutions of course further increase the number of valid moods at our disposal.

What all the above means is that the lists of causative syllogisms developed by scanning in the previous chapter can with relative ease be adapted to such changes of polarity in the items concerned. This method of substitution is a valid shortcut. There is no real need to deal with these moods separately, producing bulky spreadsheets all over again. Just change the polarities of the terms systematically as shown above, and thus re-define the forms of the premises and conclusions involved, and the job is equally well done. We can thus refer to such syllogisms as derivatives of purely causative syllogisms treated in the previous chapter.

I have not numbered the results of such multiplication of moods, but this is a job that eventual needs doing. We should anyway keep in mind the distinction between primary (independently validated) and secondary (derived) moods when we do count them.

It is probable that many, if not all, syllogism thus derived (by substitutions) can also be derived by direct reductions. I cannot predict with certainty offhand, in view of the fact that many moods have multiple (both positive and negative) conclusions or only negative conclusions. Perhaps some or all of the negative conclusions can be derived by indirect reduction (i.e. reduction ad absurdum). The matter would have to be investigated carefully in detail before we can say for sure. However, I am not about to do this job. I leave it too to others.

3. Preventive Syllogisms and their Derivatives.

But this substitutive method is not adequate for preventive syllogisms and their derivatives. The problem of negation of items is more complicated when the middle term is negated in *only one* of the premises. In such cases the conclusions are affected in ways not easily predicted by mere substitutions.

Negate middle term	Figure 1	Figure 2	Figure 3
Major premise::	Q-R	R-Q	Q-R
Minor premise:	P-notQ	P-notQ	notQ-P
Conclusion(s):	P-R + P-notR	P-R + P-notR	P-R + P-notR

Negate middle term	Figure 1	Figure 2	Figure 3
Major premise::	notQ-R	R-notQ	notQ-R
Minor premise:	P-Q	P-Q	Q-P
Conclusion(s):	P-R + P-notR	P-R + P-notR	P-R + P-notR

We can predict at the outset that it is irrelevant whether the middle term is negative in the major or minor premise. The forms involved will be named differently, according to where the middle term has been negated, but the syllogistic process and its conclusion will be the same. What is sure, however, is that the conclusions obtained will be different.[137]

---

[137] In truth, at first I was not sure of that: I assumed offhand that most premises so generated would be incompatible; but it turned out they were all compatible and most yielded conclusions. Thereafter, it seemed obvious that this should be so!

I decided therefore to explore the issue further, and developed the following tables for **3-item preventive syllogism**, with one causative premise (the major) and one preventive premise (the minor), which can all be viewed online at my website as usual. The first of these tables shows the moduses of the premises used; and the second table summarizes the next three, which detail the scanning work done in each figure. The important table is the second, viz. Table 23.2-0; please look at it carefully:

- Table 23.1 – 3-Item Preventive Syllogisms – Premises Scanned. (8 pages in pdf file.)
- **Table 23.2-0 – 3-Item Preventive Syllogisms – Summary of Conclusions Scanned. (9 pages in pdf file.)**
- Table 23.2-1 – 3-Item Preventive Syllogisms – Figure One – Scanning for Conclusions. (48 pages in pdf file.)
- Table 23.2-2 – 3-Item Preventive Syllogisms – Figure Two – Scanning for Conclusions. (48 pages in pdf file.)
- Table 23.2-3 – 3-Item Preventive Syllogisms – Figure Three – Scanning for Conclusions. (48 pages in pdf file.)
- Table 23.3 – 3-Item Preventive Syllogisms – Comparison between Causative and Preventive Syllogisms. (3 pages in pdf file.)

As can be seen, what I did here was simply change the data for the minor premise in the 3-item tables developed in the preceding chapter. The latter served as templates for the present investigation. No formulas were changed, note (which is why I have not produced a new 'formulas used' table for the tables posted here). Note too that the forms of conclusion scanned for (P-R + P-notR) are unchanged.

However, as the comparative Table 23.3 shows, the specific conclusions actually obtained are very different. What is interesting is that the conclusions here obtained are almost a mirror image of the conclusions previously obtained, except that the 'spectrum' has shifted from causative to preventive. That is, whereas with two causative premises the conclusions leaned on the side of causation, consisting roughly speaking of a mix of mainly positive causative conclusions and exclusively negative preventive conclusions – here, with a causative and a preventive premise, the conclusions lean on the side of prevention, consisting roughly speaking of a mix of mainly positive preventive conclusions and exclusively negative causative conclusions! We might express this shift by writing the conclusions in the form 'P-notR + P-R' (instead of 'P-R + P-notR') – i.e. with the preventive part first and the causative part second, though of course the order in which the parts are stated is formally irrelevant.

If you look again at the lists of conclusions and compare them, you can see that, apart from this 'spectral' shift, the conclusions are essentially the same except that the order they appear in the columns is not identical. But the differences are, as might be expected, orderly. In the first figure, the moods numbered 111-118, 141-148 reflect the moods with the same numbers (though with the spectral shift, of course – i.e. they are, as it were, laterally inversed 'mirror images'); while the moods numbered 121-128, 131-138, 151-158, 161-168, 171-178, 181-188 respectively mirror the moods 131-138, 121-128, 161-168, 151-158, 181-188, 171-178. All statistics are consequently the same, *mutatis mutandis*.

For this reason, i.e. with reference to the kind of conclusions obtained, I have called syllogism with two causative premises (and its derivatives, as above described) 'causative syllogism'; and syllogism with one premise causative and the other preventive, 'preventive syllogism'.

Note well, these labels refer to the conclusions obtained rather than to the kinds of premises (for, as we have seen in the previous section, a syllogism may have some preventive or inverse-preventive premise(s) and yet be essentially causative). Note also the distinctive format of preventive syllogism – the middle term is antithetical, being positive in one premise and negative in the other.

Just as causative syllogism has a mass of derivatives, as detailed in the previous section, so with preventive syllogism. If we take the syllogisms listed in the above tables, and negate their minor and/or major term, and/or negate both occurrences of their middle term, and/or negate their subsidiary term(s), we get a mass of derivative syllogisms as before. This means that, combining the lists of positive causative and preventive syllogisms, including all such derivatives, we truly have an exhaustive overview of syllogisms to do with causation in the largest sense.

The above listed tables concern only 3-item syllogism, of course. Concerning **4-item preventive syllogism**, we can already predict that the results will be very similar. Just as 3-item preventive syllogism yields a shift in conclusions from the causative to the preventive side, so with 4-item preventive syllogism we can expect such a shift. Similarly, the vertical order of the conclusions will be somewhat changed, though in an orderly manner as before. The 4-item results are sure to be analogous, because they have to be consistent with 3-item results. The following CONJECTURAL table shows the results one may expect offhand, based solely on analogy to the results given in Table 23.3:

- Table 23.4 – 4-Item Preventive Syllogisms – CONJECTURAL predictions of conclusions. (6 pages in pdf file.)

Please note well that this table is conjectural, and not to be relied on! The only reason I have put it here is to give the reader an idea of the sorts of results to expect. But it has yet to be proved by the scanning method. The reason why we cannot be sure of this analogical argumentation is that there may be unexpected twists when the fourth item is involved – e.g. it may be that a surprising change of polarity occurs in such cases. The scanning work must eventually be done if we wish to be exhaustive in our research. But I leave it till later or for others.

As regards the number of valid moods, we can predict that the collections of causative and preventive syllogisms will be numerically the same. By the way, regarding syllogisms with a premise (or two) that compounds causative and preventive propositions: such compound premises need not be viewed as necessitating further research, but can be treated as 'double syllogisms'. That is, the conclusion(s) of the compound premise(s) will be the sum of the conclusion(s) of the two (or more) causative and preventive syllogisms involved.

The same comment can be made here as in the previous section regarding the possibility of deriving syllogisms by direct or indirect reduction. It may even be that the above tabulated preventive syllogisms can be so derived from causative syllogisms; I have not tried. As far as I am concerned, matricial analysis, scanning and substitution are independent sources of validity: there is no formal need for the traditional methods of Aristotelian logic. However, such research would be interesting to pursue for its own sake, as it might well reveal the interdependence of all syllogisms in the various figures and with various polarities of terms. Intuitively, this seems obvious to me; but it has to be proved. We would then know what the shortest possible list of causative syllogisms includes.

4. Syllogisms with Negative Premise(s).

All the above concerns positive premises, note well. We have yet to consider what happens when one or both of the premises, be they causative and/or preventive, is/are negative. This is what we will briefly look into now. My intent here is only to sketch the way, without actually doing the whole job.

I have in the past, in phase II, briefly considered the issue. In Table 15.9, I list various negative propositions or compounds with negative elements that may appear as premises, without working out their moduses[138]. Then, in Table 15.10, I consider first figure syllogisms with both premises negative and elementary – negations of **m**, **n**, **p** abs or **q** abs and find (using the 'manual' method) no conclusions from them.

Obviously, to be thorough in our research, we must systematically list all possible moods and then (mechanically) scan for the conclusions they imply in the full range of causative and preventive forms. In positive syllogism, we considered moods with both premises positive and among the eight forms **mn**, **mq**, **np**, **pq**, **m**, **n**, **p**, **q** (these forms being ordered from the strongest determination to the weakest, and numbered 1-8, even though the first four are compounds of the latter four). Our first job, therefore, would be to consider the negations of these eight compound and elementary forms, i.e. **not-(mn)**, **not-(mq)**, **not-(np)**, **not-(pq)**, **not-m**, **not-n**, **not-p**, **not-q**.

This I have done for 3-item causative syllogism, for all three figures. I started with the spreadsheets used to produce Tables 22.6-1, 22.6-2, 22.6-3. Using these as templates, I duplicated them 3 times each, and negated the premises successively in them: first the major premise, then the minor premise, then both the premises. The formulas were so easily adapted that I do not need to produce a listing of them for you. Lastly, I recalculated everything. The reliability of the results obtained is confirmed by their symmetry. The findings are summarized in the first table, and the detailed scanning work on which it is based is shown in the next nine tables. These tables can all as usual be viewed online at my website.

- **Table 23.5-0 – 3-Item Negative Causative Syllogisms – Summary of Conclusions Scanned. (3 pages in pdf file.)**
- Table 23.5-1 – 3-Item Negative Causative Syllogisms – Major Premise Negated – Figure One – Scanning for Conclusions. (32 pages in pdf file.)
- Table 23.5-2 – 3-Item Negative Causative Syllogisms – Minor Premise Negated – Figure One – Scanning for Conclusions. (32 pages in pdf file.)
- Table 23.5-3 – 3-Item Negative Causative Syllogisms – Both Premises Negated – Figure One – Scanning for Conclusions. (32 pages in pdf file.)
- Table 23.5-4 – 3-Item Negative Causative Syllogisms – Major Premise Negated – Figure Two – Scanning for Conclusions. (32 pages in pdf file.)
- Table 23.5-5 – 3-Item Negative Causative Syllogisms – Minor Premise Negated – Figure Two – Scanning for Conclusions. (32 pages in pdf file.)
- Table 23.5-6 – 3-Item Negative Causative Syllogisms – Both Premises Negated – Figure Two – Scanning for Conclusions. (32 pages in pdf file.)

---

[138] I have in phase III worked out the 3-item moduses for the possible premises listed in this table, namely in Table 18.4.

- Table 23.5-7 – 3-Item Negative Causative Syllogisms – Major Premise Negated – Figure Three – Scanning for Conclusions. (32 pages in pdf file.)
- Table 23.5-8 – 3-Item Negative Causative Syllogisms – Minor Premise Negated – Figure Three – Scanning for Conclusions. (32 pages in pdf file.)
- Table 23.5-9 – 3-Item Negative Causative Syllogisms – Both Premises Negated – Figure Three – Scanning for Conclusions. (32 pages in pdf file.)

The summary in Table 23.5-0 should certainly be looked at. What we learn from it is that 24 moods out of 192 (12.5%) in each figure do yield a conclusion. It is always a single conclusion, and it is always negative and causative – i.e. neither positive nor preventive. 12 of these conclusions occur in syllogism with the major premise negative; and the other 12 of them occur in syllogism with the minor premise negative; no conclusions emerge when both premises are negative[139]. Of the said sets of 12 conclusions, 5 are **not-m**, 5 are **not-n**, and the other 2 are **p** abs and **q** abs. This is true in each figure, though the order of appearance varies.

This exploration in negative syllogism does not cover the whole field (see Table 15.9 again), but it is of course a good start. We can later push further afield. Note that **not-(mn)** means '**not-m** and/or **not-n**', which allows for three consistent alternative outcomes '**m** + **not-n**' or '**not-m** + **n**' or '**not-m** + **not-n**'; and similarly for the other negated compounds **not-(mq)**, **not-(np)**, **not-(pq)**. Thus, when we validate a conclusion for a negated compound, we also validate it for its three alternative outcomes; though this does not guarantee that the alternatives would not, when taken specifically, yield a more specific conclusion. Their syllogisms still need to be separately investigated.

Moreover, all this must be done for preventive as well as causative premises, as we did on the positive side. We can expect negative preventive syllogism to resemble negative causative syllogism, except that the negative conclusions obtained will be shifted to the preventive side and vertically displaced as before. Moreover, all this must be done for 4-item as well as 3-item syllogisms (and eventually also 5-item syllogisms). However, there is no cause for despair; the project is within the realm of the possible. It is big, but the scanning method is clearly known and perfect in its results. Furthermore, we (or at least I) have applicable templates – we can use the spreadsheets already developed for positive syllogisms, merely modifying them as appropriate for the various negative syllogisms.

Logicians are duty bound to solve all conceivable syllogisms. But I am not going to do all the work. I think I have done enough. I have laid the foundations of and greatly developed this important department of logic. It is a big field and there is room in it for other workers. I invite those interested to do their bit, and claim their share of the territory in history books on the subject. Look upon the task as an exercise. The best way to learn and fully understand is by doing – i.e. by taking up the challenge of actual research work.

---

139 Note that the investigation in Table 15.10 taught us the same lesson for two negative elementary premises, **not-m**, **not-n**, **not-p**, **not-q**; but here the scope is larger, since we also prove the inconclusiveness of two negative premises, when *one or both* of them are negative compound premises, **not-(mn)**, **not-(mq)**, **not-(np)**, **not-(pq)**.

5. Causal Logic Perspective.

Since I am here more or less putting an end to my share of the research, I would like to mention a further perspective. The logic of causation, or more briefly put 'causative logic', is only one department of the logic of causality, or 'causal logic'. Two other departments of causal logic seem to me very important fields that logicians must develop further. They are the logic of volition, or 'volitional logic' and the logic of influence, or 'influential logic'.

In my work *Volition and Allied Causal Concepts* (2004), I discuss the topic of volition philosophically and explain its relation to causation and influence respectively. I there debunk various widespread misunderstandings concerning these concepts and show how they can be credibly defined and correctly understood. I will not go into details here, wishing only to add a few general comments relevant to the present context.

Volition is a form of causality very different from causation. It cannot be reduced to causation. Volition cannot be fully represented by any particular modus or combination of moduses. However, volition is not entirely separate from or devoid of causation. That is to say, causation often (indeed, always) sets limits to the natural possibilities inherent in volition. Volition is never without limit (at least not that of humans or animals); it is always delimited, within a given 'realm of the possible'. 'Freedom of the will' does not signify 'unlimited freedom, period' – but 'full freedom *within* certain naturally (i.e. causatively) set limits'. This must always be kept in mind, if we do not want to fall victim to fallacious skeptical arguments. Volition goes beyond causation, into specific areas that causation does not entirely rule. But the relationship between the agent of volition, i.e. the conscious being (to whatever degree) doing the willing, and the immediate causal product of volition, that i.e. which the agent actually wills, is not devoid of causation. Specifically, we can say that without this agent – or, in some cases, another like him – the thing willed would not have occurred. Thus, we can say that the agent is causatively related to the thing willed at least in this respect – i.e. in being *a necessary cause* (or one possible instance of *a kind of* thing that is a necessary cause) of the thing emerging from the act of will.

'If not for this agent willing it, this willed thing would not have occurred'. Therefore, the agent is not only the willer of that thing, but a causative of it – specifically a necessary cause of it. However, his relation to the thing in such cases is not like that of a lifeless 'complete cause'. We cannot express the relation between the agent and his will's immediate products with reference to the moduses of complete causation. If we seek a modus to formally apply to causation it can only be (as discussed in the present volume, chapter 19) the 'last modus' – i.e. modus 16 in a 2-item framework, or more deeply modus 256 in a 3-item framework, and so forth. But this modus is too vague; it is not exclusive to volition, so it does not tell us anything about it.

Note that I have simplified the matter a bit, when I suggested that the agent (or some such agent) is necessary to the immediate effect of the will. In some cases, it is true, the same effect might equally well have been produced through entirely causative (i.e. non-volitional) means. In such cases, the agent is may be said to be only one of the many possible 'sine qua nons' to the *kind of* effect concerned. Nevertheless, with reference to that given *instance* of the effect concerned, the agent is definitely the one and only necessary cause. If someone or something else had caused a similar effect, it would have caused a similar but distinct instance of the effect. The instantiation is entirely identified to the actual cause.

Thus, though volition may involve some causation in that the agent is a necessary cause of what he or she wills, its relation cannot be confused with complete causation. Volition is not complete causation – it *replaces* the natural function of complete causation with a quite distinct form of causality. That is why we must name it distinctively and study it separately. Through an act of volition, the agent does the same job as a complete causative of 'bringing' into actuality something that was previously inactual and that would have otherwise probably remained inactual; but that does not qualify it as a complete causative. That is to say, we cannot say of it 'if the agent, then necessarily the effect' – because the agent may choose not to produce the effect, unlike a lifeless causative. Where volition is applicable, we can only say 'if the agent, possibly the effect'.

Let me now remind you of the basic insights regarding '**influence**' in my theory of volition. Influence is a causal relation – it is the relation between various objects (natural or endowed with volition) and what the agent it (i.e. the influence) impinges on wills. Influence is the intermediary between the domain of causation and that of volition. However, it radically differs from causation in this: something is an influence on volition only through *the intermediary of consciousness*. It must be cognized somewhat to qualify as an influence.

A causative may well affect a volition, by virtue of setting limits to its operation. Every volition is affected by some causatives, in this respect. But influence is something else entirely. An object, whether mental or physical, can only influence us if we are (to whatever degree or extent or depth) aware of it. Once we are aware of it, it becomes one of the factors affecting the act of will proper, i.e. the effort of the agent concerned. How does an influence affect a volition? It naturally makes it easier (positive influence) or more difficult (negative influence) for the agent to will that will. The agent must put less or more effort, respectively, to achieve the same result, i.e. the volition concerned. This is clearly very different from mere causation; it is a much weaker (and more personal) causal relation.

The other important point to keep in mind concerning influence is that it is never 100% determining. This is a requirement of consistency. If the will is free, it cannot be extremely influenced – some leeway must remain for it. If the will is truly determined by some factor external to the agent of will, then what occurs is not influenced volition but mere causation. The postulate of freedom of the will is that though volition may indeed be influenced it is never overpowered by influences, but retains some degree of liberty to go ahead in one direction or the other or to abstain from such courses of action. Generally, the reason why skeptics concerning freedom of the will err is that they fail to make this distinction between causation and influence.

All right. Why have I reviewed these basics concerning volition and influence in the present context? Simply because I want to briefly discuss **the confluence of causative logic and the logics of volition and influence**, and thus set the stage for causal logic in its widest sense. In *Volition and Allied Causal Concepts*, I lightly touched upon the formal logic relating to the concepts there studied; but there is still of course much, much work to do in that field. In the present volume, we have gone into causative logic in great detail, though there is still room for further research.

What I want to point out here is the need for a 'formal logic' style study of the matter. That is, one capable of drawing exact conclusions from *syllogisms involving a mixture of causative premises with volitional and/or influential ones*. What conclusions can be drawn from different combinations of these forms (and other related factors)? This is a big field that yet needs intensive study.

To give a formal example: if agent A willed W (under influences X, Y, Z) – and W is a complete cause (or partial cause with certain complements) of some phenomenon P – can A be said to have willed P? Offhand, I would say: no. Only if P was ***intended*** (i.e. thought about as a goal of W) would I say it was 'willed'. If P was thought of as a very likely or inevitable effect of W, P may only be said to have been ***incidental*** to A's action; and if it was neither intended nor thought of as possible or probable effect, then P was ***accidental*** to A's action. Note that the thought of P is more *influential* in intention than it is in incident – and plays no role in accident.

To give a material example: when I drive my car to the supermarket, it is my intention to get there and shop there. I know this will cause some pollution, but that is not my intent; indeed, I wish I didn't, but not enough to stop me driving; so this is incidental to my driving. I could of course knock my car into something on the way; but this is not very likely and I do my best to avoid it; so such occurrence would be an accident. (Incidents and accidents can also of course be positive events, note.)

Broadly speaking, then, given that A wills W and W causes P; what ought we to conclude regarding the causal relation of A and P? What we need are precise, complex, formal, accurate, consistent answers to all such questions. Only then would we be able to claim to have really mastered the whole field of causal logic. I leave it at that for now. Thank you for your attention.

Chapter 24. A PRACTICAL GUIDE TO CAUSATIVE LOGIC.

In the present, final chapter, I will try and provide readers with a practical guide to the logic of causation. That is, after, all the purpose of the whole exercise. This book was written over a period of many years (on and off) more as a research report than as a text book. Most readers, I assume, are not very interested in the details of how I got to such or such a result, but just want to learn how to reason correctly with causative propositions. The validations are more of concern to logicians. Lay people want practical guidance. Thus, do not expect here a systematic summary of all findings, but rather a highlighting of some of the main points.

1. What is Causation?

What is causation? This term refers to a concept – an abstraction through which we can order empirical facts in a way that makes them more comprehensible to us and helps us makes predictions. Like every reasonable concept, causation does indeed signify an existing fact – namely the fact that sets of two or more facts are often evidently related in the ways we call causation. Causation refers to certain observable or induced or deduced *regularities in conjunction or non-conjunction* between two or more things. By 'things' (or preferably, henceforth, 'items') here, understand any domain of existence: material, physical, bodily, mental, abstract, spiritual; any category of existent: substance, entity, characteristic, quality, change, motion, event, action, passion, dynamic, static, etc. – anything whatever.

As with all concepts, the concept of causation varies somewhat from person to person, and over time in each person. At one end of the spectrum, there are people for whom the concept of causation is a vague, subconscious notion, which often produces erroneous judgments. At the other extreme, there are those who clearly understand causation and use it correctly in their thinking. The purpose of causative logic, i.e. of the present detailed theory of causation and its relevance to thought, is to improve people's understanding and practice.

Causation can thus be defined, broadly – and more and more precisely, as our study of it proceeds. But can causation as such be 'proved' to exist? Yes, indeed. Causation relies first of all on the admission that there are *kinds* of things. For, generally, we establish causation (as distinct from volition, which is indeterministic causality) not for individual items, but for 'kinds', i.e. for sets of things that resemble each other in some way. When we say that X causes Y, we mean that instances of the kind X are related in a certain way to instances of the kind Y.

Now note this first argument well: if there were no kinds, there would be no causation. That is, if nothing could be said to be 'the same' as anything else, kinds would not exist and causation could not be established. But if we claim "Nothing is the same as anything else in any respect", we are engaged in an inextricable self-contradiction, for that very statement is full of assumption regarding the existence of kinds. Therefore, such a claim is logically untenable, and we must admit that kinds exist, i.e. that our concepts have some empirical basis.

Now, causation refers to the *possibility or impossibility of various combinations* of things (or their negations). For example, to say that X is never found in conjunction with not-Y and that not-X is never found in conjunction with Y, is a statement of 'complete necessary' type of causation. We can certainly argue, regarding a particular pair of items X and Y (e.g. irrational behavior and mental suffering), as to whether or not they indeed fit in this relational format; merely asserting it as fact does not of course make it fact.

But no one can logically deny that there exist some pairs of things in this world that do indeed fit this pattern of relation. It would mean that we deny that there are possibilities and impossibilities of conjunction. Note this second argument well: if we claim "No conjunction of things is possible", we are saying that the conjunctions implied by this very statement are impossible; and if we claim "No conjunction of things is impossible" we are saying that contradictions are possible. All the more so, if we claim that nothing is possible or that nothing is impossible, we are involved in logically unacceptable self-contradictions. When a thesis is self-contradictory it must be abandoned, and replaced by its contradictory thesis.

Therefore, the definitional bases of causation as such – i.e. the fact that there exists the modalities of possibility and impossibility, and thence of necessity and unnecessity – and the fact that some conjunctions in the world are bound to be related by one or the other of these modalities (nothing else is even conceivable) – are indubitable. Thus, causation, which refers to different combinations and permutations of such modalities of conjunctions, is indubitable. There are no ifs and buts about it.

Why, then, you may ask, are the likes of David Hume or Nagarjuna, and all their modern followers and imitators, so convinced of the illusoriness of causation? The answer is that they are clearly not committed to reason or logic, but merely express their cognitive or psychological problems; or they are not very intelligent. Nagarjuna relied heavily on fallacious reasoning to support his alleged critique of causation. Hume's search for an empirically observable phenomenon of 'connection' or 'bond' was a red herring; it implied that causation is something concrete, i.e. tangible or otherwise materially detectable. No wonder he could not find it! No: to repeat, causation is an abstraction, through which we order our empirical observations and predict similar events of the same sort[140].

Hume admits as much when he defines causation as 'constant conjunction' between things. However, that definition is flawed inasmuch as it draws attention to only the positive side of causation; it ignores the crucial negative side (the constant conjunction between the negations of the things). Hume also ignores the different determinations or degrees of causation. And in attempting to 'explain away' causation by referring it to habitual associations of ideas, he contradicts himself – since such explanation is itself an appeal to causality; i.e. it purports to tell us 'why' we assume causation. Causation is formally the same whether it is assumed to occur in the material surrounds or in the mind, i.e. whether it correlates things or ideas. The fact that causation is usually induced by means of generalizations does not allow us to equate it to association of ideas. And anyway, association of ideas can occur even where causation is doubted; so these concepts cannot be the same in our minds.

As shown above, the concept of causation rests on two pillars, two fundamentals of human knowledge. The one is the fact of similarity and the other is the fact that conjunctions may be possible or impossible.

---

140 To give an example: a subcategory of causation in physics is the concept of 'force'. This is in no way thought of as something substantial – yet we consider it to be a reliable scientific reference, because it is an abstract inductive postulate through which we are able to order and predict various physical phenomena. Even if a particle theory of force is developed, it depends on the causative understanding that such particles obey certain abstract laws of behavior.

You can deny that two or more particular objects are similar, but you cannot deny that there are somewhere similar objects and that we are able to identify them in principle. You can deny that two or more particular objects are sometimes or never conjoined, but you cannot deny that there are somewhere objects that are sometimes or never conjoined and that we are able to identify them in principle. When I say "you cannot deny", I mean you cannot do so without self-contradiction – i.e. you cannot do so with the sanction of logic, i.e. you do so against logic.

Ontologically, causation occurs because *not* everything is possible in the world. If nothing was impossible, everything could proceed every which way. The limitations that exist in Nature constitute obstacles in its free flow, and 'force' it to flow along specific routes. Nature's course is determined by where it cannot go, rather than by where it must go. The stream of events follows the groove formed by the limits set.

There are as many modes of causation as there are modes of modality. Rational argument refers to the logical (*de dicta*) mode of causation. Extensional causation is based on extensional modality. Natural, temporal and spatial causation likewise are based on these (*de re*) modes of modality. It is logically inconsistent to admit one mode of causation (e.g. the logical) and refuse to admit the others (e.g. the natural mode). There is formally no reason to discriminate between them.

In conclusion, causation is a mental overlay through which we order observed reality. But this overlay does not force reality into any arbitrary patterns; it is not an invention of ours. It is merely an acknowledgement that certain patterns do observably occur, and our task in causative reasoning is to identify when they do occur as well as possible. The overlay is not a distortive filter or a hindrance to knowledge. It is based on experience of the world and helps us to more correctly and profoundly discern and understand the world, and thus also to better predict and deal with it.

The concept of causation has no doubt a long history, dating from the beginnings of humanity, if not earlier still in its wordless animal ancestors. Certainly, the moment our ancestors thought or said "because…" or "therefore…" they displayed their belief in or knowledge of causation. The *study of* the concept is a much later development, of course, which coincides no doubt with the dawning of philosophy, especially in ancient Greece. But it is, I think, in modern times that people began to look for applications of causation in a very conscious manner. I refer of course to the advent of modern experimental science in Europe.

Two important philosophical figures in this context were Francis Bacon and John Stuart Mill. Not because they discovered causation theoretically or the ways to find it in practice, but because they sought to verbalize causative logic. However, neither of these thinkers asked all the right questions or gave all the right answers. Surprisingly, no one made a big effort to follow up on their work, discouraged perhaps by the skepticism instilled by David Hume. It is not until the present study of causation that we have a full analysis and practical guide to causative reasoning, a truly formal logic of causation. This is really a historic breakthrough.

2. How is Causation Known?

We have in the previous section explained that causation is an 'abstract fact' and established that it is knowable by humans. Our definitions of the various types and degrees of causation provide us with **formal criteria** with which we are able to judge whether causation is or is not applicable in given cases. But to affirm that causation *as such* is definable and knowable does not tell us just how to know it *in particular cases*.

Can we perceive causation? Not exactly, since it is not itself a concrete phenomenon but an abstract relation between concrete phenomena (and more broadly, other abstractions). It has no visual appearance, no color, no shape, it makes no sound, and it cannot be felt or tasted or smelled. It is an object of conception.

Can it then be known by direct conceptual 'insight'? This might seem to be the case, at first sight, before we are able to introspectively discern our actual mental processes clearly. But eventually it becomes evident that causation must be based on concrete experience and logical process. We cannot just accept our insights without testing them and checking all the thinking behind them. The foundation of causative knowledge – i.e. of knowledge about causation between actual things – is evidently induction.

That is to say, quite common and ordinary processes like generalization and particularization or, more broadly, adduction (the formulation and empirical testing of hypotheses). These processes are used by everyone, all the time, though with different degrees of awareness and carefulness. The bushman who identifies the footprints he sees as traces of passing buffalo is using causative logic. And the scientist who identifies the bandwidth of rays emanating from a certain star as signifying the presence of certain elements in it is using the same causative logic. The bushman is not different from or superior or inferior to the scientist. Both can make mistakes, if they are lazy or negligent; and both can be correct, if they are thorough and careful.

How is a given causative relation induced? Take for instance the form "X is a complete cause of Y". This we define as: "If X, then Y; if not X, not-then Y; and X and Y is possible". How can these propositions be established empirically? Well, as regards "X and Y is possible", all we need is find one case of conjunction of X and Y and the job is done. Similarly for "if not X, not-then Y"; since this means "not-X and not-Y is possible", all we need is find one case of conjunction of not-X and not-Y and the job is done.

This leaves us with "If X, then Y" to explain. This proposition means "X and not-Y is impossible", and we cannot by mere observation know for sure that the conjunction of X and not-Y never occurs (unless we are dealing with enumerable items, which is rarely the case). We must obviously usually resort to generalization: having searched for and never found such conjunction, we may reasonably – until and unless later discoveries suggest the contrary – assume that such conjunction is in fact impossible. If later experience belies our generalization, we must of course particularize and then make sure the causative proposition is revised accordingly.

Another way we might get such knowledge is more indirectly, by adduction. The assumption that "X and not-Y is impossible" might be made as a consequence of a larger hypothesis from which this impossibility may be inferred. Or we may directly postulate the overall proposition that 'X is a complete cause of Y' and see how that goes. Such assumptions remain valid so long as they are confirmed and not belied by empirical evidence, and so long as they constitute the most probable of existing hypotheses. If contrary evidence is found, they are of course naturally dropped, for they cannot logically continue to be claimed true as they stand.

Another way is with reference to deductive logic. We may simply have the logical insight that the items X and not-Y are incompatible. Or, more commonly, we may infer the impossibility of conjunction – or indeed, the whole causative proposition – from previously established propositions; by eduction or syllogism or hypothetical argument or whatever. It is with this most 'deductive' source of knowledge in mind that the complex, elaborate field of causative logic, and in particular of causative syllogism, is developed. This field is also essential to ensure the internal consistency of our body of knowledge as a whole, note well.

Additional criteria. It should be added that though causation is defined mainly by referring to various possibilities and impossibilities of conjunctions – there are often additional criteria. Space and time are two notable ones. Two events far apart in space and time may indeed be causatively related – for example, an explosion in the Sun and minutes later a bright light on Earth. But very often, causation concerns close events – for instance, my eating some food and having a certain sensation in my digestive system. In the both these cases, the effect is temporally after the cause. In the latter case, unlike the former, the cause and effect are both 'in my body'.

Between the Sun's emission of light and its arrival on Earth, there is continuity: the energy is conserved and travels through all the space from there to here, never faster than the speed of light, according to the theory of relativity. But what of recent discoveries (by Nicolas Gisin, 1997), which seem to suggest that elementary particles can affect each other instantly and at a large distance without apparent intermediary physical motion? Clearly, we cannot generalize in advance concerning such issues, but must keep an open mind – and an open logic theory. Still, we can say that in most cases the rule seems to be continuity. When we say 'bad food causes indigestion', we usually mean that it does so 'within one and the same body' (i.e. not that my eating bad food causes you indigestion).

As regards natural causation, we can formulate the additional criterion that the cause must in fact precede or be simultaneous with the effect. But this is not a universal law of causation, in that it is not essential in logical and extensional causation. In the latter modes, the causative sequence may be reversed, if it happens that the observer infers the cause from the effect. Although, we might in such cases point out another temporal factor: when we infer (even in cases of 'foregone conclusion'), we *think of* the premises before we think of the conclusions. That is to say, there are two temporal sequences to consider, either or both of which may be involved in a causal proposition: the factual sequence of events, and the sequence of our knowledge of these events.

Similarly, quantitative proportionality is often indicative of causation; but sometimes not. Although it is true that if the quantity of one phenomenon varies with the quantity of another phenomenon, we can induce a causative relation between them; *it does not follow* that where no such concomitant variation (to use J. S. Mill's term) is perceived, there is not causation. In any case, the curve quantatively relating cause and effect may be very crooked; 'proportionality' here does not refer only to simple equations, but even to very complicated equations involving many variables. In the limit, we may even admit as causative a relation for which no mathematical expression is apparent. An example of the latter situation is perhaps the quantum mechanics finding that the position and velocity of a particle cannot both be determined with great precision: though the particle as such persists, the separate quantities p and v are unpredictable (not merely epistemologically, but ontologically, according to some scientists) – which suggests some degree of natural spontaneity, in the midst of some causative continuity.

Thus, we *must* stick to the most general formulations of causation in our basic definitions, even as we admit there may be additional criteria to take into consideration in specific contexts. It follows from this necessity that we can expect the logic of causation certain inferences (like conversion, or those in second and third syllogism) where what is initially labeled a cause becomes an effect and vice versa. Keep this in mind[141].

Laws of causation. We should also here mention the cognitive role of alleged laws of causation. We have already briefly discussed laws relating to space and time.

In times past, it seems that some degree of sameness between cause and effect was regarded as an important law of causation. Upon reflection, the proponents of this criterion for causation probably had in mind that offspring have common features with their parents. But apparently, some people took this idea further and supposed that the substance (and eventually some other characteristics) of cause and effect must be the same. But though this criterion may be applicable to biology or other specific domains (e.g. the law of conservation of matter and energy in physics could be so construed), it is not generally regarded as universal. Formally, I see no basis for it.[142]

The law of causation most often appealed to (at least in Western thought) is that 'everything has a cause'. But though it is evidently true of most things that they have causes, and the belief in this law often motivates us to look for or postulate causes (i.e. even if none is apparent, we may assume one to exist), we have not in our study found any *formal* grounds to affirm such a law as universal. Admitting the fact of causation does not logically force us to admit its universality. This does not prove that it is not empirically universal; and it does not prevent us from formulating such universality as an adductive hypothesis. In any case, today, as evidenced by quantum physics and big-bang cosmogony, it seems generally assumed by scientists that this law is indeed not universal (which does not mean it is not very widely applicable).

I wonder anyway if it was ever really regarded as universal. I would say that in the 19th Century, this law was assumed universal for physical phenomena – but not necessarily for mental phenomena; human volition was generally taken to be an exception to the rule, i.e. freedom of the will was acknowledged by most people. Paradoxically, in the iconoclastic 20th Century, while the said law of causation was denied universality for material things, every effort was made to affirm it as regards human beings and thus forcefully deny freedom of the will[143]. Intellectual fashions change, evidently. But as far as I am concerned, while I admit the

---

141 It is interesting to note here that J. S. Mill's definitions of causation use the expression: "*is the effect, or the cause,…* " – meaning he had in mind the general forms.

142 If we want to go more deeply in the history of 'laws of causation', we would have to mention, among others, the Hindu/Buddhist law of karma, according to which one's good and bad deeds sooner or later have desirable or undesirable consequences, respectively, on oneself. It is the popular idea that 'what goes round must come round'. Though I would agree this is sometimes, frequently or even usually empirically true, we must admit that it does not always seem confirmed by observation – so it is at best a hopeful generalization (to a life after this one) intended to have positive moral influence. In any case, I see no formal basis for it. The same can be said concerning reward or punishment by God – though it might well be true, it is not something that can readily be proved by observation or by formal means; an act of faith is required to believe in it (I do, on that basis). In any case, the latter can hardly be called a 'law of causation', since the free will of God is thought to be involved in bringing about the effect.

143 Actually, both these changes were (I suggest) consciously or subconsciously motivated by the same evil desire to incapacitate mankind. Their proponents effectively told people: "you cannot control matter (since it is ultimately not subject to law) and you cannot control yourself (since you have no freewill) – so give up trying". People who believed this nonsense (including its advocates) were *influenced* by it to become weaker human beings. Virtue was derided and vice was promoted. We see the shameful results of this policy all around us today.

possibility that this law may not-be universally true of matter, I have no doubt that it is inapplicable to the human will[144].

Another alleged law of causation that should be mentioned here (because of the current interest in it, in some circles) is the Buddhist notion that 'every thing is caused by everything'. As I have shown in the present volume, this idea of universal 'interdependence' is logically untenable. It is *formally* nonsensical. Indeed, if you just think for a moment, you will realize (without need for complex formal analysis) that to affirm interdependence is to deny causation, or at least its knowability. Every concept relies on our ability to distinguish the presence and absence of the thing conceived; if it is everywhere the same, it cannot be discerned. I think the Buddhist philosopher Nagarjuna can be said to have realized that; and this would explain why he ultimately opted for a no-causation thesis. However, that does not mean that causation can logically be denied: as already explained earlier, it cannot.

Well, then. Are there any 'laws of causation'? Of course there are, a great many! Every finding concerning the formal logic of causation in this volume is a law of causation, a proven law. For instance, the fact that not all positive causative syllogisms yield a positive conclusion of some sort is an important law of causation, teaching us that a cause of a cause of something is not necessarily itself a cause of that thing.

3. A List of the Main Causative Arguments.

I have in previous chapters developed deduction of causation in considerable detail, but mostly in terms of propositional symbols. This form of expression is gibberish to most people, and so useless. I will therefore here list some of the essential arguments in ordinary language, i.e. in plain English. Hopefully, by studying these validated and invalidated arguments, everyone can improve their causative reasoning.

My 'Practical Guide to Causative Logic' would consist of the following tables, which may as usual be seen and freely downloaded, in .pdf format, at my website (www.thelogician.net):

- **Table 24.1 – Practical Guide to Causative Logic – List of Forms, their Oppositions and Eductions. (4 pages in pdf file).**
- Table 24.2 – Practical Guide to Causative Logic – Merged List of 3- & 4-Item Causative Syllogisms (Symbolic). (6 pages in pdf file).
- **Table 24.3 – Practical Guide to Causative Logic – Merged List of 3- & 4-Item Causative Syllogisms (Textual). (36 pages in pdf file).**
- Table 24.4 – Practical Guide to Causative Logic – Abridged List of 3- & 4-Item Causative Syllogisms, including only Positive Conclusions. (16 pages in pdf file).

---

[144] I argue this issue elsewhere, in my *Volition and Allied Causal Concepts*. It should be mentioned that an analogue to the law of causation is often postulated, consciously or not, for the mind. We tend to think that every act of volition has a cause, in the sense of *being influenced or motivated*, by something or other. Though largely true, this assumption taken literally would exclude purely whimsical volitions; thus, I tend to doubt it, for reasons explained in my said book. In any case, do not confuse this 'law of influence' with the 'law of causation' here discussed. These are very distinct forms of causality, which cannot be lumped together.

The first table consists of four parts. The first page shows the basic definitions of the generic determinations of causation. The second page lists the various forms of causative propositions (causative, preventive, inverse causative and inverse preventive) that emerge from these definitions by making various changes of polarity. The third page clarifies their main oppositions – i.e. what each form implies, denies or neither implies nor denies. And the fourth page clarifies their main eductions, i.e. inversions, conversions and contrapositions. These lists permit the reader to interpret causative propositions and understand how they interrelate individually.

The second table is valuable, though not of interest to people who have not gotten used to the symbols. It is a needed technical preparation for the third table, in that it merges into one table all the 3- and 4-item causative syllogisms previously listed separately. The results thus here collected are then converted to text, using various devices like concatenations, find/replace and ad hoc macros. This processing produced the third table.

Let us look more closely at Table 24.2, before further ado. This table merges Tables 22.6-0 and 22.7-0 in an appropriate order, eliminating what they had in common. The table for 3-item syllogisms, you will recall, listed 64 moods per figure; these moods were all of subfigure (a), concerning either two hard premises or a hard premise with a weak absolute premise or two weak absolute (abs/abs) premises. The table for 4-item syllogisms listed 89 moods, because it replaced all single absolute weak premises with a corresponding relative weak premise, and each abs/abs combinations with two analogous combinations abs/rel (subfigure (b)) and rel/abs (subfigure (c)); since the latter doubling of moods occurred 25 times per figure, the number of moods went from 64 to 89. When we add the 64 '3-item' moods to the 89 '4-item' moods, we do not get 153 moods but only 144 moods (per figure). The reason for that is that the '4-item' listing includes some entries that are really 3-item moods – i.e. the moods without any weak premises, namely the 9 moods **mn/mn**, **mn/m**, **mn/n**, **m/mn**, **n/mn**, **m/m**, **n/n**, **m/n**, **n/m**.

Having explained all that, let us now look at the **statistics** implied by this new merged list of moods. In each figure, we find 27 moods (19%) without causative or preventive conclusions. They are the 9 moods numbered 44, 47, 48, 74, 77, 78, 84, 87, 88, in subfigures (a), (b) and (c). These moods may be referred to as 'invalid' – that is to say, any causative or preventive conclusion proposed for them is invalid. This leaves us with **117 valid moods (81%) per figure**, i.e. moods that yield one or more positive and/or negative, causative and/or preventive conclusion(s). For each figure: 41 moods (28% of total) yield positive conclusions, all causative, whether elementary or compound; 63 moods (44%) yield one, two or three negative causative conclusion(s); 97 moods (67%) yield some causative conclusion(s), positive and/or negative; 76 moods (53%) yield one to four negative preventive conclusion(s); no moods yields any positive preventive conclusion. Of course, some of these conclusions overlap.

We could count each conclusion obtained from a given mood as constituting a separate valid syllogism, and thus greatly increase the number of valid syllogisms! But I prefer to regard all the conclusions obtained from each mood as together constituting 'the (compound) conclusion'.

Table 24.3 offers the reader a complete list of all 3- and 4-item causative syllogisms in plain English, for all three figures. That is to say all the syllogisms with positive causative premises of any sort. That amounts to 144 moods per figure, including the moods which yield no causative or preventive conclusions. The following example shows how the data is presented:

Mood 122 (b) - premises: **mq/mq** (abs / rel S)

Q is a complete and contingent cause of R

P is a complete and (complemented by S) contingent cause of Q

Positive conclusion(s): **mq** abs

P is a complete and contingent cause of R

Negative conclusion(s): causative: **not-q** rel to notS; preventive: none

P (complemented by notS) is not a contingent cause of R

The mood concerned is first defined numerically and symbolically; then the major and minor premises are verbally listed; then the positive conclusion(s), if any, are given, both symbolically and verbally; then the negative conclusion(s), causative and/or preventive, if any, are given, both symbolically and verbally. Note that conclusions are divided into positive and negative ones. The fourth table is an abridged version of the third, showing only moods with positive conclusions. The important conclusions for ordinary discourse are the positives, although the negatives are also useful information (e.g. in consistency checking or to construct 'ad absurdum' reductions). A general finding is that the positive conclusions (if any) of causative syllogisms are always causative, whereas the negative conclusions (if any) may be causative or preventive.

The domain of causative syllogism (in the broadest sense) is of course much larger than the moods here listed. Here, we have only shown syllogisms (valid and invalid) *from positive causative premises with positive terms*. Syllogisms involving conflicting middle terms (and hence a mix of causative and preventive premises) are not included. Nor are syllogisms involving one or more negative premise(s). Nevertheless, the syllogisms here listed are the most typical and commonly encountered. For a larger perspective, see earlier chapters.

It must be stressed that the results presented here are exhaustive and certain. They are exhaustive in the sense that all conceivable conclusions, of any causative or preventive form, positive and negative, have been tested and either validated or invalidated. They are certain, in that everything is calculated by means of spreadsheets (totaling over 72'000 pages!) and found consistent with previous findings by other means. The actual validation and invalidation work is not shown here, but is open to scrutiny in previous chapters.

This is the first time anyone has worked out and published these syllogisms, which are crucial to both ordinary and scientific thinking processes.

4. Closing Remark.

I here bring to an end my account of phase III of *The Logic of Causation*, having considerably rationalized and expanded the research project, and indeed brought it to a successful conclusion. It is perhaps not the very end of the matter, but the most important work is done. I still may, in the coming months or years, G-d willing, try to further the research. I still dream of producing software capable of receiving actual data input and dishing out the best inductive and deductive conclusions from it (this would hopefully solve 5-item syllogism). But if my effort or my life should cease now I would feel I have already fulfilled my self-appointed mission.

I hope only that other people, reading this research report, realize its great originality and importance to logic and philosophy and to all the special sciences; and that they make the effort to study, assimilate and expand its findings, and to pass on its teaching in universities and other forums.

With heartfelt thanks to G-d,
For all his constant kindness to me.
Avi Sion
Geneva, July 2010.

DIAGRAMS

TABLES

In Phase I

In Phase II

In Phase III

Only some of the tables and details of some others are published in this volume. The full set of tables is posted in the website www.TheLogician.net at:

http://www.thelogician.net/4_logic_of_causation/4_lc_tables.htm

Table 22.7-13B – 4-Item Causative Syllogisms – Scanning Preventive Conclusions to Moods 131-138. (1735 pages in pdf file.)
Table 22.7-14A – 4-Item Causative Syllogisms – Scanning Causative Conclusions to Moods 141-148. (1720 pages in pdf file.)
Table 22.7-14B – 4-Item Causative Syllogisms – Scanning Preventive Conclusions to Moods 141-148. (1735 pages in pdf file.)
Table 22.7-15A – 4-Item Causative Syllogisms – Scanning Causative Conclusions to Moods 151-158. (1032 pages in pdf file.)
Table 22.7-15B – 4-Item Causative Syllogisms – Scanning Preventive Conclusions to Moods 151-158. (1041 pages in pdf file.)
Table 22.7-16A – 4-Item Causative Syllogisms – Scanning Causative Conclusions to Moods 161-168. (1032 pages in pdf file.)
Table 22.7-16B – 4-Item Causative Syllogisms – Scanning Preventive Conclusions to Moods 161-168. (1041 pages in pdf file.)
Table 22.7-17A – 4-Item Causative Syllogisms – Scanning Causative Conclusions to Moods 171-178. (1720 pages in pdf file.)
Table 22.7-17B – 4-Item Causative Syllogisms – Scanning Preventive Conclusions to Moods 171-178. (1735 pages in pdf file.)
Table 22.7-18A – 4-Item Causative Syllogisms – Scanning Causative Conclusions to Moods 181-188. (1720 pages in pdf file.)
Table 22.7-18B – 4-Item Causative Syllogisms – Scanning Preventive Conclusions to Moods 181-188. (1735 pages in pdf file.)
Table 22.7-21A – 4-Item Causative Syllogisms – Scanning Causative Conclusions to Moods 211-218. (1032 pages in pdf file.)
Table 22.7-21B – 4-Item Causative Syllogisms – Scanning Preventive Conclusions to Moods 211-218. (1041 pages in pdf file.)
Table 22.7-22A – 4-Item Causative Syllogisms – Scanning Causative Conclusions to Moods 221-228. (1720 pages in pdf file.)
Table 22.7-22B – 4-Item Causative Syllogisms – Scanning Preventive Conclusions to Moods 221-228. (1735 pages in pdf file.)
Table 22.7-23A – 4-Item Causative Syllogisms – Scanning Causative Conclusions to Moods 231-238. (1720 pages in pdf file.)
Table 22.7-23B – 4-Item Causative Syllogisms – Scanning Preventive Conclusions to Moods 231-238. (1735 pages in pdf file.)
Table 22.7-24A – 4-Item Causative Syllogisms – Scanning Causative Conclusions to Moods 241-248. (1720 pages in pdf file.)
Table 22.7-24B – 4-Item Causative Syllogisms – Scanning Preventive Conclusions to Moods 241-248. (1735 pages in pdf file.)
Table 22.7-25A – 4-Item Causative Syllogisms – Scanning Causative Conclusions to Moods 251-258. (1032 pages in pdf file.)
Table 22.7-25B – 4-Item Causative Syllogisms – Scanning Preventive Conclusions to Moods 251-258. (1041 pages in pdf file.)
Table 22.7-26A – 4-Item Causative Syllogisms – Scanning Causative Conclusions to Moods 261-268. (1032 pages in pdf file.)
Table 22.7-26B – 4-Item Causative Syllogisms – Scanning Preventive Conclusions to Moods 261-268. (1041 pages in pdf file.)

Works by Avi Sion:

ISBN 978-1495221101

Made in the USA
Charleston, SC
19 April 2014